After the Three Italies

RGS-IBG Book Series

The *Royal Geographical Society (with the Institute of British Geographers) Book Series* provides a forum for scholarly monographs and edited collections of academic papers at the leading edge of research in human and physical geography. The volumes are intended to make significant contributions to the field in which they lie, and to be written in a manner accessible to the wider community of academic geographers. Some volumes will disseminate current geographical research reported at conferences or sessions convened by Research Groups of the Society. Some will be edited or authored by scholars from beyond the UK. All are designed to have an international readership and to both reflect and stimulate the best current research within geography.

The books will stand out in terms of:

- the quality of research
- their contribution to their research field
- their likelihood to stimulate other research
- being scholarly but accessible.

For series guides go to www.blackwellpublishing.com/pdf/rgsibg.pdf

Published

Putting Workfare in Place
Peter Sunley, Ron Martin and Corinne Nativel

After the Three Italies: Wealth, Inequality and Industrial Change
Mick Dunford and Lidia Greco

Domicile and Diaspora
Alison Blunt

Geographies and Moralities
Edited by Roger Lee and David M. Smith

Military Geographies
Rachel Woodward

A New Deal for Transport?
Edited by Iain Docherty and Jon Shaw

Geographies of British Modernity
Edited by David Gilbert, David Matless and Brian Short

Lost Geographies of Power
John Allen

Globalizing South China
Carolyn L. Cartier

Geomorphological Processes and Landscape Change: Britain in the Last 1000 Years
Edited by David L. Higgitt and E. Mark Lee

Forthcoming

Living Through Decline: Surviving in the Places of the Post-Industrial Economy
Huw Beynon and Ray Hudson

The Geomorphology of Upland Peat
Martin Evans and Jeff Warburton

Publics and the City
Kurt Iveson

Driving Spaces
Peter Merriman

Geochemical Sediments and Landscapes
Edited by David Nash and Susan McLaren

Fieldwork: A Historical Geography of Science Outdoors
Simon Naylor

Natural Resources in Eastern Europe
Chad Staddon

Peoples/States/Territories
Rhys Jones

Climate and Society in Colonial Mexico
Georgina H. Endfield

Mental Health and Social Space
Hester Parr

Consuming Ethics: Markets and the Globalisation of Care
Clive Barnett, Nick Clarke, Paul Cloke and Alice Malpass

After the Three Italies

Wealth, Inequality and Industrial Change

Michael Dunford and Lidia Greco

Blackwell
Publishing

BLACKWELL PUBLISHING
350 Main Street, Malden, MA 02148–5020, USA
9600 Garsington Road, Oxford OX4 2DQ, UK
550 Swanston Street, Carlton, Victoria 3053, Australia

First published 2006 by Blackwell Publishing Ltd

1   2006

Library of Congress Cataloging-in-Publication Data

Dunford, Michael.
　After the three Italies : wealth, inequality and industrial change /
Michael Dunford and Lidia Greco.
　　　p. cm. — (RGS-IBG book series)
　　Includes bibliographical references and index.
　　ISBN-13: 978-1-4051-2520-8 (hardback : alk. paper)
　　ISBN-10: 1-4051-2520-9 (hardback : alk. paper)
　　ISBN-13: 978-1-4051-2521-5 (pbk. : alk. paper)
　　ISBN-10: 1-4051-2521-7 (pbk. : alk. paper)
　1. Italy—Economic conditions—1994- —Regional disparities.
2. Regionalism—Economic aspects—Italy. 3. Industries—Italy. 4. Income
distribution—Italy. I. Greco, Lidia, 1969- II. Title. III. Series.

HC305.D76 2005
338.945—dc22

                                                                    2005006161

A catalogue record for this title is available from the British Library.

Set in 10/12 plantin
by SPI Publisher Services, Pondicherry, India
Printed and bound in India
by Replika Press

For further information on
Blackwell Publishing, visit our website:
www.blackwellpublishing.com

Dedication

For Lotte and Robin (Michael Dunford)
To my return to Italy (Lidia Greco)

Contents

Figures

Tables

Series Editors' Preface

The RGS/IBG Book series publishes the highest quality of research and scholarship across the broad disciplinary spectrum of geography. Addressing the vibrant agenda of theoretical debates and issues that characterise the contemporary discipline, contributions will provide a synthesis of research, teaching, theory and practice that both reflects and stimulates cutting edge research. The Series seeks to engage an international readership through the provision of scholarly, vivid and accessible texts.

Nick Henry and Jon Sadler
RGS-IBG Book Series Editors.

Preface and Acknowledgements

This book is one of the results of a research project (L213252028) funded by the UK Economic and Social Research Council under its 'One Europe or Several?' programme. The project was entitled Regional Economic Performance, Governance and Cohesion in an Enlarged Europe. The aim of the project was to discover whether, as EU integration and enlargement moved ahead, Europe's regional economies were moving closer together, or whether new geographical differences and inequalities were being created. To this end the research concentrated on developed and less-developed areas in four countries: England, Italy, Poland and Slovakia. Aggregate data on the performance of regional economies were collected in each country. A plant-level survey of 482 establishments was also completed, as were 165 interviews with firms and regional development institutions.

The research profited greatly from the support of the Programme Director, Helen Wallace (European University Institute, Florence), the contributions of the main researchers, Jane Hardy (University of Hertfordshire), Ray Hudson (University of Durham), Al Rainnie (Monash University), David Sadler (University of Liverpool) and Adrian Smith (Queen Mary, University of London), and the energy and enthusiasm of two research fellows, Alex Bolland and Brian Haywood (University of Sussex). As far as the Italian research was concerned, a major contribution was made by Ray Hudson who took part in the Italian field work and whose reflections helped shape this volume. In Italy, Gioacchino Garofoli (Università degli Studi dell'Insubria), Anna Giunta and Flavia Martinelli (Università della Calabria), Alfredo del Monte (Università di Napoli Federicio II) and Vittorio Nicolardi (Università degli Studi di Bari) made helpful and sometimes vital contributions. A number of other people played helpful roles in setting up and carrying out surveys and interviews: Emanuela Altomare,

Cesare Benzi, Matteo Boemi, Oronzo Catucci, Lesley Dinnen, Fabrizio Guzzo, Caterina Marini, Alessandro Muscio, Michele Santangelo and Vincenzo Sileo.

The idea of writing this particular book emerged in the final year of the project. At that time Lidia Greco took part in some of the interviews in southern Italy, and the two authors co-drafted a report on the Italian research for the End of Project Workshop held in Sussex in April 2002. At the same time both authors had worked for more than a decade on aspects of Italian regional and urban development and shared a concern about the adequacy of the perspectives and interpretations that dominated much of the international literature.

The aims of this volume are threefold. The first is to provide an updated picture of Italy's regional inequalities and to develop more generally applicable methods for documenting and analyzing them. To this end Italy's economic performance, its recent development relative to other European countries and the rest of the world, its internal differentiation and its persistent regional inequalities are all analyzed. The second is to examine the changing territorial division of labour in a number of economic sectors. A disaggregation of trends in regional development permits the identification of the contribution to regional performance of firm performance, which is itself examined in the light of corporate strategies that unfold in the context of wider product and value chains. The third is to help develop a critique of neoliberal theories of convergence and spatial development, to provide a new synthesis of theoretical ideas in economic geography, and to contribute to revitalized political economy approaches to territorial development.

As far as the examination of Italian regional inequality is concerned, it is important to recall that the overturning in the 1990s of the post-WWII political order, the restructuring of interest group representation, the implementation of a series of administrative and policy reforms and the process of European integration created hopes of more successful economic and political development. In the field of regional studies, the exceptional growth of industrial districts in the centre and northeast of the country, innovative regional policy reforms, which replaced the top-down approach of the previous 40 years with a more democratic and participative bottom-up approach, and the evidence of economic vitality in some parts of the south renewed the hope that Italy's deep-rooted regional divisions might diminish.

A careful analysis of quantitative and qualitative data suggests however a more qualified picture. Although territorial inequalities have started to diminish in the recent past, wide territorial inequalities remain. These inequalities are primarily macroterritorial and are determined by a divergence in employment rates rather than in productivity. The scale of contemporary inequality is in part a consequence of the fact that in 1974–96 Italy experienced sharp increases in territorial inequality, reversing earlier processes of catch-up and convergence: the economics and politics of excellence rewarded and strengthened successful regional economies, while unsuccessful economies were weakened and discouraged, often on the grounds that the result would be a faster increase in aggregate wealth.

At the same time the Italian economy has lost momentum. Although Italy continues to occupy an important economic position vis-à-vis other advanced European countries, it confronts a number of serious challenges. Tight Euro-zone financial controls and strict constraints on macroeconomic management have constrained growth, while the end of currency devaluations relative to its European partners has exposed some of the structural weaknesses of Italy's economic apparatus. At present, profitability and growth are suffering from the country's specialization in traditional sectors and its dependence on slow-growing markets.

What this analysis of regional inequality and economic performance highlights is the existence of a wide range of regional trajectories on the one hand and the interdependence and unevenness of Italy's economic geography on the other. In much of the recent geographical literature, however, these central features of Italy's economic geography are frequently overlooked and are no longer considered as something requiring explanation. Often this literature is characterized by region-centred approaches and a concern with what happens inside regional economies. Consequently, little attention is paid to the study of unequal economic development, socioeconomic inequalities and the role of power relations in shaping the space economy. Instead, accounts of the Third Italy, and also of the development of southern districts, addressed just one aspect of the structure and development of the space economy. Additionally, some of this literature implicitly assumes that areas that are more developed offer role models for areas that are less developed. A similar idea is an explicit aspect of neoclassical convergence models. As these models, with their expectations of catch-up, prevailed until recently, the economic literature also was difficult to reconcile with the reality of widening inequalities.

The existence of these gaps between models/theories and reality suggests that there is a need for a more systematic account of regional inequalities that considers the relationships between development and underdevelopment. If development and underdevelopment are interconnected, it follows that, instead of a single pathway of modernization and development, there are multiple pathways, and that, alongside mechanisms that equalize development, there are mechanisms that make it more unequal. More radical versions of this thesis suggest that, in capitalist societies, development and underdevelopment are two sides of the same coin. An implication of these ideas is that the central object of analysis should be the shifting map of areal differentiation and the shifting mosaic of uneven development.

In this volume, the analysis of regional inequality accordingly opens the way to the elaboration of a more comprehensive approach to regional economic development. To this end elements of different theoretical approaches are synthesized into an alternative and new conceptual framework. The aim is to disclose the nature of variations in regional development as the outcome of complex and geographically differentiated political and economic processes.

More specifically, the framework seeks to identify the connections between comparative regional development, the underlying territorial division of labour and the mechanisms that shape it. The idea is that one can connect the changing trajectories of Italy's regional economies and the associated changes in inequality with the changing position of Italian enterprises in European and international divisions of labour, and that one can combine quantitative evidence of regional performance with quantitative and qualitative evidence of firm performance and industrial change.

In the perspective that we suggest, the profit-seeking firm is the central economic actor in the explanation of industrial change. Nonetheless, the firms' external environments and their network relations with other firms are also considered. Networks exist at a regional level. In contrast to the new regionalism, however, more attention is paid to interregional relationships and international networks. In addition, whilst acknowledging the fundamental profit-seeking nature of enterprises, the approach adopted recognizes that institutional variables, primarily the institutional environment in which companies operate, affect corporate economic performance. This institutional environment comprises not just other firms but also a set of institutional, political and cultural conditions.

To summarize, this volume offers, first, an analysis of regional performance. In seeing regional performance as the outcome of the shifting relative weight of centrifugal (equalizing) and centripetal (unequalizing) forces, it provides an alternative to standard theories of catch-up. Second, it examines the role of firm performance in shaping regional performance. In doing so, it examines some of the micro-foundations of regional performance. Third, it centres explanations of the geography of industrial change on an analysis of profit-seeking in the context of wider value chains and institutional configurations. In doing so, it aims to insert a recognition of the role of profit-seeking into a theory of corporate conduct that is less rudimentary than those of mainstream economics.

As with any joint work, there was a division of labour in the preparation of this volume. Michael Dunford was primarily responsible for chapters 2–4 and 6–8. Lidia Greco was primarily responsible for chapters 5 and most of 9. The conclusion and this preface were jointly written. All the chapters were discussed together and developed or edited in the light of these discussions.

The authors wish to thank a number of people who played important roles in the preparation and publication of this volume. Franco Chiarello, Nick Henry, Ray Hudson, Benito Giordano, Roger Lee, Diane Perrons, Francesco Prota and three anonymous referees provided valuable comments on the script and/or were involved in helpful discussions of different parts of this work. Hazel Lintott (University of Sussex) drew many of the illustrations. Debbie Seymour carefully copyedited the script, while Angela Cohen and Jacqueline Scott saw the book through its various stages of production. The authors owe, finally, a special debt of gratitude to Ray Hudson for his role in initiating the collaboration that made this book possible.

Abbreviations

ASI	Area di Sviluppo Industriale (Industrial Development Area)
CEEC	Central and Eastern European Countries
CGIL	Confederazione Generale Italiana del Lavoro (General Confederation of Italian Workers)
CIG	Cassa Integrazione Guadagni (Temporary Layoff Wage Compensation Fund)
CIGS	Cassa Integrazione Guadagni Straordinaria (Special Layoff Wage Compensation Fund)
CIPE	Comitato Interministeriale per la Programmazione Economica (Interministerial Committee for Economic Planning)
CIPI	Comitato Interministeriale per la Programmazione Industriale (Interministerial Committee for Industrial Planning)
CNA	Confederazione Nazionale Artigiani (National Craft Confederation)
CSF	Community Support Framework
CSI	Consorzio di Sviluppo Industriale (Agency for Industrial Development)
CISI	Centro Integrato per lo Sviluppo dell'Imprenditorialità (Integrated Entrepreneurship Development Centre)
CISL	Confederazione Italiana Sindacati dei Lavoratori (Italian Confederation of Workers' Unions)
DC	Democrazia Cristiana (Christian Democratic Party)
DS	Democratici di Sinistra (Left Wing Democrats)
DIA	Direzione Investigativa Antimafia (Central Anti-Mafia Investigation Agency)

DPS	Dipartimento per le Politiche di Sviluppo e Coesione (Treasury Department for Development and Cohesion Policies)
EC	European Community
ECE	Eastern and Central Europe
ECSC	European Coal and Steel Community
EDF	Electricité de France
EEA	European Economic Area
EFTA	European Free Trade Association
EI	Intervento Straordinario (Extraordinary Intervention)
ENI	Ente Nazionale Idrocarburi (National Hydrocarbons Agency)
ESF	European Social Fund
EU	European Union
EU15	First 15 European Union Member States
EVC	European Vinyls Corporation
FDI	Foreign direct investment
FDY	Fully drawn yarn
FI	Forza Italia (Go-Ahead Italy)
FIAT	Fabbrica Italiana Automobili Torino
FIM	Federazione Italiana Metalmeccanici (Italian Metal-Mechanical Workers' Federation)
FIOM	Federazione Italiana Operai Metalmeccanici (Italian Federation of Metalworkers)
FISMIC	Federazione Italiana Sindacati Metalmeccanici ed Industrie Collegate (Italian Federation of Unions for Metal-Mechanical and Related Industries)
FOY	Fully oriented yarn
GATT	General Agreement on Tariffs and Trade
GDP	Gross domestic product
GEPI	Società di Gestione delle Partecipazioni Industriali (Agency for the Management of Industrial Shareholdings)
GRP	Gross Regional Product
GSP	Generalised System of Preferences
ICE	Istituto Commercio Estero (Institute for Foreign Trade)
IMI	Istituto Mobiliare Italiano (Italian Institute for Short-Term Credit)
INPS	Istituto Nazionale di Previdenza Sociale (National Social Security Agency)
IRI	Istituto per la Ricostruzione Industriale (Institute for Industrial Reconstruction)

ISTAT	Istituto Centrale di Statistica (Central Statistical Office)
LN	Lega Nord (Northern League)
MAP	Mobilità di Accompagnamento alla Pensione (early retirement mobility allowances applicable to employees who would otherwise be eligible to receive a pension within a maximum of three to four years)
MDI	Diphenylmethane di-isocyanate
MSI	Movimento Sociale Italiano (Italian Social Movement)
NAFTA	North Atlantic Free Trade Agreement
OECD	Organisation for Economic Cooperation and Development
OEM	Original Equipment Manufacturer
OPEC	Organisation of Petroleum Exporting Countries
OPT	Outward processing traffic/trade
PDSI	Partito Social-Democratico Italiano (Italian Social Democratic Party)
PET	Polyethylene terephthalate
PIT	Progetti Integrati Territoriali (Integrated Territorial Projects)
PLI	Partito Liberale Italiano (Italian Liberal Party)
POR	Programma Operativo Regionale (Regional Operational Programme)
POY	Partially oriented yarn
PPS	Purchasing power standards
PP.SS.	Partecipazioni Statali (State Holding Companies)
PRI	Partito Repubblicano Italiano (Italian Republican Party)
PSI	Partito Socialista Italiano (Italian Socialist Party)
PTA	Purified terephthalic acid
PVC	Polyvinyl chloride
R&D	Research and development
SATA	Società Automobilistica Tecnologie Avanzate (Advanced Technology Automotive Company)
SEECs	South-East European Countries
SIR	Società Italiana Resine
SME	Small and medium-sized enterprise
SPI	Società per la Programmazione Industriale (Agency for Industrial Planning)

TDI   Toluene di-isocyanate

UGL   Unione Generale del Lavoro (General Union of Workers)
UIL   Unione Italiana Lavoratori (Union of Italian Workers)
UILM   Unione Italiana Lavoratori Metalmeccanici (Union of Italian Metal-Mechanical Workers)
UTE   Unità Tecnologica Elementare (Elementary Technological Unit)

VCM   Vinyl chloride monomer

WMAD   Weighted mean absolute deviation
WTO   World Trade Organisation
WWII   World War II

Chapter One

Introduction: Growth, Inequality and the Territorial Division of Labour

Areal Differentiation and Development Models

Italy is a country with long-standing and deep-seated regional inequalities. Until the 1980s, most attention was paid to the development divide that separates the Mezzogiorno from the centre-north (Figure 1.1 noting that the Mezzogiorno comprises the southern regions of Abruzzo, Molise, Basilicata, Puglia, Campania, Calabria, Sicilia and Sardegna and, in some definitions, the south of Lazio and Figure 1.2 which identifies the underlying relief including the Alps in the north, the Po valley and the Appenine chain which stretches through the centre of Italy). Of course there are differences within each territorial entity no matter how small it is, and every place is ultimately unique. The reason for grouping places into regions and making generalizations about them is that, in spite of their differences, the places grouped together have certain characteristics in common. The areas comprising the Mezzogiorno, for example, all shared characteristics of economic backwardness, even if economic structures, the degree of backwardness and the particular causes of relative underdevelopment differed.

In 1977, the publication of Bagnasco's *Tre Italie* (*Three Italies*) resulted in a further distinction between the metropolitan economies of the northwest and the rest of the centre-north (centre-northeast). The northwest was made up of four regions: Piemonte, the Valle d'Aosta, Lombardia and Liguria. Centred on the three cities of Turin (Piemonte), Milan (Lombardia) and Genoa (Liguria) that comprise the apexes of Italy's industrial triangle, these areas were the heartland of Italy's largest industrial enterprises. The centre-northeast, or Third Italy, conversely was made up of regions that were more rural, whose leading urban centres were smaller and which were characterized by economies in which SMEs played a larger role (Trigilia, 1992; Garofoli, 1991; Becattini, 1990).

Figure 1.1   Italy's territorial units and regions.

The distinction between the Three Italies is evident in Figure 1.3. This figure records demographic growth on the horizontal axis and manufacturing jobs per resident on the vertical axis for the Census years from 1951–2001 for ten groups of regions. The groups comprise regions with similar trajectories. All three metropolitan areas in the northwest (Piemonte, Lombardia and Liguria) started out with a large number of manufacturing jobs per resident. The centre-northeast had a moderate share, while the Mezzogiorno and Lazio had a small share. The metropolitan areas saw significant population growth until 1971 or 1981 but then decline set in. The areas of the Third Italy, conversely, saw very limited population growth but very strong increases in manufacturing jobs per resident. In the

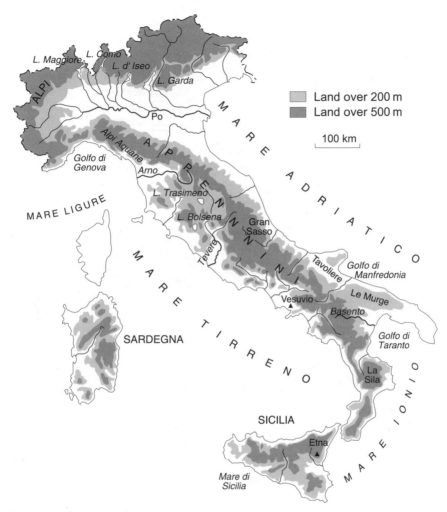

Figure 1.2   Italy: main relief.

south, several distinctions can be made. In a number of small regions that initially lost population, there was significant manufacturing job growth. Calabria lost ground marginally on both indicators. The regions containing the large cities of Naples and Bari, the islands and Lazio saw strong population growth but relatively small increases in manufacturing jobs per resident.

As further regional differences were identified, additional geographical distinctions were made. The Third Italy was itself divided into the anticlerical and socialist/communist 'red' provinces of central Italy and

Manufacturing jobs per '000 residents, 1951–2001

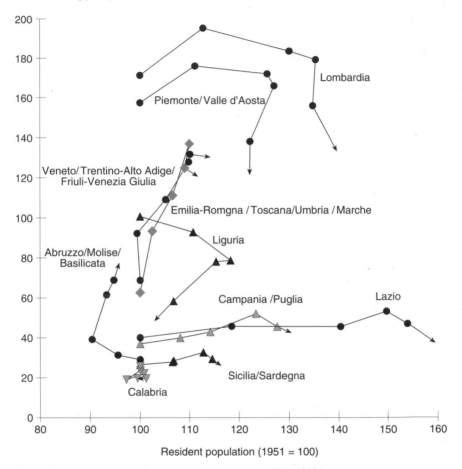

Figure 1.3   Manufacturing job and population trajectories, 1951–2001.

the formerly Austrian, Catholic 'white' provinces of the northeast. Areal differentiation was also discovered in the Mezzogiorno. Arlacchi (1983), for example, identified three Calabrias: an area of great estates and class conflict in the area around Crotone; an area of small, self-sufficient peasant farming around Cosenza; and a zone of export-oriented citrus fruit farming controlled by Calabria's traditional Mafia around Gioia Tauro. Other distinctions cut across these divisions: as the district model of small-scale entrepreneurship and flexible specialization took hold in formerly weakly industrialized areas along the Adriatic coast, Fuà (1983), for example, identified the Adriatic Way.

The identification of these areal differences was closely associated with successive analyses of Italy's remarkable modernization after World War II (WWII). In the 30-year Golden Age (1945–75) Italy, along with other relatively developed capitalist economies in Europe, east Asia and countries settled by Europeans, experienced unprecedentedly fast economic growth. Average growth rates were fastest in a series of initially less-developed economies, of which Italy was a striking example. As a result these less-developed countries started to catch up with the world's leading economy (the United States) and other initially more-developed countries. High rates of growth coincided therefore with strong economic convergence. In the mid-1970s, however, the situation changed. After a sharp and generalized decline in profitability that started in the mid-1960s, average growth rates declined. Although they increased in the late 1980s and in the years from 1993 until the aftermath of the economic crises of 1997–8 and the fin-de-siècle collapse of the new economy bubble, rates of growth remained comparatively low for the next 30 years.

In this context of slower growth, the rise of the Third Italy had a major international impact. One reason why was that the relative dynamism of the regional economies of the centre-northeast enabled Italy to resist the growth slowdown to a greater extent than other capitalist (mixed economy) metropoles until the early 1990s. In retrospect it is clear that the rise of some of these regional economies antedated the turning point that occurred in the mid-1970s. What struck observers, however, was the fact that the regional economies of the centre-northeast grew quite rapidly, as the metropolitan economies of the northwest lost momentum, and as the south stagnated and came to be identified with economic inefficiency, corruption and crime.

A great deal of attention was subsequently paid to these models and their role in shaping Italy's performance in the 1970s and 1980s. As well as the literature on the Three Italies, there were studies of Italy's industrial districts (Garofoli, 1991; Dei Ottati, 1995), and of the models afforded by particular regional economies, such as the classic Emilia-Romagna model. In political science, the thesis of Putnam, Leonardi and Nanetti (1993) concerning the role of civic cultures in shaping economic performance was originally developed in relation to the sharply contrasting economic performance of the centre-north and the Mezzogiorno. This research stimulated a large literature on the role of social capital and 'good governance' and their implications for the destinies of different economies. In the 1990s the resulting regional cleavages played an important part in reshaping the Italian political scene, especially with the rise of the LN as the representative of the northern petty bourgeosie (Agnew, 2002).

The second reason for international interest in the Third Italy lay in two arguments advanced by Piore and Sabel (1984). The first was that the growth slowdown was associated with a crisis of the Fordist industrial model (defined as mass production of standardized goods with dedicated machines and unskilled workers). The second was that the developmental success of the Third Italy was a result of a model of industrial development that offered an alternative to the mass production system and a way out of the crisis of Fordism. Markets for mass-produced goods, they argued, were saturated in advanced countries, while consumers were interested in goods that were more diversified and had a higher design content. In this situation smaller, more flexible and specialized enterprises, that made more diversified goods and services, with skilled craft workers and flexible equipment, had a competitive advantage. (At the same time the instability of demand gave integrated systems of highly specialized and adaptable SMEs, without major overheads, advantages over large firms). Piore and Sabel argued that the combination of a demand-side response, in the shape of customized products and changes in product mix, and a supply-side response, in the shape of the employment of skilled workers and flexible machines, amounted to the re-emergence of a craft model of development that they called 'flexible specialization'. In Italy, West Germany and Scandinavia, they argued, specialized industrial districts and regional economies made up of networks of such SMEs and supporting local collective services and community networks were dynamic and successful. The success of these areas offered, they continued, the prospect of the end of mass production, its replacement by a model of 'flexible specialization', a reconsolidation of regional economies as integrated units of production, and the creation of a new post-Fordist world. Indeed, they argued, if development does finally proceed along this path, the 1970s and 1980s will be seen in retrospect as a major turning point in the historical process of industrialization: whereas the nineteenth century saw mass production destroy craft production, the 1980s may represent a second industrial divide in which industrial societies started to abandon mass production and returned to nineteenth-century craft methods.

In the wider geographical literature there were parallel developments. On the one hand, some research concentrated on the economic foundations of agglomerated economic activities in traditional sectors as in the Third Italy and in innovation-oriented new technology industries. In this research emphasis was placed on the role in local development of the geography of economic transactions (Scott, 1988) and the supply-side determinants of creativity, knowledge, innovation and learning (Castells and Hall, 1994). On the other there was a development of region-centred approaches to local economic development. These approaches emphasized the socioinstitutional and associational foundations of regional economies and, in particular, of those considered successful. Offering explanations of economic performance that rely primarily on institutional factors (Amin and

Thrift, 1994), these studies paralleled the Putnam thesis on the centrality of the cultural/institutional determinants of economic action. Called the new regionalism, and associated with the subsequent 'cultural turn' in economic geography, this research enabled economic geography to widen its vision to include cultural and institutional dimensions of economic life. In doing so, it added to earlier 'structuralist' explanations of regional uneven development, including political-economy approaches, which had overemphasized economic causes, and were too deterministic, paying insufficient attention to social and political action (Hudson, 2003). The institutional and cultural foundations of local and regional economies do shape the possibilities for, and impose constraints on, development. They also help explain, for instance, why economic activities agglomerate and grow in certain places rather than in others. A paradoxical consequence of this approach is, however, that some economic geographers have increasingly downplayed the significance of the economy in their own field (Martin and Sunley, 2001). Although culture and institutions are important dimensions and determinants of socioeconomic activity and development, they are not the only determinants, nor are they usually the most important. As Sayer (1997) argued, economic forces are especially important in contemporary life. The importance of economic analysis should not therefore be underestimated.

The new economic geography of the economists (henceforth called, after Martin and Sunley (2001), geographical economics) offered yet another way of seeing the processes that shape economic landscapes. Within this tradition, emphasis is on the interaction of economies of scale, input-output relations and markets in creating economic agglomerations. The models associated with this tradition have, however, several limitations. From a conceptual perspective, these models extend the formal mathematical modelling of early location theory to develop equilibrium solutions to new maximization problems. All are essentially concerned with the allocation of resources, or with the long-run supply side determinants of growth and development. What these models assume is is a smooth and automatic process of adjustment to economic shocks: resources released as a result of structural change are quickly re-employed. Although the starting point is profit-seeking enterprises, the underlying concepts of corporate strategy and of the relevant external environment are narrow ones, while the derived economic landscapes are abstract and highly simplified (see Chapter 2 and Perrons, 2004). Not surprisingly, these models do not ask whether different/similar mechanisms of growth operate differently/similarly in different regions. Essentially, these approaches neglect the significance of the characteristics of real places and the place-specific characteristics of growth and development. Whereas the new economic geography gives too much weight to the cultural/institutional determinants of economic action, the new geographical economics pays too little attention to the specific characteristics of regional economies and the significance of institutional and cultural factors in shaping economic landscapes.

After the Three Italies

In the last section we identified several sets of ideas that drew attention to, and helped explain, certain overlooked and/or emerging realities in Italy. In addition, we indicated that these ideas were of wider international significance. More specifically, three developments were highlighted. First, the literature on the Three Italies identified aspects of areal differentiation that had been overlooked and that were of increasing national and international importance. Second, the literature on flexible specialization and industrial districts identified an increase in the importance of a model of industrial organization involving small-scale industries in semi-rural and small-town environments in the centre-northeast of Italy and elsewhere in the developed and developing world. Third, region-centred approaches, drawing on earlier work on industrial districts, were thought to offer ways of explaining the structure and trajectory of these areas.

At an early stage, however, the generality of the flexible specialization thesis was challenged. In many sectors, it was argued, mass producers can take advantage of flexible production techniques, and oligopolistic groups will dominate the scene. As Coriat (1991) showed, to succeed flexible specialization requires that economies of scope are greater than economies of scale. This condition will hold if overall demand is static, and if demand is irregular, or products are subject to rapid obsolescence. In these conditions, an enterprise that can change almost instantaneously from one product to another will have the edge over one that seeks to realize scale economies. Furthermore, in the conditions of monopolistic competition that characterize these sectors, the more dynamic, flexible firms can actively pursue strategies of product differentiation so as to maximize monopoly rents. Strategies of flexible specialization offer, therefore, not a new macroeconomic model of development but a principle for success in certain conditions of demand and competition. The model is in fact most appropriate in particular sectors (manufacture of clothes, knitwear, textiles, ceramics, and shoes where fashion and/or seasonal factors lead to rapid product changes), in conditions of unequal income distribution, and in economic conjunctures in which instabilities predominate. Indeed, argued Coriat, its orientation towards elite markets and its monopolistic character qualify the ideal properties its advocates attribute to it.

Similarly Bagnasco and Oberti (1998) were able to point to the existence of several worlds of production (see also Bagnasco, 1977; Storper and

Salais, 1997). One of these worlds was centred on the SME systems that grew from the 1960s onwards, mainly in parts of centre-northeast Italy outside of the main areas of modern industry. Characterized by strong trade, craft and sharecropping traditions, high rates of of family saving and a dense tissue of small towns, these areas saw a strong development, across three decades, of microenterprises in traditional industries, and sometimes of related machine tool industries. These SME systems were a result of several mechanisms: a residual mechanism involving the survival of SMEs in segmented markets where there was little competition from modern enterprises; a specialization mechanism involving concentration on traditional activities where scale economies were relatively unimportant; and a diffusion mechanism involving a decentralization of employment to escape social protection legislation and the industrial conflict characteristic of large plants. The world that resulted survived and prospered in the shadow of the Fordist industries in the northwest and the underdeveloped and subsidized south.

A second was the world of large-scale industries which were restructured in the 1980s and 1990s. Although this transformation enabled some of them to retain their economic importance, a sharp decline in direct employment reduced their direct impact on society and its class structure. In these sectors fixed capital investment remained large, requiring organizational stability and a foreseeable future. In the absence of these conditions, short-term profitability was difficult to achieve.

A third world was associated with a diversified service sector. This sector included traditional activities that were modernized but in which significant self-employment survived, and a more dynamic set of information, communication, entertainment, advertising, leisure, health, education and financial services. In these sectors organizational structures and economic logics differed profoundly in particular from the world of large-scale industry. Fixed capital investments were small, while production was of immaterial goods. Enterprises were flexible and could adapt rapidly to market fluctuations. For simple tasks, these enterprises employed part-time workers or staff on fixed-term contracts. For professional work, individualized semi-professional relations were preferred. In these circles, what prevailed was a market logic that prioritized short-term profitability and a culture centred on laissez-faire, individualism and speculation.

SME systems were in other words just one of several worlds. In the conditions prevailing in the 1970s and 1980s this world was one that achieved dynamic growth and played a leading role in sustaining Italian economic expansion. As a result, it did warrant special attention, but not to the exclusion of other worlds.

The importance of transcending existing concepts of the Three Italies stems not just from the existence of other worlds of production. As early as the 1980s, it was also evident that the development path of which flexible specialization was just a part was going hand in hand with sharp increases in territorial inequality. In Italy, after 1974, there was a reversal of earlier processes of convergence and catch-up, which could also be observed more widely in the capitalist world: successful regional economies were rewarded and strengthened, while unsuccessful economies were discouraged and relatively weakened, always on the grounds that the result would be a faster increase in aggregate wealth than a more equitable model of development would offer. Surprisingly, the geographical literature, perhaps due to its region-centred approach, had moved away from the study of unequal economic development, socioeconomic inequalities and the role of power relations in shaping the space economy. Within economics what prevailed were, until recently, expectations of convergence that were difficult to reconcile with the reality of widening inequalities. A major feature of the economic geography of Italy and the developed world was, accordingly, either overlooked or not explained.

The significance of the uneven development of different worlds of production has increased in recent years. One reason why is that, after three decades of growth, the relative economic performance of the Third Italy deteriorated sharply, as did that of Italy as a whole. At the root of these growth slowdowns were the interaction of specialization in the traditional 'Made in Italy' sectors, a changing external environment and changed conditions in areas of growth. As far as the external environment was concerned, three factors were important. First, globalization and political and economic transition in the former Communist world resulted in the emergence of new competitors. Second, the growth of demand remained weak. Third, the fixing of Italy's exchange rate relative to those of other Euro zone countries prevented further devaluations. As a result, one of the main mechanisms permitting the adaptation and development of the 'made in Italy' industries was closed off. Alongside these external factors there were several internal factors. First, the adaptive capacities of SMEs were limited by their small size, limited research and development and limited access to external finance. Second, congestion and other external diseconomies had a strong negative impact on company performance, while collective efforts to support innovation, training and knowledge-related activities were not sufficiently effective. Third, a number of social factors that had played a major role in aiding industrial development were under strain. One was the erosion of traditional family solidarities. Another was the expansion of education which made the young reluctant to follow in their parents' footsteps. As a result, there were labour shortages and problems of entrepreneurial succession. In these circumstances, the adaptation and survival of districts requires a profound transformation of the district model. At present, one of the main responses is for individual enterprises to delocalize tasks to CEECs and SEECs. As a result, the enterprises that lack the resources required to internationalize are abandoned. If, however, entire districts are to survive, what are required are the internationalization of entire SME systems and a strengthening of their presence on international markets (Bersani and Letta, 2004).

Accounts of the Third Italy should be seen, as they were in Bagnasco's study, as just one part of a more systematic account of the structure and development of the space economy. A more systematic account requires a consideration of the interdependence of multiple worlds of production and the relationships between development and underdevelopment. A view that we shall defend is, accordingly, that, instead of a single path of modernization and development, there are multiple, interdependent, changing and context-dependent pathways, and that, alongside mechanisms that equalize development, there are mechanisms that make it more unequal. More radical versions of this second thesis suggest that in capitalist societies development and underdevelopment are two sides of the same coin (Hymer, 1975). An implication of this second set of ideas is that the central object of analysis should be the shifting map of areal differentiation and the shifting mosaic of uneven development.

In recent years, a shift of attention has started to occur as the relative fortunes of different parts of Italy have changed. Amongst other developments, there was a renewed focus on the south. A number of new empirical studies concentrated on southern industrial districts. Centred on traditional craft activities, some of these areas achieved remarkable growth (see Cersosimo and Nisticò, 2001; Viesti, 2000; Meldolesi, 1996). To contest the simple opposition between a successful Third Italy and a uniformly stagnant south, economists, geographers and sociologists were able to re-emphasize the fact that the Mezzogiorno manifested striking internal differences. Areas of economic crisis and slow growth coexisted alongside dynamic industrial districts, specialized primarily in export-oriented manufactures (Trigilia, 1992). At the same time, some critical voices considered the enthusiasm that the identification of areas of growth in the south generated as excessive and as overlooking some of the critical problems southern development confronted. More generally it was clear that improvements in the relative economic performance of the south during the second half of the 1990s were due to the fact that national economic stagnation had a smaller impact on the Mezzogiorno than on the centre-north. Nonetheless, the Mezzogiorno is again a widely debated topic. As Petrusewicz (2001: 63) recently argued:

> The changing Mezzogiorno, the Mezzogiorno at work, Apulia's tiger economy, Basilicata's administrative efficiency, the cultural revival, the 'Bassolino phenomenon', Gioia Tauro harbour: these are the terms that begin, just

begin, to sketch out a new discourse on the Italian south. Southern municipal administrators – Antonio Bassolino of Naples, Gerardo Bianco of Catania, Leoluca Orlando of Palermo, Giacomo Mancini of Cosenza – are national figures and constitute a leading component of the new movimento dei sindaci (mayors' movement). The winner of the Nobel Prize for literature, Dario Fo, says he prefers Naples to Milan, because the former invests in culture, while the latter invests in the fashion business. . . .

The so-called new meridionalismo ('southernism') of the 1980s and 1990s, in an abundant and lively literature, has questioned many stereotypes and commonplaces in the Mezzogiorno's history, sociology, anthropology and economy. An even newer revisionist wave is provocatively tackling the questions of southern identity . . . , 'civicness' . . . , culture . . . , and economic dynamism . . . . Much of this scholarship makes use of new methodological approaches – orientalism, discourse analysis, deconstruction – contributing in the process to the rejuvenation of Italian social sciences.

A New Economic Geography of Uneven Development

Accounts of the Third Italy and of flexible specialization always were partial accounts of economic and social reality. These aspects of reality were, however, given special emphasis as they were thought to point towards a new economic order. The world that these discourses sought to explain has, however, changed, while some features of the old world including the combination of relatively slow growth and increasing inequality were overlooked. In this situation, the challenge that arises for geographical scholarship is to describe and explain the coexistence of sharp cyclical movements and relatively slow average growth, of remarkable economic modernization alongside relative stagnation, of shifting patterns of areal differentiation and of changing trends in social and territorial inequality in a changing international geopolitical context (for one attempt see Perrons, 2004). Involved is a shift in geographical perspectives on industrial change and regional development. Such a shift is made possible by the scope for synthesizing elements of different approaches that have emerged in recent years, in the belief that the complexity of the phenomena under investigation requires an approach that recognizes the complementarity of different narratives.

Accordingly, the first aim of this volume is to offer a way of synthesizing elements of existing narratives precisely to develop new ways of analysing the shifting geographies of economic and industrial performance. Empirically, we shall analyse the changing trajectories of Italian regional economies relative to the rest of the European Union, the changing position of Italian enterprises in European and international divisions of labour and their impacts on Italy's regional economies. The aim is to demonstrate that variations in regional development are the outcome of complex and geographically differentiated politicoeconomic processes. More specifically, our aim is to identify the connections between comparative regional development, the underlying territorial division of labour and the mechanisms that shape them. The approach adopted considers the firm as the central economic actor in explanations of industrial change. Considerable attention is also paid, however, to the impact of a firm's environment. One aspect of this environment is a firm's network relations with other firms. Networks exist at a regional level. In contrast to the new regionalism, however, the approach adopted in this volume pays greater attention to the embeddedness of firms in international networks. In addition, whilst emphasizing the fundamental profit-seeking nature of enterprises, it recognizes that an institutional environment also affects corporate economic performance, and that this institutional environment comprises not just other firms but also a set of institutional, political and cultural conditions (Usui and Colignon, 1996; Powell and DiMaggio, 1991).

A second aim of this volume is to suggest that what has to be explained is the coexistence of increases in wealth, shifting trends in inequality and profound and interdependent contrasts in territorial development. Commencing at a macroeconomic scale, the aim is first to measure growth and trends in regional inequality. These trends in growth and inequality are the aggregate result of the changing trajectories of Italy's regional economies and of their underlying development models. Making sense of the whole requires, therefore, a study of the parts, where the parts include not just the district and flexible specialization model of the Third Italy but also the mixed fortunes of Italy's metropolitan economies, the recent upturn in relative development, after years of falling behind, of the Italian south, and the striking internal differentiations within the south that separate growing districts from earlier poles of development. The next step, however, is to disaggregate what is happening at a regional scale and to examine the impact on relative development of the changing social and territorial division of labour and the changing position of Italian enterprises located in particular regional economies in European and international divisions of labour. Most attention will be paid to large industrial corporations and to capital- and technology-intensive industries, including the remarkable reshaping of Italy's economic landscape as a result of the privatization of the state holding companies (PP.SS). One reason why is that little recent attention has been paid to these sectors, although it must be emphasized that the examination of particular industries is designed primarily to illustrate a method of analysis of more general applicability.

Connecting microevolutions, regional development models and wider trends is important as it makes it possible to place the former in context and to reveal the variety that the latter conceals. This task is a difficult one.

There are, however, ways of making progress. What part does the evolution of an industry play in creating the aggregate wealth of a locality? That question can be answered by disaggregating aggregate trends. Aggregate trends in relative development are ultimately the sum of millions of often strongly differentiated, place-specific, microeconomic changes or constancies. Aggregate trends, reflected in aggregate data, are in other words the result of the underlying changes/inertia in the social and territorial division of labour that result from millions of actions of varying size and impact, although the speed and direction of change of the latter are affected by interaction effects and the system-wide results of adding up the parts. These underlying changes involve the destruction of the old, the rise of the new and the associated changes in social and territorial inequality. As we shall show, a new international division of labour is emerging. Old industries that paid high wages to relatively unskilled people are subjected to strong cost competition, move offshore or go to the wall. New industries and new technological and organizational models provide high incomes for symbolic analysts (Reich, 1991) most often in rich areas. New low-paid services create jobs for a large tranche of people (including immigrants) who work but are poor. To these changes one could add the erosion of the status and rewards of welfare professionals/public administrators, although it is not dealt with in this volume. To put it another way, trajectories of accumulation and their impact on the changing volume and composition of the demand for labour and the rewards paid for different kinds of work play a key role in explaining trends in comparative development.

After identifying the ways in which a particular sector is expanding or contracting, detailed analysis of its components can be carried out to see to what extent what is happening to the parts is the same as what is happening to the whole and can help us understand it. What are the parts? In this volume the parts are the enterprises. Instead of the territorial entities of the new regionalism, the point of departure is the profit-seeking firm. The approach will differ, however, from that of methodological individualism. Attention will be paid, insofar as the individual enterprise is concerned, to a combination of value chain approaches to upgrading and analyses of locational behaviour. Actions also depend, however, on the systems of which actors are a part. Emphasis will accordingly also be placed on the nature and role of the context of action set by interfirm relations and firm-political/cultural environment relations and governance mechanisms. Qualitative and quantitative case-study material from the ESRC research project and the authors' other research will help in this respect. As Sayer and Morgan (1985:153) argued, 'by looking at firms in ... causally relevant [contexts] ... and examining what they actually did, the logic or structure behind ... [seemingly] ... inexplicable patterns in the aggregate data becomes clear'. A similar point was made by Hudson (2003:745) when he wrote that 'the precise effects that . . . [broad structural relations] do or do not have are contingent matters and must be revealed by careful, theoretically-informed empirical research, typically via an intensive case study (of factories, firms, or regions, for example)'. In the spirit of these arguments this volume will show that the decision-making processes of companies depend on a complex interaction of principally economic but also political and cultural factors at a range of scales.

Although developed in relation to Italy-centred evidence, and concerned to provide new insights into Italian development, the result will be of wider significance. In particular, this volume aims to provide (1) a new way of conceptualizing industrial change and regional development, that synthesizes some of the major insights of recent research, (2) a way of connecting aggregate/meso trends with micro developments and (3) a framework applicable to the analysis of place-specific change in other parts of the world.

The Structure of the Book

After an account in Chapter 2 of recent theories of comparative development and territorial economic organization, Chapter 3 presents a conceptual framework for the analysis of regional economic performance and the changing social and territorial division of labour. Chapter 4 documents trends in economic growth and inequality in Italy. Trends in inequality have received far too little attention in the geographical literature on the new regionalism, yet increased inequality is paradoxically a major feature of the new economic geography. At this stage attention must also be paid to the changing contours of the Italian political scene and Italy's political economy that were instrumental in shaping recent trends in growth and inequality (Chapter 5). The next step is a disaggregation of these trends in order to identify the performance of individual regional economies and of some components of aggregate demand (Chapter 6) and the contribution to regional performance of particular industries (Chapter 7). An important aim of Chapter 7 is to start to explore the relations between regional performance and industrial/firm performance. To this end a disaggregation of regional performance into its sectoral components is complemented by an initial analysis of industrial performance and its regional impacts.

At this point, we had already started to concentrate on certain industries, regional economies and economic zones. The first decision was to use two industries to exemplify the approach we propose: the motor vehicle and chemical industries. The second was to examine their restructuring in four intersecting worlds: core EU Member States in which Italian goods and services are sold and in which Italian enterprises invest/acquire assets; the transition economies of which some are closely integrated into interregional and international divisions of labour and flows of goods, people and capital with Italy; the Mediterranean area with which Italy has close relations and which faces risks of marginalization as Europe's centre of gravity moves eastwards; and the wider world where relevant, as for example in the globalization strategies of Italian corporations. The third was to concentrate an analysis of territorial impacts on (although not exclusively) the regional economies of Lombardia, Piemonte, Puglia and Basilicata.

The reasons for choosing these industries were threefold. The first was a desire to concentrate on economic activities that serve wider markets rather than on the recycling of income within regional economies. The second was to choose sectors that are redeploying at an international scale. The third was to choose sectors in relation to which the authors have detailed survey and interview data. Insofar as the approach is fruitful, it can be extended to other sectors. As far as the regions are concerned, the choices were a result of three considerations. First, the volume is concerned with the new inequalities generated through the integration of the most and the least developed parts of Italy into the international division of labour. Second, these areas have received scant attention in the recent past, in spite of the fact that the two northern metropolitan regional economies account for a large share of Italy's economic activity.

The next step is the development of explanations of the restructuring of our chosen industries with a view to explaining those aspects of the territorial division of labour that make significant contributions to regional economic performance. To explain these aspects of the territorial division of labour and their impact on inequality we examine, in Chapter 8 (motor vehicles) and 9 (chemicals), corporate profit, upgrading and internationalization strategies of Italian enterprises. The central argument is that accumulation is leading to changes in the demand for labour. In core countries/regional economies these changes reduce the demand for unskilled labour and raise the wages of symbolic analysts. In areas into which economic activities are delocalized there are increases in the demand for relatively unskilled labour. These trends have sharply differentiated regional consequences, adversely affecting for example places in which old industries that paid high wages to relatively unskilled people are destroyed and advantaging the new centres of the knowledge economy which includes core strategic functions in old and new industries. These and other conclusions are summarized in Chapter 10.

Chapter Two

Convergence, Divergence, Regional Economic Performance and the New Economic Geographies

Analyzing regional economic performance

Analyses of regional economic performance tend to fall into two groups. In the first there is the quantitative economic geography research which seeks to measure and identify the sources of differential regional performance (see, for example, Dunford, 2002; O'Leary, 2001; Fagerberg, Verspagen and Caniëls, 1997). In the second there are qualitative studies of regional economies deemed successful to different degrees (such as Cooke and Morgan, 1998; Dunford and Hudson, 1996). To the latter one can add the more quantitatively-oriented research of some political scientists interested in the impact of institutional performance and the character of civil society on the performance of regional economies (see, for example, Putnam, Leonardi and Nanetti, 1993).

As far as possible, studies of regional performance should involve measurement. Measurement is not, however, unproblematic, as it requires the identification of sensible territorial units, and the existence of accurate estimates of meaningful indicators. Questions about regional performance also imply, however, a concern with the qualitative characteristics of places, and an assessment of whether places, and the ways of work and life of their inhabitants, are getting more similar, whether differences are reproduced, or whether new differences are created. In this case, the major difficulty lies in deciding what does and does not matter in the face of the impossibility of systematically enumerating every facet of every regional economy.

In this volume the starting point is GDP, as GDP is a measure of the new wealth and income created in a particular territory in a particular period. (Note however that wealth is a means to an end rather than an end in itself. The real aim of development should be to improve human well-being, which itself depends on the distribution of wealth and income and the ways in which wealth and income are used). A community's GDP can be interpreted and measured in three ways:

1   From the point of view of production, GDP is a measure of aggregate value added:

$$Gross\ Domestic\ Product \equiv \sum_{i=1}^{n} Value\ added_i \qquad (1)$$

Geographically it is a reflection of the geography of the production of goods and services for regional and wider markets.

2   As the new wealth created in production is distributed to those who participate in economic life, GDP is also a measure of primary incomes. From this distributional perspective, GDP is divided into profits (comprising the income of business, interest and rents paid plus depreciation) and wages paid to regional and extra-regional subjects:

$$Gross\ Domestic\ Product \equiv Gross\ operating\ surplus + Compensation\ of\ employees$$

$$\equiv \sum_{i=1}^{n} Gross\ operating\ surplus_i$$

$$+ \sum_{i=1}^{n} \sum_{j=1}^{m} Employment_{ij} \times Average\ wage_{ij} \qquad (2)$$

where there are $i = 1, 2, \ldots, n$ industries and $j = 1, 2, \ldots, m$ occupations. Of these wages and profits, a part accrues to local residents and a part to nonresidents. In this formulation, regional GDP and its evolution depend on three elements: the distribution of income between wages and the gross operating surplus; regional profiles of employment by industry and occupation; and the associated distribution of wages by industry and occupation.

3   The primary distribution of income is subsequently modified by state-administered redistribution. The result is the secondary distribution of income. These secondary incomes are saved and/or spent, and thereby permit the establishment of claims over the goods and services produced. Measurement of these expenditures (conventionally divided into consumer and government expenditure, investment/savings and net exports) offers a third way of measuring GDP.

In this volume most attention will be paid to the production and distribution of new wealth and income and the underlying profiles of employment and earnings.

In a proximate sense, an area's wealth and income depend on the interaction of a four sets of factors.

1   The first is its natural and created resource endowments. Included are its natural resources, infrastructures, population, skills, degree of development of the forces of production, etc.. An area's resource endowment is therefore **a cause of development**, specialization and the division of labour.

2   The second is (a) the degree of mobilization of an area's resources and human potential and (b) the effectiveness, efficiency and productivity with which they are used. In turn these drivers depend on:

   i   growth and development strategies of enterprises and other economic actors, and the consequent development of learning, innovation and technological and productive capabilities,
   ii   institutional structures, capacities and performance, systems of regulation, and social relations of production (including relations of trust), and
   iii   the wider national and international context and the rules of the game: in a cold climate (a zero-sum game) the most successful prosper, but, even if everyone emulates the winners, not all can win; in a warm climate (a positive sum game) the opportunities for all are greater.

3   The third is the extent to which an area's inhabitants can command and control the resources of others, the degree of unequal exchange and the extent of transfers of wealth.

4   As resources are created, conserved and reproduced in and through development, **resource endowments are finally also a consequence of development**: many of the resources on which development depends are endogenous rather than exogenous, are a result of growth and investment and are greatest in areas that are already most developed. Development processes are therefore cumulative.

These propositions derive from the existing literature. The aim of this chapter is to review this literature and to identify contributions to explanations of differences and trends in regional value added, earnings and employment. The chapter will start with a brief examination of the role of convergence (centrifugal) and divergence (centripetal) tendencies in shaping the relative performance of regional economies. In subsequent secetions the connection between quantitative assessments of comparative regional economic performance and the changing territorial division of labour, implicit in Equation 1, will open the path to an identification of the contributions of the new economic geographies of the geographers (the new economic geography) and the economists (the new geographical economics) to the study of location and the territorial division of labour.

Convergence or divergence

Economic analyses of regional performance and growth are concerned with regional differences in the new wealth and income produced in a particular period, and, especially, with the way in which comparative levels of development change over time. In a political context these analyses are also concerned with where regional economies are heading and with the identification of actions that have shaped their past trajectory, determine their current evolution and will permit them to arrive at a preferred future state (Fagerberg, 2003).

Neoclassical convergence models

Theoretical models of relative regional economic performance produce conflicting answers. In the recent period of hegemony of neoliberal economic ideas and programmes, models that predict catch-up have predominated, especially in discussions of globalization. The expectation in this case is that convergence will result from the alignment of technologies, lifestyles, consumption patterns and government policies, in part as a consequence of the discipline exercized by global financial markets and dominant economic and political interests.

The theoretical rationale for these expectations is provided by a range of neoclassical growth, regional development and trade models. These models all rest on assumptions of (1) constant returns to scale, implying that costs of production do not change as the volume of output changes, (2) perfect competition, with its implication that the economy is made up of small enterprises none of which can have any impact on market prices, and (3) instantaneous re-employment of resources, implying that no resources remain unemployed.

The first is the Solow (1956) growth model, and the growth accounting exercises to which it gave rise. This model rested on the idea that regional economic performance depends on (1) growth in the volume of capital and labour inputs, of which an area's resource endowment is made up and (2) increases in productivity, due to technical change. The Solow model predicts that the growth path of a country or region will converge on a steady state. As long as labour is mobile (or regional rates of population growth are identical) and technological knowhow is public, this steady state growth rate is the same for every country or region, while in this steady state per capita variables such as per capita output do not vary. According to these models, the existence of similar conditions in the economies under

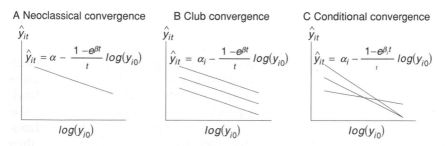

Figure 2.1 Convergence in reduced form neoclassical growth models.

consideration would result in a convergence in per capita income. The reason why is that economies that are less developed are expected to grow/approach the steady state faster than economies that are more developed, while in the steady state per capita variables do not change.

Empirically this claim is associated with the β convergence hypothesis (Barro and Sala i Martin, 1992). According to this hypothesis, the slope parameter, $(1 - e^{\beta t})/t$, of regressions of per capita GDP growth in t years, $\hat{y}_t$, on the logarithm of initial GDP per head, y_0, will be negative (Figure 2.1A). As Figures 2.1B and 2.1C suggest, and as we shall indicate, this strong claim was later replaced by alternative theories of conditional and club convergence that predict cross-sectional heterogeneity.

Two factors underpin the expectation of faster growth of less-developed economies. The first is the existence of diminishing returns to capital and constant returns to scale. At any given level of technological development, these conditions imply that: (1) increases in per capita output get smaller as the amount of capital per worker increases, (2) increases in productivity cease once a steady-state/equilibrium amount of capital per worker is reached, (3) increases in per capita output associated with incremental investments are greatest in areas/enterprises that are the least mechanized and (4) countries that differ only in initial levels of per capita GDP will converge on the same level of GDP per capita. The second is the view that technology is a public good available at no cost to everyone, and that the diffusion of technology and of knowledge from advanced to less-developed enterprises and areas will close technology and productivity gaps.

The second set of models are neoclassical models of regional economic development. These models suggest that the flow of capital to low income areas and of labour to high income areas will result in an equalization of income per head (Borts and Stein, 1964). In this case the idea is that the movement of labour from low-to high-wage areas will raise wages in the former and reduce them in the latter. Capital conversely will move in the opposite direction as the rate of profit is expected to be higher in

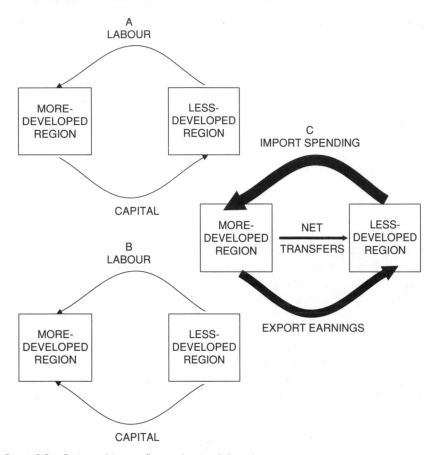

Figure 2.2 Factor and income flows and regional dynamics.

less-developed areas (Figure 2.2A). Movement will continue, it is argued, until there are no further differences in rates of return in different areas.

As with growth models, the assumption of constant returns to scale is critical. If, instead, increasing returns prevail, the returns to capital may well be greater in more than in less-developed areas, and any net flow of capital may take place in the opposite direction from less to more-developed areas (Figure 2.2B). Indeed, precisely this argument was one of the main strands of alternative theories of circular and cumulative causation (Kaldor, 1973; Myrdal. 1957). According to these theories, comparative development depends on the relative weight of unequalizing centripetal forces (increasing returns and backwash effects) and equalizing centrifugal forces (spread effects).

Criticisms of these neoclassical models are not confined to their account of the nature of market mechanisms but also point to other mechanisms that they overlook. Any redistribution that does occur may result, for example, not from market forces but from net flows of public expenditure not considered in these models and put in place specifically to counteract the unequalizing effects of market mechanisms (Figure 2.2C).

The third set of models are neoclassical factor proportions explanations of specialization and trade. These models predict an equalization of factor prices independently of factor mobility (see Armstrong and Taylor, 2000: 122–31). According to these models, areas specialize in activities that make intensive use of resources that are relatively abundant. Increases in the demand for relatively abundant resources and reduced demand for relatively scarce resources will raise the incomes of the former and reduce the incomes of the latter.

One of the difficulties with these models that all provide strong support for market ideologies is that empirical research has lent little support to their expectations of an equalization of development (see for example the Italian evidence presented in Chapter 4). Empirical research into the convergence hypothesis shows, for example, that there are differences in regional steady state growth rates and permanent differences in income per head. As a result alternative theories of conditional and club convergence (Figure 2.1B and C) were introduced. In the case of club convergence differences in performance are considered to stem from the fact that economies fall into different groups/clubs because of differing initial conditions, while conditional convergence models attribute differences to the cross-area heterogeneity of control variables such as rates of accumulation or the quality of modes of governance. Giannetti (2002), for example, shows that technology diffuses but the capacity of economies to benefit from technology spillovers depends on their initial specialization, skills and knowhow.

Cumulative causation and endogenous growth

The next step was the development of new models of endogenous growth. These models led to radically different conclusions, although the conclusions were not new, as they had been anticipated in earlier theories of cumulative causation. Associated principally with Rohmer (1990), endogenous growth models made technical progress endogenous and suggested that improvements in knowledge would lead to a continuation of investment and growth in the areas that are most advanced, helping to explain why economically less-developed areas fail to catch up (for a clear account see Fagerberg, 2003). These models rest on the idea that learning and the development of successive generations of equipment and goods and services are associated with cumulative improvements in human know-ledge and technology, and with cumulative increases in human skill and knowhow. At first these improvements were considered to amount to spillovers that led to increasing returns to scale at the level of the economy but constant returns at the level of the firm. As a result economists could continue to assume that firms operated as if they inhabited a world of perfect competition. At a later stage allowance was also made for econ-omies of scale at the level of the firm and for imperfect competition. This change reflected a final recognition of the fact that firms that carry out organized research and development can recoup the costs incurred by securing a temporary monopoly and restricting temporarily the diffusion of the innovations that result.

According to these models, growth can stem from societal spillovers associated with the general advance of knowledge, and from the results of partially appropriable research and development. An important feature of this approach is the reconceptualization of the nature of knowledge. First, it is recognized that knowledge is a nonrival good that can be used simultan-eously by different people, and over and over again by the same people at close to zero marginal cost. Second, it is recognized that knowledge is a partially excludable good. Third, investments that increase human know-ledge yield increasing returns at the level of the firm. There are two reasons why. First, in so far as enterprises can limit the diffusion of new ideas, they can spread the fixed costs of developing new products and technologies over a large volume of sales, yielding strong decreases in average costs as output increases. Second, investments that increase knowledge nonetheless generate strong externalities. The reason why is that knowledge cannot be perfectly patented or kept secret (and is therefore only a partially excludable good): once it is known that something can be done, other firms can seek to replicate it, so that new knowledge has a positive effect on the production possibilities of other enterprises.

Applied to the study of comparative development, this movement away from constant returns to scale and perfect competition to a world of increasing returns and imperfect competition can imply cumulative caus-ation and divergence rather than convergence, as new investments in places and enterprises that are already advanced create new development gaps (see Dunford, 2002, for a fuller account).

A particularly striking application of these ideas is found in the work of Quah (1996; see also the discussion in Perrons, 2003). Quah predicts the emergence of twin peaks in the distribution of regional income/household income in what he calls weightless (superstar) economies in which value is embodied in immaterial things (see Figure 2.3 which portrays the transition from an economy in which there are a lot of people with middle incomes and few rich and few poor people to a twin peak model). One cause of the

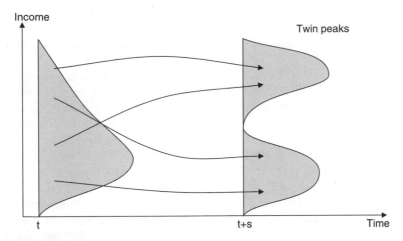

Figure 2.3   Twin-peaks.
Source: Quah, 1996.

twin peak model lies in the supply side effects identified earlier. For Quah, in weightless economies ideas are themselves commodities. The immaterial goods that result are distinct in that they do not have to be transferred from one person to another but are simply replicated or copied at a marginal cost that is close to zero (and where, if equipment is required to use these goods, the upfront costs are small). Alongside this supply side factor there are two demand-side factors. First, the market for many immaterial goods is extremely large and often global in character (lifting the constraints on the division of labour posed by the extent of the market). Second, consumers often prefer the products/services of the famous to those of people who are less famous but whose talents, skills and abilities may differ only marginally from those of the most famous. The combination of a very high level of demand for the goods/services offered by superstars and their very low costs of reproduction creates very high incomes in a world of winners and losers and wide inequalities. These mechanisms clearly have an impact on territorial development as immaterial goods and services are produced somewhere just as the people whose incomes are derived from them must live and work somewhere. What is more, Quah also suggests that the twin-peak model may be applicable to regional economies (see Perrons, 2004).

As is clear, these competing models lead to the identification of two sets of mechanisms. On the one hand, there is a set of centrifugal forces, of which one of the most important is the transfer of technological and organizational knowledge to less-developed areas or less advanced enterprises. On the other hand, there is a set of centripetal forces that arise as enterprises and areas that are developed create new sources of competitive advantage through, in particular, further investments in knowledge and skills. At any point in time the relative performance of different regional economies depends on the relative weight of these two sets of forces.

These developments in economics were reflected in the increasing concern in the geographical literature with the role not of resource endowments but of knowledge and innovation (Figure 2.4), although the source of inspiration in geography was more often the approach of evolutionary economists such as Schumpeter and Freeman. Geographers increasingly argued that regional growth depended on innovation and learning, on the existence of policies that support investment and raise research and development especially in technologically progressive sectors, and on the diffusion of knowledge and its determinants. A great deal of emphasis was placed on the degree to which the speed and extent of diffusion (whether through emulation, social interaction or technology transfer) is hampered by distance, encouraged by proximity, constrained by national frontiers, perhaps to the disadvantage of small countries, and shaped by the impact of institutional and cultural factors on absorptive capacities. A limitation of this research was however an excessive concern with the role of technologically progressive sectors and small start-ups, and too little concern with the outcomes of the research and development activities of large groups and the role of large enterprises in developing industrially innovative products and processes.

Territorial Divisions of Labour

The changing map of development is a reflection of the changing geography of the production of goods and services and the changing distribution of earnings and income. Explaining the changing map of relative development therefore implies explaining what economic geographers often call the territorial division of labour. In our view the territorial division of labour identifies who does what, where and when, what rewards they receive and in what relationships they stand to other people and economic activities in other places.

Geography and management studies

In the past geographical analyses of the territorial division of labour tended to concentrate on abstract models of economic landscapes. Examples include the work of von Thünen, Weber, Hotelling, Christaller and Lösch (for a summary, see Haggett, 2001). Subsequently, more attention was paid

Table 2.1 The hierarchical division of labour within the multinational corporation and uneven territorial development

Level	Function	Locational requirements	Locations
1	Strategy	Close to capital markets, media and government	World's major cities: New York, London, Paris, Berlin, Tokyo, Moscow, Beijing
11	Coordination	Close to sources of white-collar labour, communications and information	Large cities, regional capitals.
111	Day to day operations	Responsive to the pull of labour, markets and materials	Widely dispersed

Source: based on Hymer, 1975.

to dynamic concepts of space and place with the development of core-periphery and growth pole theories (Perroux, 1970; Boudeville, 1966). In the 1970s the emphasis was on conceptually informed classifications (see Table 2.1 and Massey, 1979; Lipietz, 1977; Hymer, 1975). The merit of the latter approach lay in the elaboration of historically-specific generalizations and in the concern, also found in the core-periphery tradition, with explaining differences in levels of development and relative development trajectories. Conversely, the advantage of the earlier locational tradition is that it considered more systematically, though incompletely, the factors that explain the character of economic landscapes.

In the more recent past, there has been a remarkable diversification of approaches involving work in geography, institutional economics, sociology and to a lesser extent political science. The result is a rich array of conceptual models designed to characterize geographical systems of production, explain their structure and account for their relative economic performance. A fundamental aim of this literature is to identify and explain the characteristics and the diversity of economic landscapes and their evolutions, drawing largely on qualitative, case study methodologies. What has resulted is the identification of a range of spatial categories and a number of core causal mechanisms (Figure 2.4).

Amongst the former are: industrial districts (Becattini, 1991; Marshall, 1961); areas of specialized production, local productive systems and system areas (Garofoli, 1991); localized industrial systems (Courlet and Pecqueur, 1992); clusters (Porter, 1990); new industrial spaces (Scott, 1988); technopoles and science parks that comprise innovative enterprises, research centres and universities; worlds of production (Storper and

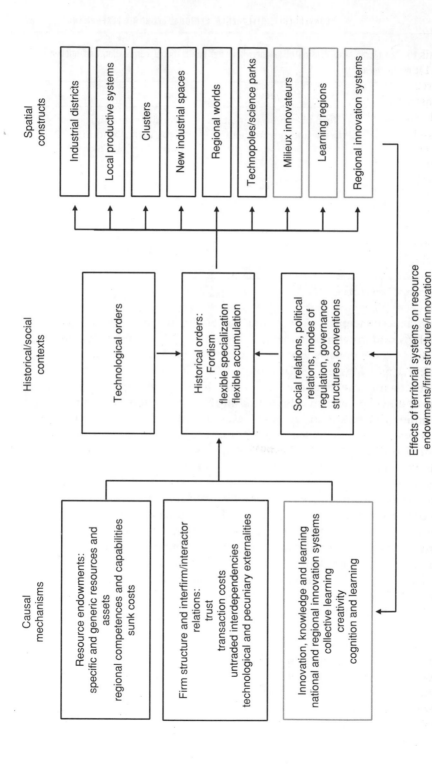

Figure 2.4 Spatial structures and causal processes in the new economic geography.

Salais, 1997); milieux innovateurs (Aydalot, 1986; Crevoisier and Maillat, 1991); regional innovation systems (Cooke and Morgan, 1998); and learning regions (Maskell and Malmberg, 1999; Hudson, 1999; Morgan, 1997). Industrial districts are, for example, areas with a large number of specialized SMEs in a single industry. Garofoli's categories denote industrial districts whose development, while externally-oriented, is increasingly self-centred, as opposed to extraverted. As one moves from one category to the next, development is increasingly determined, therefore, by endogenous as opposed to exogenous variables. For Courlet and Pecqueur local productive systems can comprise large and/or small enterprises, and are not confined to a single sector, yet involve specialization, dense interfirm relations and a synthesis of market coordination and cooperation. Cooperation, in their view, is important and involves partnership, joint development, strategic alliances, and joint ventures. Scott's new industrial spaces are a consequence of a spatial reagglomeration of economic activities as a result of a transition to a new model of development. This new model of development is characterized by two trends. The first is vertical disintegration, which is itself due to the interaction of scale and scope economies and the structure of transaction costs. The second is spatial agglomeration, which is a result of the impact of spatially dependent transaction costs and external economies. (Economies of scale exist if costs diminish as the volume of output increases, while economies of scope[1] exist if the joint costs of providing related products and services are smaller than the costs of providing these products and services separately).

To explain these spatial structures, and to explain their reciprocal impact on development, three main groups of causal mechanisms are examined, with the relevant mechanisms varying across the different spatial constructs. First, there are theories rooted in a traditional concern with resource endowments that emphasize the role of generic and specific assets, the competence and capabilities of an area's inhabitants and the role of sunk costs (Courlet, 2001; Pecqueur, 2000; Clark and Wrigley, 1997). In many of these cases the emphasis is on the development of the milieu rather than on the development of enterprises. Second, there are theories that emphasize the role of industrial organization, industrial strategies and interfirm relationships. At the centre of these approaches are emphases on the impact of externalities and spatially- and nonspatially-dependent interfirm and interagent relationships, with authors drawing on neoclassical theories of transaction costs (Scott, 1988), or concepts of untraded interdependencies (Storper, 1995), trust (Humphrey and Schmitz, 1998), network organization and governance, institutional thickness (Amin and Thrift, 1994) or the degree of local rootedness/embeddedness of social interaction. These last four themes also reflect a concern, characteristic of the first approach, with the features of the local milieu in which interagent relationships develop. To them one can add the more quantitatively-oriented research of some political scientists interested in the impact of institutional performance, the character of civil society and social capital on the performance of regional economies (see, for example, Putnam, Leonardi and Nanetti, 1993). Third, there are theories that emphasize the role of innovation, systems of innovation, knowledge, individual and collective learning and creativity.

As a result of interpretations of theories of regulation, institutional economics, or evolutionary sociology, these causal mechanisms are seen as operating in a series of specific historical contexts and as socially mediated elements of a range of historical development models (Dunford, 1990; Peck and Tickell, 1995). These contexts and models derive from a synthesis of changing technological, economic, institutional and cultural factors. Storper and Salais' (1997) regional worlds are, for example, possible action frameworks, comprising economic and noneconomic forces such as institutions, cultures and social practices that take the form of conventions, defined as largely implicit and routinized rules of action designed to deal with coordination problems.

A limitation of this rich, recent geographical literature is that relatively little attention is paid to the analysis of firm strategies as a whole or to the systemic processes or structural factors that affect the behaviour of capitalist enterprises. This omission is in part the outcome of the fact that these approaches were themselves developed as alternatives, rather than as complementary, to the political economy tradition (Hudson, 2000; Massey, 1995; Lipietz, 1977) in which this volume is rooted. The framework that we shall develop, which will give a greater degree of emphasis to the profit and upgrading strategies of enterprises, will however seek to integrate the insights that recent work has provided into the roles of industrial organization, innovation and the milieu in which economic activities develop.

Alongside this geographical literature, there are several other approaches. In management studies, for example, Porter (1990) argued that neoclassical factor endowment theories are too simple. In seeking competitive advantage, he argued, national and regional economies can actively create sophisticated resource endowments. According to his diamond model, this created competitive advantage depended on four interacting factors: firm strategy, structure and rivalry; factor conditions; demand conditions; and related and supporting industries. A subsequent five forces model was developed to explain the structure and trajectory of an industry in terms of the interaction of rivalry amongst existing enterprises, the power of buyers and suppliers, and the threats posed by new entrants and substitute goods or services (Porter, 1998).

The new geographical economics and territorial divisions of labour

More recently, there has been renewed interest in comparative regional economic development and in economic geography amongst economists. In this case, there are two starting points. The first is the theory of economic growth and its application to the question as to whether areas that are economically less developed will catch up with areas that are economically more advanced. The second is the new geographical economics. Essentially, the new geographical economics is centred on the theory of the location of profit maximizing firms often in a world of increasing returns and imperfect competition (Fujita, Krugman and Venables, 1999). The aims of the new geographical economics are severalfold. First, it seeks to explain the geographical concentration of the activities of individual enterprises, the agglomeration of different economic activities and their consequences for the distribution of the shares of aggregate economic activity across different localities. Second, it aims to explain geographical specialization, which refers to the fact that within certain localities some activities are relatively strongly represented. Third, it examines the impact of economic integration and globalization on the territorial distribution of economic activities and on core-periphery structures. The means are the construction of artificial models, drawing on analyses of the impacts of increasing returns, forward- and backward-linkages, external economies, endogenous growth, product differentiation, external diseconomies, etc. to generate imaginary economic landscapes. A further aim is to compare these imaginary worlds with the real world with a view to arguing that correspondence between them is indicative of the identification of the generative mechanisms. The methodological approach is similar. therefore, to that of earlier location theorists such as von Thünen, Weber, Lösch and Christaller and the subsequent quantitative research tradition that dominated urban and regional geography in the 1960s and early 1970s (Martin, 1999).

At the root of the new geographical economics was the development of tractable models of competition in a world of increasing returns. What made these models possible was the development by Dixit and Stiglitz (1977) of a formal model for the theory of monopolistic competition. The theory itself was developed originally in the 1930s by Robinson and Chamberlain. In a world of monopolistic competition firms can set their own prices. Firms are faced, in other words, with downward-sloping demand curves, such as the curve DA in Figure 2.5, and can increase sales by reducing prices. Marginal revenue (MR) is therefore less than price. To maximize profits, firms select a price (pA) and output rate (qA) for which

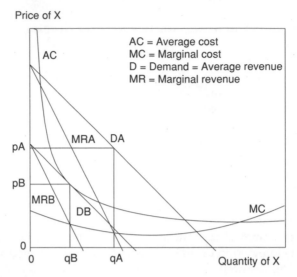

Figure 2.5 Monopolistic competition.

$MC = MR$. At this profit maximizing position, price exceeds marginal costs (MC), giving firms an incentive to try to sell more by shifting the demand curve to the right, perhaps through advertising. In Figure 2.5, the firm is on the downward sloping part of its average cost curve, so that it has excess capacity, in the sense that output is not sufficient to minimize average costs. Total revenue, equal to pAqA, exceeds total costs, so that the firm earns excess profits. Such firms have an incentive to prevent the entry of rivals. The existence of high profits will however encourage other firms to enter the market, perhaps to make alternative varieties of the product. If other firms do enter, they will reduce the market share of existing firms: as new firms enter, the demand curve for the initial firm in Figure 2.5 moves leftwards, and, it is assumed, will continue to move leftwards until it reaches DB where it is tangential with the firm's long run average cost curve, at qB and pB. At this point $MC = MR$ and there are no excess profits.

The Dixit-Stiglitz model was the starting point for the increasing returns/imperfect competition revolution in economics. Four developments were involved: new theories of industrial organization; new trade theory; new growth theory; and, finally, new geographical economics theories.

In new geographical economics theories, the spatial organization of production is a result once again of the interplay of centripetal and centrifugal forces. Centripetal tendencies depend, first, on the interaction between scale economies, transport costs and the size and relative location of output and input markets (Figure 2.6, in which the dots denote people, there are three goods and services denoted by the shaded triangles, squares and hexagons, with the hexagons representing a consumer good, and the scale of production is represented by the size of the shape).

Essentially, footloose consumer good manufacturers wish to concentrate production in a few locations because of scale economies at the plant level even if the distribution of population is relatively even, yet seek also to locate these plants near their customers because of transport costs (Figure 2.6A). As specialization and the complexity of the division of labour increase, access to customers and suppliers is greatest where other firms locate because of market-size/home market effects, though the strength of this centripetal force will vary with the significance and density (measured perhaps by the share of intermediate goods in production) of forward and backward linkages with other firms and industries (Figure 2.6B).

A second set of factors that encourages agglomeration is the development of a wide range of nonpecuniary (technological) external economies (non-market size effects) found in existing agglomerations (Figure 2.6B). Emphasized in the literature on industrial districts rather than geographical economics models, these technological externalities include: scientific, technological and knowledge spillovers, often as a result of personal interaction; spillovers from local public goods; dense and sophisticated labour markets; and the fact that the presence of one producer has the effect of directly reducing the costs of other producers, as a result, for example, of the ability of producers to share infrastructures and specialized providers of inputs.

A third set of factors that intervene to shape the process of agglomeration/concentration is the mobility of labour and differential population growth which affect the distribution of the population and the geography of expenditure (Figure 2.6C). As agglomerations emerge, for example, and as the geography of job opportunities and the range of goods and services/quality of life change in their favour, people move to them. Mobility alters the distribution of expenditure and market demand, adding further to the size of the market in areas of net population growth (the expenditure shifting aspects of migration). At the same time, it increases the supply of labour, checking the growth of wages (the production shifting aspects migration).

Concentration is not just a result of scale economies. If plant size is held constant, agglomeration can occur through changes in the number of plants. An increase in the number of plants will occur if there are a number of varieties of a particular good or service and if, in this situation, two conditions prevail. The first is that all varieties of differentiated goods and services are demanded. The second is that there are substitution effects, as consumers choose between domestically produced varieties and more expensive imported varieties, encouraging local production. In large markets

A Dispersed rural population, yet scale economies lead to relatively few production sites for industrial goods: how few depends on the trade-off between scale economies and transport (iceberg effect) costs.

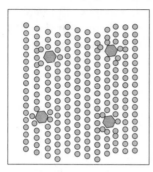

Average cost

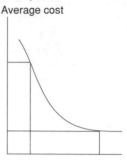

Quantity of good produced

B Specialization, division of labour and colocation near suppliers and customers. External economies also encourage proximity.

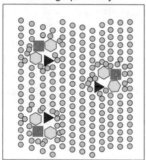

C Mobility reinforces agglomeration as does market access: suppliers of differentiated products in dense markets.

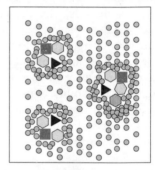

D Differential transport costs lead to relative growth of increasing returns activities and population in transport hubs.

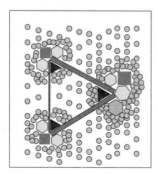

E Endogenous growth/accumulation and competition differentially increase/ decrease size of economic activities. Dispersion due to diseconomies/reduced transport/factor costs.

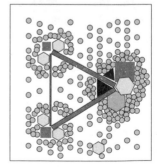

Figure 2.6 Explaining economic landscapes.

the share of imported varieties is smaller, and so there is greater substitution in favour of domestically produced varieties (Figure 2.6C).

A further mechanism derives from a set of self-reinforcing market access advantages stemming from the differential development of transport networks. In Figure 2.6D the width of the lines is inversely proportional to transport costs. Costs are low on wide routes making the eastern city a transport hub, as it is cheaper to get from it to the other two cities than it is to travel between them. One reason why might be the coexistence of small differences in traffic and either increasing returns in the transport sector or differences in the quality of intercity transport infrastructures. Other things being equal, the hub will be more attractive to increasing returns activities than the other two cities, setting in motion a self-reinforcing process: a city in which economic activities are concentrated will occupy a central position in the transport network reinforcing its advantages as a production location and creating a core-periphery pattern (Krugman, 1998).

All of these factors (the interaction between scale economies, transport costs and the size and location of output and input markets; nonpecuniary external economies; the mobility of labour; the manufacture of different varieties of goods and services; and the impact of transport networks) help explain concentration, mainly through their impact on the location of factors of production and the nature and differentiation of market demand. To them should be added the dynamic effects of their interaction and the ways in which accumulation and growth reinforce the cumulative process (Figure 2.6E). Countries, regions and cities with large markets invest more, further enlarging those markets, while investment in research and development and in people augment the assets and resources that underpin spatial concentration. Spatial concentration is in other words also a consequence of mechanisms of endogenous growth discussed earlier (Krugman, 1998), while endogenous growth and in particular the innovations it generates can reinforce agglomerations (see, for example, Martin and Ottaviano, 2001).

The strength of this circular and cumulative agglomeration process will depend on the relative importance of these centripetal forces and a set of countervailing centrifugal forces. The latter stem from three sets of factors. The first is the relative immobility of land-based agricultural activities, natural resources and the population they support. This immobility creates an off-setting incentive to locate establishments in these areas where there are few local competitors. The second is the ordinary operation of factor markets: wages and rents are higher in the neighbourhood of existing centres of economic activity; high costs can encourage activities to locate in areas where costs are lower. The third is external diseconomies (Figure 2.6E).

There are several reasons why these models of cumulative regional economic development are important. The first is that they offer explanations (scale economies, market size effects, the density of forward and backward linkages, externalities, differential transport investment and mobility) of the geographical concentration of industries and of the agglomeration of economic activities and people. Agglomeration is an important feature of economic landscapes. Explaining why agglomerations occur, what their characteristics are, how many agglomerations there are and where agglomerations are located is therefore an important task.

Agglomerations differ also, however, in their per capita incomes. A second contribution of these and related models lies in their contribution to explanations of these differences. The identification of inequality-generating mechanisms derives, for example, from the integration of these ideas with ideas about endogenous growth. If growth, for example, depends on the accumulation of knowledge and skill, if income depends on the embodiment of knowledge in economic activities, and if the effective use of knowledge depends on the past development trajectories of a locality and its neighbours, income differences can emerge. Moreover, if richer areas find it easier to re-equip and to invest in new knowledge, differences could increase.

A third contribution lies in attempts to explain the impact of globalization and integration on industrial specialization and location. This contribution is related to the second. The reason why is that it also deals with the evolution of inequality and the rise and fall of core-periphery structures. In these models globalization and economic integration are understood as involving the reduction of transport and trade costs (Krugman and Venables, 1996). Suppose initially that trade and transport costs are very high. In these conditions centrifugal forces are stronger than centripetal forces, and economic activity will tend to be dispersed, as producers locate near their markets. Consider the simple two region model represented by the so-called tomahawk diagram in Figure 2.7 (in which solid lines are stable and the broken lines unstable equilibria). At the right hand end of the horizontal axis denoting high trade and transport costs the share of industry in each of two regions is 0.5. As trade and transport costs decline, so does the relative magnitude of the centrifugal forces, until a break point is reached at which agglomeration forces outweigh centrifugal forces, industrial activities concentrate in one of the two regions, and the initial symmetry is broken. If trade and transport costs continue to fall, the agglomeration will survive until a critical sustain point is reached, at which centrifugal forces would once again outweigh centripetal forces, and economic activities would disperse, in this case as distance is virtually irrelevant in cost terms. A striking feature of this model is the fact that, once established, agglomerations are usually able

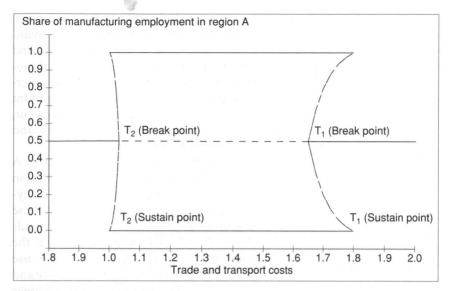

Figure 2.7 Core-periphery bifurcations.

to survive even under conditions, represented by trade and transport costs that lie between the break and the sustain points, that would not cause them to form in the first place. A second feature is the existence of bifurcations with qualitative behaviour changing abruptly when the quantitative balance of forces passes some critical level. (In the real as opposed to this fictitious world, of course, there are high degrees of inertia, due to role of sunk investments that generate enduring differential advantages/disadvantages).

The implication of this argument is that there is an inverted U-shaped relationship between integration/globalization and agglomeration. Initial reductions in high trade and transport costs cause agglomeration and reinforce core-periphery structures. Once trade and transport costs are fairly low, further reductions cause dispersion, initially of industries that are relatively labour-intensive and have relatively weak interindustry linkages. To put the point another way, reducing the costs of transactions between core and peripheral regions gives peripheral producers better access to markets in the core, but also gives core producers better access to peripheral markets, and the net effect for peripheral producers will be negative at intermediate trade and transport costs but positive if these costs become insignificant. (The idea that deconcentration will finally prevail is questionable as it rests on the idea that what matters is absolute rather than relative trade and transport costs and that there is limited inertia).

A fourth feature of these new models of location and development is the recognition that the geography of economic space is a result of a dialectical, historical process in which causes are consequences and consequences are causes (see Dunford and Perrons, 1983) and in which humanly-created resource endowments are an endogenous consequence of an area's development path. As Krugman (1999) recognizes:

> explanations of economic location are almost always historical, while history tends to have a 'one damn thing after another' character. If you try to explain why a particular region is home to a particular industry, you usually end up explaining it largely by describing the sequence of events that caused the industry to be there.

The reason why is that what happens at each stage is in part a result of what happened earlier. Moreover, as Krugman concludes in relation to the New England case, 'the one overwhelming lesson of New England's economic history is that, while each successive stage reflects forces that are obvious in retrospect, attempts at prediction nearly always get it completely wrong'.

An important consequence is that these models offer an alternative to traditional approaches to the study of trade and specialization. Traditional explanations were cast in terms of factor intensities, rooted in the Heckscher-Ohlin view that places specialize in those activities that use intensively resources that are relatively abundant in that place, and Ricardian views concerning the role of technologically induced differences in productivity. At any moment in time there is some sense in the factor proportions idea that differences in resource endowments will, for example, cause peripheral areas to specialize in agricultural production, raw material exports and relatively unskilled manufacturing, while core areas concentrate on knowledge- or skill-intensive activities. Cumulative causation models rest however on the idea that resource endowments must themselves be explained and are, in fact, created historically. Most of the resources on which development depends are a result of previous human activity. Even in the case of natural resources, the productivity with which they are exploited depends on historically created capabilities, although natural features may act as the seeds around which cumulative processes unfold.

The new geographical economics has, in short, considerable merits. First, it draws on and develops the earlier locational tradition in geography to identify possible answers to why questions. Second, it attempts to derive spatial organization from analyses of profit-seeking. Third, it acknowledges the importance of path dependence and of Myrdal's cumulative causation thesis. The arguments advanced nonetheless remain abstract, and little progress has been made in identifying the relative importance of different causal mechanisms. What is more, these models have several limitations. The first is that the models considered so far are all essentially concerned with the allocation of resources or with the long-run supply side determinants of growth and development. What these models assume is a smooth and automatic process of adjustment to economic shocks in which resources released as a result of structural change are quickly re-employed. The demand for labour depends, however, on the demand for the output workers produce and the real wage, while the demand for output depends on aggregate demand. In the real world, as opposed to the fictitious worlds these models depict, adjustment can involve significant relative declines in aggregate demand that can have a powerful negative impact on output and therefore on employment: a recession will impede the re-employment of resources in new spheres of economic life, invalidating many of the normative conclusions of orthodox neoclassical models. The second is that their concept of corporate strategies and of the relevant external environment is a narrow one. As Neary (2001:549–50) remarked, 'except for the fact that it incorporates increasing returns, the new economic geography [geographical economics] has industrial organization underpinnings which are very rudimentary', also noting the failure to model explicitly the transport and logistic system. Third. in their search for parsimony these theorists produce models that are too deterministic in that they overlook forces that play an important role in shaping economic landscapes (see, for example, Pinch and Henry, 1999).

Conclusions

Contemporary trends towards greater economic integration and globalization are often, although not always and not at all scales, leading towards a widening of territorial inequalities, as a result of the reshaping of the social and territorial division of labour. In this chapter we have suggested that the new economic geographies of the geographers and the economists can be considered as ways of identifying centripetal and centrifugal forces in the geography of economic activities. Construed in this manner, the recent theoretical literature differs sharply from earlier theories of convergence in that it suggests that divergence can also occur as relative development depends on the relative weight of centripetal forces related to scale economies and market access on the one hand and and centrifugal forces related to the relocation of factor-cost sensitive activities in low cost peripheral areas on the other. This literature accordingly points to relationships between relative macro development trajectories of territorial entities and the underlying dynamics of the territorial division of labour. To make further progress, however, a fuller conceptualization of the territorial division of labour and the changing functional roles of establishments in different kinds of localities is vital.

Chapter Three

Theorizing Regional Economic Performance and the Changing Territorial Division Of Labour: Value Chains, Industrial Networks, Competition and Governance

Introduction

Many of the analytical ideas present in endogenous growth and new economic geography/geographical economics literatures can figure in a more comprehensive conceptual framework that seeks to identify the connections between comparative regional development, the underlying territorial division of labour and the mechanisms that shape them. The aim of this chapter is to develop such a framework.

The framework we propose is outlined in Figure 3.1. This figure has two important characteristics. First, it identifies a reverse causal chain in which successive downward steps identify factors that underlie or explain elements at each higher level, and whose traversal in the opposite upwards direction permits a movement from microlevel activities to meso- and macrolevel phenomena. Second, it synthesizes literature dealing with (1) corporate strategies and commodity and value chains (see Smith et al., 2002), (2) production networks, customer-supplier and employer-employee relations and relations to competitors and rivals and (3) and the regulation and governance of economic change. These three elements are represented by the three main boxes in the lower part of Figure 3.1, which are themselves derived from the distinction between the internal resources and strategy of an individual enterprise (what takes place within the boundaries of the firm which themselves shift as a result of mergers, acquisitions and alliances) and the external environment of the firm (the external

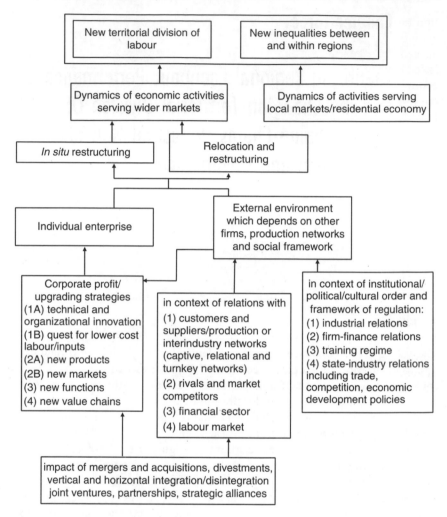

Figure 3.1 Territorial divisions of labour.

resources and constraints embodied in the relation of the firm with other firms and with noncorporate governance systems).

Basic and Nonbasic Industries

To examine the relationships between regional performance and the territorial division of labour it is helpful to draw on a long-established distinction between basic and nonbasic industries. Basic industries are economic activities found in a locality that predominantly serve nonlocal markets and accordingly generate net exports, a net transfer of wages and, if the industry is locally controlled, profits to their home area. Nonbasic industries are economic activities that predominantly serve local markets. These activities comprise a residential economy serving the residential population of a locality. This residential economy includes a wide range of neighbourhood services financed either through government transfers or through the local recycling of incomes. In developed economies a large share of employment and of recent employment growth is of the nonbasic/residential economy kind, with a large share of jobs for women and in low paid personal services (Perrons, 2003; 2001). Indeed, the predominance of nonbasic jobs is one of the main reasons why there is such a strong correlation between the geography of recent employment growth and the distribution of the population.

Clearly, the regional location of nonbasic employment is closely related to the location of the markets or users that these economic activities serve, even if some of these sectors source globally. In the case of basic activities, on the other hand, on which this volume henceforth concentrates, location depends on the interaction of factors identified in the new and old economic geographies. Market size and location play a role, but so too do scale economies and endogenous growth, the strength, intensity and geography of vertical linkages, external economies, and the interaction of factor intensities and resource endowments.

Explaining the Dynamics of Activities Serving Wider Markets

The trajectories of regional economies depend not just on the location/ relocation of economic activities but also on their *in situ* evolution and the consequent changes in output, employment and income. To explain these trajectories it is therefore essential to examine the strategies of firms and the conditions that explain them.

The starting point of such an analysis is the profit strategies of firms and the productive strategies they put in place to achieve their profit goals (Figure 3.2). Enterprises develop strategies to deal, first, with the imperative that capital invested earn at least the average rate of profit and, second, with two uncertainties. The first is the uncertainty as to whether the goods and services produced will be sold (see Boyer and Freyssenet, 2000: 6–25) and in particular whether the costs an enterprise incurs will be recovered.[1] There are several ways of dealing with this situation. The first is to identify goods and services for which there is effective demand. Once these products are identified, an enterprise seeks a competitive advantage in satisfying

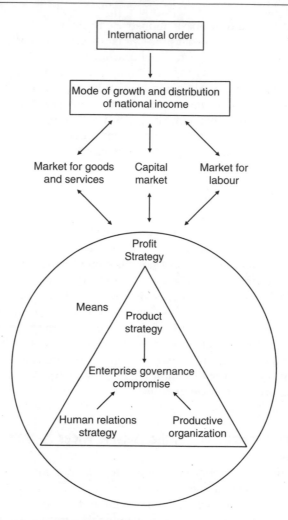

Figure 3.2 The productive model and its context.
Source: elaborated from Boyer and Freyssenet, 2000:24.

the desires of consumers through their price, quality, diversity, novelty and availability. This type of competitive advantage can derive from the products themselves, the means of production used, the organization of work or the wages paid to the workforce. It can also derive, however, from collective conditions of production (infrastructural conditions, taxation regimes, the quality of training systems, innovation support, etc.) provided in a particular area perhaps though public action. In yet other cases it can derive from strategies that transfer the risk to others (which we shall consider later).

The second is through the establishment of a mode of growth that makes the evolution of demand more predictable or that limits interfirm or inter-worker competition. As this second path, involving the creation of a more organized model of capitalism, indicates, the evolution of enterprises depends not just on internal relationships within an enterprise but also on its external relationships with its wider environment (Figure 3.2). These environmental aspects of change will be considered in the next two sections.

The second uncertainty is the uncertainty as to whether the workforce will produce goods and services of the right quality, in the right quantity and at the right time. Also in this case there are two broad paths. The first is to establish relations of trust with wage earners, exchanging autonomy and career progress for improvements in the performance of the enterprise. The second is to design machine systems and methods of work organization to channel and discipline the activity of the workforce in ways considered compatible with the profit strategy of the enterprise. Historically, it is this second path that has prevailed.

This profit strategy, involving as it does a quest to secure satisfactory returns and to keep ahead of competitors, may be construed as entailing attempts to upgrade. Construing it in this way enables a connection to be established with the value chain literature, which itself is a development of the commodity chain and input-output literatures (Dietzenbacher and Lahr, 2004; Gereffi, Korzeniewicz and Korzeniewicz, 1994). Similarly categories applied in the new geographical economics literature can be recast in this framework. An advantage of such recasting is that it permits the examination of 'what' questions concerning the profit strategies of firms as well as 'where' and 'when' questions. Close relationships also exist with other traditions including Schumpeter's concept of innovation which, in its second and wider formulation, included: (1) the introduction of a new good or a new quality of a known good, (2) the introduction of a method of production or a mode of commercial management new to the sector, (3) the opening up of a new market whether or not it previously existed, (4) the conquest of a new source of raw materials or semi-manufactured goods, and (5) the establish-ment of a new organizational model in any industry, including the creation or dismantling of a monopoly position (see Galvão, 1998).

Attempts to upgrade involve attempts to exploit specific sources of profit or of surplus profit/economic rent and can be classified into four types (identified as 1A and 1B, 2A and 2B, 3 and 4 in Figure 3.1).

The first path involves changes that reduce costs relative to prices and increase market share. **Cost reduction** can be achieved in a number of ways. *In situ* it can result from the introduction of new productivity-increas-ing technologies, new methods for handling material and information flows or the reorganization of operational structures (1A). One way of achieving such an advantage is through the realization, where they exist, of **economies of scale** involving the spreading of fixed costs over a larger volume of output to reduce unit costs and **economies of learning**. In this way a link is first established with the new geographical economics models discussed earlier, although these theories should be widened to include consideration of dynamic learning and experience curve effects.[2] Another is productive **flexibility** which permits a rapid adjustment of costs to changes in demand. Alternatively cost reduction can result *in situ* from the intensification of work and the reduction of idle time. Cost reduction can also occur through the transfer of certain operations to areas where wage costs are lower, or the searching out of cheaper sources of energy, materials and components (1B). (Again these strategies are related to the factor cost arguments embodied in the Heckscher-Ohlin models of trade and location). Other ways include the transfer of costs onto society as a whole through, for example, recourse to state-financed early retirement schemes.

More generally, these distinctions are closely related to the analysis of the labour process represented by Figure 3.3 where T is the apparent duration of work, t_v is the uniform time in which value is added, and t_n is time socially necessary to produce goods whose value is equivalent to the wage paid. An employer's surplus is equal to the difference between t_v and t_n. This surplus can be increased in three ways. An employer can reduce the difference between T and t_v either by intensifying work and reducing the porosity of the working day, or by extending T by for example introducing shift work. Alternatively, an employer can reduce t_n relative to t_v through a reduction in the wage bill, perhaps as a result of increases in productivity in the wage goods sector, or the acquisition of cheaper wage goods.

The second path involves the introduction of **new commercially relevant products**, increases in the **variety** of products to capture new segments of market demand, improvements in the **quality** or design of

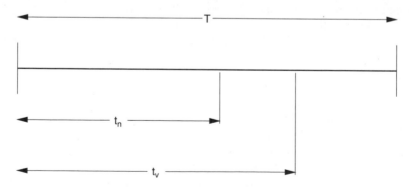

Figure 3.3 The labour process and the working day.

existing products (2A), and entering new markets (2B) which permit short- or long-term increases in the prices the products command and/or sales expansion, increases in market share and increases in the utilization of existing capacity. Markets are opened up in a range of different ways which include politically mediated processes of integration, liberalization and privatization. (New geographical economics models highlight the impact of the size, diversity and location of these markets for the location of the activities that supply them, while new products result from research and development and marketing initiatives which are implicit in the endogenous growth literature).

The third path involves changes in the **relative weight of different functional roles** within the sector or value added chain to increase the share of functions commanding higher returns (Humphrey and Schmitz, 2002). Examples include concentrating on knowledge intensive activities or on marketing and distribution in the same value chain, or concentrating on what are conceived as core competences and working with specialist suppliers of functions considered noncore.

To these adaptations can be added a fourth path involving **disinvestment** in one sphere of activity **and a transfer of resources** to new lines of activity and **from one chain to another**. Acquisitions and sales or openings and closures, for example, enable companies to modify their portfolio of plants. These activities enable companies to specialize in new activities and alter the geography of their operations. At a global level any consequent changes in the profile of activities should result in increased output per head for the economic system as a whole. In some cases these new activities involve a movement into new sectors/chains. A classic example is Nokia, which from 1865 until 1960 was a manufacturer of paper, rubber and cables.

All of these pathways to upgrading amount to ways of increasing the volume of profits (value added less wages) and/or the rate of profit defined as:

$$
\begin{aligned}
Rate\ of\ Profit &= \frac{Profits}{Capital\ Advanced} \equiv \frac{Profits}{Value\ Added} \times \frac{Value\ added}{Capital\ Advanced} \\
&\equiv \frac{Profits}{Value\ Added} \times \left(\frac{Value\ Added}{Employment} \times \frac{Employment}{Capital\ Advanced} \right)
\end{aligned}
$$

$$\equiv Profit\ Share \times Productivity\ of\ Labour \times Capital\ per\ Person\ Employed$$

$$(1)$$

In each case the change may occur *in situ* (expansions, contractions and restructuring) or may also involve locational shifts (openings, closures and relocations). Market-seeking investments of the type made by western companies in the CEECs were driven, for example, by a desire to expand sales by acquiring distribution and production capabilities in the CEECs and perhaps also to increase the rate of utilization of existing capacity.

These profit and upgrading strategies can be connected to the literature on convergence and the new geographical economics. For example, technology transfer that raises relative productivity in less-developed areas contributes to catch-up, though profits repatriation will affect the territorial distribution of income from inward investment. Conversely, the relative upgrading of core economies has centripetal effects, as can measures that put in place transport improvements or widen market access and permit the realization of scale economies from fewer production locations.

In devising a profit strategy, no enterprise ignores any of these sources of profit or pathways to upgrading. At any point in time, however, not all, and not all combinations, are equally feasible. Choices depend on market and wider social conditions, and vary over the course of time. In the motor vehicle sector, for example, at least six different historical combinations can be identified: a quality strategy; diversity and flexibility; volume and diversity; permanent cost reduction; and innovation and flexibility (Boyer and Freyssenet, 2000: 18).

Corporate profit strategies and their implications for *in situ* and locational change at the level of the enterprise result in outcomes that affect regional performance. At a regional scale, for example, upgrading can be defined as an increase in the wealth and employment created in a particular area, or in the wealth appropriated by the inhabitants of a certain area relative to others. Enterprise and regional upgrading are not the same. On the one hand, enterprise upgrading is designed to increase the volume or rate of profit rather than aggregate wealth. On the other, arriving at regional wealth involves adding up the wealth that all of an area's establishments create. The impact on regional wealth of the upgrading of an individual establishment depends therefore not just on changes in the wealth it creates but also on the magnitude of the direct and indirect effects it has on other parts of the regional economy. Included, for example, is the magnitude of displacement effects, increased/decreased orders for local suppliers and competition for workers.

The implementation of a profit or upgrading strategy requires the existence of appropriate and coherent means/instruments/resource endowments (a productive model) (Figure 3.2). To take the motor vehicle sector as an example, a volume and diversity strategy requires multipurpose machines and multiskilled workers, while a volume strategy requires standardized processes of production and unskilled workers. In practice, the means at an enterprise's disposition are a path-dependent result of the specific demands of successive choices and may not prove mutually coherent.

Boyer and Freyssenet (2000) suggest that a productive model involves three elements (Figure 3.2). The first is a product strategy that involves the identification of markets and market segments, target sales volumes, the range and design of products, the quality of products and planned margins. The second is a productive strategy which deals with the means and methods chosen to achieve the product strategy. The productive strategy includes decisions about the organization of product design, manufacturing and marketing, the choices of production techniques, the spatial organization of activities, sourcing strategies and management criteria. The third is a human relations strategy covering systems of recruitment, job classifications, wage determination, and modes of representation of worker interests.

Assembling the material and intellectual means also entails the establishment of a compromise (a governance strategy) with all of the major actors. Of these actors some are internal (owners, managers, workers and unions plus new staff recruited from outside to support changes of direction) while others are external (suppliers, joint venture partners, companies acquired, shareholders, etc.).

These concepts offer some of the foundational elements of an account of regional performance and the territorial division of labour. Although the starting point is the inherited territorial division of labour, its trajectory depends on:

1   the profit and upgrading strategies of individual enterprises;
2   the appropriateness and coherence of the resources the enterprise can mobilize and the governance compromise it seeks to put in place in order to implement its chosen profit strategy;
3   the ways in which they impact upon the impact on the territorial division of labour.

Territorial development and regional economic performance depend on the nature of these strategies, their degree of success and, equally, the extent to which they fail, as can happen if the profit strategy is misjudged, or the means are not appropriate and coherent, or a suitable governance strategy proves elusive.

Attention has so far been concentrated, however, on the internal determinants of the growth of the firm. Growth was seen to depend on the capacity of a firm to earn profits and compete for market share. This capacity was itself seen to depend on its internal resources (its inventiveness, material resources, economies of scale and scope, accumulated knowledge, skill, competences and experience) and on its organizational capacity and collective effort which shape its ability to see and take advantage of its product development and marketing, production and human resource possibilities.

Enterprises and Their Environment: Establishing the Frontiers/ Boundaries of the Firm

Companies are never self-sufficient. Accumulation and growth depend also on interactions across enterprise boundaries with other economic actors (suppliers or customers, competitors, market organizations, universities, political institutions, research institutes, financial institutions and market research and consultancy groups) and with the macroeconomic environment. These frontiers that separate intra- and extrafirm relations and that shape the number, size and organization of firms depend on a number of factors.

Of particular importance are the degrees of vertical and horizontal integration (see the lowest box in Figure 3.1). Standard (Coase-Williamson) analyses of integration and disintegration involve a consideration of the interaction of three factors. The first is economies (and diseconomies) of scale, which influence the size and number of firms, and exist if costs diminish (increase) as the volume of output increases. The second is economies (and diseconomies) of scope, which exist if the joint costs of providing related products and services are smaller (higher) than the costs of providing these products and services separately. The third is economies (and diseconomies) of diversification, which depend on the advantages and disadvantages of a firm diversifying into unrelated processes and products (conglomerate strategies). Surprisingly little attention is paid to the revenue implications: as the goal of capitalist production is not to minimize costs but to maximize the rate of profit and/or gain market share and accumulate, strategies of integration and disintegration will also depend on their revenue and risk implications, especially where enterprises have an influence over prices.

Economies of scope are connected with the range of products a firm makes. This range depends in part on the degree of vertical integration upstream into activities supplying inputs and downstream into activities involving further transformations of the commodities it produces. In these cases the scope of a firm depends on a series of 'make or buy decisions' and on choices between markets and hierarchies as coordination arrangements. These decisions depend on relative internal and external production costs and the costs of organization of economic activities. Organization costs comprise (1) the costs of internal coordination, which increase as a firm integrates upstream and downstream, and (2) transaction, or external coordination, costs, which increase as vertical integration declines and the volume of market transactions or transactions across a firm's boundaries increases (Figure 3.4). Scope economies can also derive from horizontal

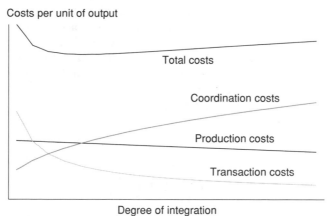

Figure 3.4 Strategies of integration and the structure of costs.

integration. Examples include situations where there are flexible machines that can be set to produce a mix of products or a range of different designs and that make joint production cheaper than separate production, and situations where the expenses associated with the development of, for example, a brand can be spread across different products such as, for example, clothing and shoes. Conglomerate strategies, which also influence the boundaries of firms, differ in that the aim is to diversify with a view to spreading risks, moving into more profitable activities or increasing group size.

As this discussion of vertical and horizontal integration makes clear, the determination of the limits and therefore also of the structure of enterprises depends on the character of the wider economic environment. Chandler's (1977) 'modern corporation', for example, realized economies of scale and scope by controlling value chains through the vertical integration of suppliers into the multidivisional enterprise. The viability of a high degree of internalization and vertical integration and of administrative as opposed to market coordination depended, however, on the size and stability of markets, and, therefore, on general conditions relating to the model of growth or regime of accumulation. Conversely, more recent literature has suggested that, in conditions of slower growth and greater instability, the competitive edge lies with networks of small firms related via arm's length market transactions (Scott, 1988; Humphrey and Schmitz, 2002).

In this section we have outlined some ideas that help establish the frontiers of enterprises. The enterprises that result are, however, never self-sufficient. Their performance and the performance of the wider economic system depend not just upon their capacity to mobilize their own constantly evolving resource endowments to pursue their own development strategies but also upon relationships across enterprise boundaries with their external environments. To examine these relations, the external environment of an individual enterprise can be divided into two parts. The first comprises a set of interfirm relations with customers, suppliers and rivals that we shall consider in the next section, along with employer-employee relations that we shall not consider explicitly. The second comprises an enterprise's relations with the wider social environment and growth model, that we shall consider in the next section but one.

Enterprises and their Environment: Interfirm Relations

Firms and production networks

A firm's interactions with other enterprises influence and augment its resources, capabilities, structure and strategy. Of these external relations and interactions, the forward and backward linkages and the externalities examined in new geographical economics models are a subset. In the contemporary industrial order the acquisition and internal absorption of external resources and knowledge, cost-reducing external economies and the risk reduction derived from sharing and networking are perhaps especially important due to the tendency for enterprises to concentrate on core competences and give greater weight to external economies derived from interfirm divisions of labour and network relations with other actors.

In the literature a number of different forms of interfirm relations, with different strengths and weaknesses, are identified. One possibility is reliance on arm's length market transactions. Another is vertical near-integration where independently-owned firms are organized via stable market and nonmarket (partnership and joint production/development) relations into interdependent hierarchical systems/networks. A range of types of vertical near-integration can be identified according to the relative weight of horizontal relationships of partnership and vertical customer-supplier relations (Figure 3.5). A third involves vertically integrated corporations of the kind Chandler identified whose interrelated activities are organized hierarchically (Leborgne and Lipietz, 1988). Similar distinctions are made by Humphrey and Schmitz (2002) who argue that one can distinguish market (arm's length), network and quasi-hierarchical relationships.

Generally speaking arm's length market relations are held to be effective for standard products where coordination requirements are limited, where information about costs, standards, reliability and supplier performance are readily available, and perhaps where specialization offers scale advantages

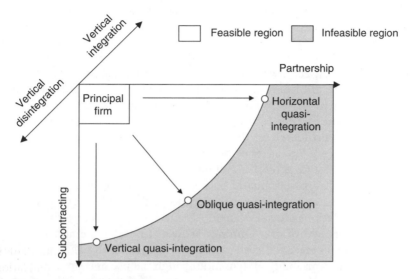

Figure 3.5 Strategies of integration.
Source: elaborated from Leborgne and Lipietz, 1991.

(implying that production costs would rise, in Figure 3.4, as integration increases). Vertical near-integration is appropriate where exchanges are more frequent, regular, time-critical, complex or customized, and involve greater asset specificity or complementary capabilities. Vertical integration is favoured, finally, where more customized products, greater asset specificity, difficulties of monitoring supplier performance and greater risks of opportunistic behaviour raise transaction costs relative to internal coordination costs.

These distinctions are useful, although the underlying conception of market relations is in some respects questionable. Aglietta and Brender (1984:31–53) have convincingly suggested, for example, that standard economic representations of market exchange are deficient in several different ways. First, markets are characterized by relatively stable relations of exchange, communication and distribution amongst active economic agents and in particular amongst functionally distinct financiers, producers, distributors and workers (with different sources of income: interest, profits, dealers' margins and wages). Second, markets are generally characterized by a relative stability of prices. Third, market exchanges of money for goods and services, and goods and services for money, are relations of complementarity, as when a retailer is able to offer a producer an outlet for his/her products. Fourth, market exchange is also a relation of conflict over the distribution of value added. Indeed, the prices at which market transactions take place are not so much a vehicle for the communication of information as an expression of power relations (and related information asymmetries) and a determinant of the distribution of value added between a buyer and a seller (and therefore between people and places). The importance of this approach is that it gets away from the idea that prices are simply a means of communicating information about relative scarcities and a device for ensuring an efficient allocation of resources. In its place is put an emphasis on two things. One is relative economic power in buyer-seller relationships, which is also considered in the literature on industrial organization and market structures. The other is the social determination of (1) the worth of different activities (in a market system what someone is worth is what they are paid) and (2) the command over social wealth (what someone is paid determines their command over the results of the work of others).

Interfirm relations are also shaped (due to its impact on profit strategies) by the management of risk. Dunin-Wasowicz (2002) has emphasized how, in response to increased uncertainty, enterprises are rationalizing capacities and hedging for variability of demand through the use of shared suppliers in networks. Networks combining specialized suppliers (providing nonstandardized intermediate goods) with several manufacturers allow a pooling of demand uncertainty (associated with idiosyncratic shocks) and generate economies of sharing costly productive capacity. In this situation the strength of network externalities (depending on innovation, link structures and link costs) will influence cost positions and the relative competitiveness of different locations, their capacity to capture the wealth they create and their adaptability.

Interfirm production and buyer-supplier networks exist at a range of scales, from industrial districts, to national economies, to continental trade blocs and the global system. These networks also involve a range of different relationships, and assume a range of forms (considered earlier). Berger et al. (2001) identify three specific models. The first model is of relational production networks. Associated with 'industrial districts' and the flexible specialization model, these networks are characterized by dense networks of specialized SMEs, close social and spatial proximity and long-term but flexible and noncontractual interfirm relationships. In these industrial districts interfirm relations are governed by modes of social interaction involving family relations, trust, reciprocity, reputation and peer pressure. These social relationships reduce the threat of opportunistic behaviour and offer an alternative to the internal hierarchy of the integrated firm and simple arms-length trade relations. The existence of a dense mosaic of small firms permits the manufacture of small volumes, short lead times, rapid delivery and fast adaptations to market conditions. In a world of small firms, however, there are limits to the realization of scale economies, sales may be insufficient to warrant sizeable research and development initiatives and the security of supplies may be jeopardized by high birth and death rates of firms (Dunford et al., 1993). Other research shows that the survival chances of firms located in clusters increases for firms operating in complementary industries but decreases for firms in the same industry (Staber, 2001).

The second model is of captive production networks (associated with lean production systems) where a dominant lead firm coordinates tiers of largely captive suppliers. These systems are characterized by near-vertical integration. In them the lead firms may have a financial stake in their suppliers, may play an active part in upgrading their suppliers' technologies and quality control systems, and may demand 'just-in-time' deliveries. The disadvantages are ones of dependence and rigidity due to the cost and difficulty of breaking buyer-supplier relationships, on the one hand, and the weakness of external information flows and linkages, on the other.

The third model comprises turnkey production networks (Sturgeon, 2000) associated with the 'virtual corporation' model. This model arises when manufacturers acquire specialized inputs from specialized suppliers who sell to several manufacturers or design products that are made by contract manufacturers. Turnkey networks are an alternative to vertical integration where Original Equipment Manufacturers (OEMs) make their own specialized inputs, and are distinguished from captive networks by the merchant character of turnkey suppliers. The word merchant denotes the market (cost-sensitive, short-term, and fluid) character of customer-supplier relations made possible by the existence of a large and diverse pool of customers. Turn-key suppliers provide a wide range of services including component and module design, the purchase or manufacture of components, the manufacture and assembly of subassemblies or final products, packaging, distribution and after-sales service. Often they specialize in processes, components and services that have applications in a range of sectors, products and services (across different end-markets, end-products and end-users). An advantage for OEMs is that they can switch partners and can rapidly connect and disconnect from suppliers in different places. Through the use of what are in effect shared facilities that the OEMs do not own, investment in own production capacities is reduced. This type of development of shared facilities is also an important aspect of a hedging strategy for reducing OEM risk in the face of the variability of demand and uncertainty about market trends. Another advantage is that the sharing of suppliers enables higher levels of capacity utilization and greater efficiency. What are advantages for OEMs are, however, potential disadvantages for contract manufacturers. If market demand weakens, contract manufacturers may find themselves with excess capacities, stock whose ownership is a subject of disagreement and limited visibility. The disadvantages for the OEMs are different. The first is the risk that if suppliers gain in financial strength, and increase their competence in product and process development, and if OEMs become overreliant on them, suppliers may find themselves able to alter the terms of trade and division of value added in their favour, or may even start to offer competing end-products (with clear implications for the geography of economic activities). The second lies in the risk of a leakage of market-sensitive information to rivals.

The identification of these models of interfirm production networks along with distinctions in the commodity chain literature between supplier-driven and buyer-driven chains are indicative of changes in the characteristics of the actors and of the ways in which these actors compete for shares in, and control of, value chains. These changes are closely related to contemporary features of the wider economic environment. Market volatility, short innovation cycles and high R&D costs are, for example, an important shaper of changes in production networks, as are the changing technological and organizational capabilities of companies. Computer simulation and digital codification of design specifications make it possible, for example, to outsource increasingly complex functions once performed in-house (Berger et al., 2001). At the same time the emergence of global suppliers can permit companies to produce in new parts of the world, creating new territorial divisions of labour, and altering the territorial distribution of value added, without having to rely solely on their own internal capabilities or on the existing capabilities of the areas into which they move. Instead they can rely on specialized suppliers who may relocate with them.

Competition and rivalry

As indicated earlier the capacity of a firm to earn profits and compete for market share depends not just on its capacity to organize its internal resources, but also on its ability to draw on complementary resources through interactions across enterprise boundaries with other economic actors. These actors include suppliers or customers. In this case the relationships are mainly ones of complementarity, although these relationships also involve conflicts over prices, the distribution of value added and the relative social worth of different activities.

Accumulation and growth also depend however on the capacity of enterprises to compete with other enterprises producing either similar goods and services or substitutes for them. In this case there are once again complementarities and shared interests over, for example, state rules/actions affecting enterprises in general or particular industries. Measures that open up markets or that protect industries from overseas competitors are examples. Of more importance however are competitive relationships

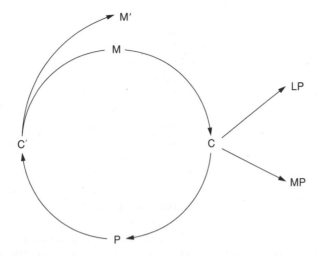

Figure 3.6 The circuit of capital.

which simultaneously (1) provide a motive to upgrade/differentiate the conditions of production and exchange more rapidly than competitors, and (2) impose on competitors a requirement to adapt/equalize the conditions of production and exchange. (These concepts of differentiation and equalization of the conditions of production and exchange are closely related to the ideas of equalizing and unequalizing tendencies introduced in our discussion of convergence and territorial development theories).

Assume, for example, that an employer producing under average conditions of production introduces a new method of production that increases the apparent productivity of labour. In these circumstances the individual value of the commodity is reduced and falls beneath its social value (see Dunford, 1988: 29–32). If the capital advanced remains the same, the increase in output comes up against unchanged conditions of demand. The reason why is that the capital advanced determines the size and composition of distributed incomes, while the incomes so formed determine the size and composition of the output that can be sold. To conquer a larger part of an unchanged market, the employer who has raised productivity must therefore lower the offer price to a level that lies above the commodity's individual value but below its social value. As the employer still earns extra profit, there is therefore a motive for each individual employer to reduce prices by increasing the productivity of labour. Once the new method has been introduced, however, the other producers whose sales may have fallen (depending on the impact of reduced prices on demand) must adapt to the new norms of production. As other producers adapt/copy the new methods, the social value of the commodity concerned is reduced, although the employer who first introduced the new method can invest the extra profits obtained to introduce yet newer methods and accumulate more rapidly than his or her competitors. (The profit and accumulation cycle is depicted in Figure 3.6 where M represents the capital advanced, C the commodity inputs made up of means of production, MP, and labour power, LP, P the process of production in which the commodity inputs are transformed into commodity outputs, C', and M' the monetary proceeds of the sale of the commodity outputs).

In practice the intensity and character of competitive mechanisms of this kind will depend on a range of economic, political and institutional factors including, for example, the degree of concentration, the nature of the competitive environment, the specificity of assets, capital requirements, the ease of entry and the costs of switching.

Modes of Governance and Growth

The evolution of enterprises and of the territorial division of labour depends, finally, on a range of other economic, political, institutional and cultural resources that enterprises do not create, and on the overall model of growth. As sociological neoinstitutionalism suggests, corporate strategies are not simply rational strategies of profit maximization. These strategies also reflect their emdeddedness in specific cultural and cognitive contexts. In situations of uncertainty and incomplete/insufficient information companies (and individuals) make choices that are influenced/constrained by what is considered appropriate and legitimate in the environment in which they operate. Considerations of this kind can help explain the persistence and inertia of institutions that are inefficient. More generally, these factors subsumed under the concept of isomorphism help account for the homogeneity of behaviour. Powell and DiMaggio (1991) show, for example, that the homogeneity of a model is likely to be determined by institutional isomorphism, legal isomorphism and mimetic isomorphism.

Analyses of the role of social and cultural conditions are associated with the idea that there are national variants of capitalism. According to this view the specialization and the trajectories of different national economies depend on nationally-distinct institutional configurations, with distinctive sets of rules, laws and institutions regulating the system of corporate governance, the industrial relations system, the educational and training system, the relations between companies, the credit and financial system, and state-industry relations, embracing macroeconomic, industrial, competition, innovation and spatial policies (Dunford, 1991; Streeck, 1989). These distinctive institutional configurations are seen as falling into a range of types identified as varieties of capitalism. An example of literature in this genre is 'Capitalism against capitalism' (Albert, 1993) in which the author distinguished 'neo-American' and 'Rhine' models of capitalism, and sought to show that the social foundations of capitalism play a major part in determining comparative economic performance. Albert argued, justifiably, it seemed in the early 1990s, in favour of the economic and social superiority of the Rhine model. As the 1990s advanced, however, it was Anglo-American economies that outperformed continental European economies. These differences in economic performance led to arguments concerning the superiority of the Anglo-American 'outsider' model of corporate control with large equity markets, dispersed ownership and active corporate control markets as compared with the 'insider' continental European model with a smaller number of quoted companies, more concentrated share ownership and a relatively small amount of takeover activity (for a discussion of these arguments, see Streeck, 2001). At the end of the decade, however, increased awareness of the extent of stock market speculation, on the one hand, and of the unexpectedly weak results of many mergers and takeovers, on the other, undermined the view that the Anglo-American model was an efficient way of allocating resources and disciplining companies.

An important question that arose in the face of system competition and increased integration and globalization is whether there is a convergence in institutional structures and organizational models towards common market-driven models. As in the case of economic convergence, there are in fact forces working in two directions at the level of the firm and at the level of their institutional environment. As firms demand more autonomy from national product, capital and labour market rules, and greater freedom to make their own choices, there is a weakening of national frameworks of regulation, implying a certain degree of convergence essentially on a neoliberal model. At the same time the movement of regulation to the international level implies a consequent dissolution of differences between national and regional systems, and a replacement of public regulation by private international product and capital markets (with their stricter criteria of profitability and shareholder value). Yet differences in the structures and strategies of firms and in national systems remain. At present, there is not a single optimal way for managing international or national operations. Instead, structure and strategy vary widely and depend in part on the history of a firm, its country of origin and its method of adaptation to the specific national contexts in which it locates. Streeck (2001) suggests that the pressure for convergence is reduced by several factors. On the one hand, regulation is in many cases a response to specific market failures or is designed to establish certain social rights and to make firms responsible not just to their shareholders but also to their employees and the general interest. Generally, firms have found ways of using these arrangements to increase their productivity and competitiveness, making them less onerous. A further erosion of this framework is not without risks that are both economic and social. At a social level, for example, it complicates the task of governments that must satisfy the needs of firms that pay less tax, choose where to pay it and make a smaller contribution to social objectives, while ensuring the social viability of a competitive market economy. On the other hand, enterprises can secure competitive advantage by exploiting the specific capabilities and advantages of different countries, essentially through specializing in activities that reflect their economic, institutional and cultural advantages, and can find complementary resources not provided by their own national institutional settings internationally. The Italian district model might survive, for example, if those parts of the value chain involving low-skilled and low-paid manufacturing jobs were integrated into the extra-Italian part of integrated production networks, while activities suited to Italian institutional conditions remained in Italy itself.

The specific pathways to development will also depend in important ways on the wider institutional context and, in particular, on the framework of rules regulating trade and competition. New trade rules, economic liberalization and the collapse of Communism have altered the conditions of access to foreign markets and investment in overseas economies. The creation of regional entities like the European Union has transformed access to markets. Institutional configurations also have an impact at the local level, as do the nature of local cultures, political systems and institutional performance (Figure 3.7 which identifies possible interactions without attributing relative causal weight to them).

Also relevant, finally, are modes of growth themselves characterized by the relative importance of different sources of income and the ways in which income is distributed (Boyer and Freyssenet, 2000:10–17). As the mode of growth varies, the risks associated with the evolution of the markets for goods and for labour differ, altering the conditions in which profit strategies unfold. Growth can depend to different extents, for example, on investment, exports or domestic consumption. In the first case a large share of income is invested in infrastructures or in capital goods. The dominance of domestic demand implies the distribution of a large share of national income in ways that increase the purchasing power of consumers. A leading role for exports implies that economic performance depends on the dynamism of international markets and the competitiveness on global markets of domestic output. Similarly, income distribution plays an important role. Boyer and Freyssenet (2000:10–17) distinguish : competitive distribution which depends on power relations and the scope for financial, commercial and real estate gains; rationed distribution characteristic of wartime economies, years of reconstruction or Soviet-type economies with high rates of investment; moderately

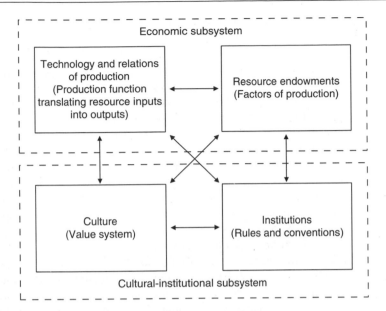

Figure 3.7   Explaining development (adapted from Hayami, 1997:11).

hierarchical coordinated/managed distribution; and unequal distribution in societies in which small oligarchies appropriate a large share of national income. The combination of sources of income and income distribution give rise to modes of growth which determine the volume and structure of the demand for goods and labour, the underlying contradictions and some of the contours of the economic activities that shape the territorial division of labour.

Conclusions

In this chapter we have outlined a framework for analysing trends in territorial inequality and the constant reshaping of the social and territorial division of labour. The creation of this framework involved a synthesis and further development of a number of different perspectives (see also Perrons, 2001). First, we argued that the evolution of the territorial division of labour depends not just on where activities are located but also on the answers to what and when questions, which implies that analyses of location should be integrated into a more general explanation of the evolution of capitalist enterprises. Second, we pointed out that these more general explanations should involve an approach which starts with the profit and upgrading strategies of individual enterprises, yet integrates this approach with analyses of their changing relations with their external environment. Third, we suggested that the conditions in which development strategies unfold and by which they are shaped are, as cumulative causation approaches make clear, the result of previous phases of development.

This framework offers ways of interpreting some of the major changes that characterize the contemporary evolution of the social and territorial division of labour. To reduce their exposure to risk, for example, formerly vertically integrated corporations are concentrating on core competences and acquiring goods and services they no longer produce from specialist suppliers. Some OEMs concentrate on research, design and distribution and entrust manufacturing operations to specialized operations. Often these specialized suppliers assume a disproportionate share of risk and are subject to strong downward pressure on costs. At the same time, there are processes of relocation driven by a range of factors. Management, research, development, design, marketing and advanced producer services requiring highly skilled white-collar workers are concentrated in core countries, while less-skilled manufacturing is located in less-developed areas, and countries where wage costs, in particular, are small. In the cores, alongside these export-oriented activities, there emerges a raft of personal services associated with low incomes and contributing to strong intraregional inequality. The delocalization of unskilled jobs to create parallel (rather than replacement) structures can help companies secure concessions from employees and governments in high cost areas and can erode existing social compromises. Most often, it results in the replacement of operations in cores by similar work in peripheries, creating new complementarities between core and peripheral locations. Sometimes there are also movements of technology and skill intensive manufacturing operations to peripheral areas not just to supply local markets but also to serve the cores, suggesting that there may be a more wholesale relocation of manufacturing to peripheral areas. What results is the location of different parts of value chains in different places in accordance with variations in costs (including transport and logistic costs), skills, the research environment and services, while the consequent weaving of different places into transnational production systems linked by global communications, transport and logistic structures creates new sets of pressures for changes in, for example, relative wages as the relocation of jobs shifts the map of demand and supply for labour of different kinds.

At a regional scale these dual processes of reorganization and relocation of economic activities are leading to radical shifts in the types of enterprise and areas of activity found in different places. As a result regional occupational and employment structures, skill mixes and wage profiles change, reshaping the map of relative regional economic development. At the same time, every enterprise and every regional economy is constantly evolving and the frontier of human development is constantly shifting. Not all enterprises and areas are, however, at different stages of the same development path, as is implied for example in some of the simple upgrading literature and in some of the literature on modernization and stages of economic growth. Instead there is a variety of development paths. For these reasons, the more-developed parts of the world or of a country do not afford a picture of the future of the less-developed parts.

Chapter Four

Growth and Inequality: The Political Economy of Italian Development

Introduction

The fundamental aims of this chapter are fourfold. The first aim is to compare Italy's economic performance with those of the first 15 European Union Member States (EU15) and to a lesser extent with their Mediterranean, Middle East and formerly Communist neighbours. The second is to identify the main features of the map of territorial development within contemporary Italy. As we shall show, there is an enduring north-south/core-periphery cleavage, although we shall also identify other important distinctions. The third aim is to ask whether Italy's regional economies tend to converge or diverge or, to put it another way, to ask whether centrifugal (equalizing) or centripetal (unequalizing) tendencies predominate. This question can be answered by examining the degree of dispersion of Italy's regional economies and by examining the evolution of individual regional economies. What we shall show is that after substantial convergence until the mid-1970s, when southern development policies were strong, divergence prevailed until the mid-1990s, and that while there was renewed convergence in the late 1990s it was associated with the comparatively weak performance of Italy in the EU15 and of the EU15 and its neighbours in the world. A fourth aim is to take the first steps towards an explanation of differences in regional development, identifying the respective roles of productive performance and demographic change in shaping the development of Italy's regional economies.

To these ends, we shall identify the size and nature of the development gaps in contemporary Italy and between Italy and its neighbours, and we shall ask how Italian regional inequalities have changed and whether convergence or divergence tendencies prevail.

Italy's Economy in its European and Mediterranean Context

Today Italy is one of the world's major industrial nations: in 1975 it was a founding member of the Group of Six (G6) leading industrial countries with France, Germany, Japan, the UK and USA.[1] Although its aggregate Gross Domestic Product (GDP) is amongst the largest in the world, in per capita output terms Italy occupies a lower position in international league tables, due to the high scores of a number of small economies. Italy is nonetheless one of the core high-income EU Member States, with, in 2001, a GDP per capita, measured at Purchasing Power Standards (PPS), of more than 300 per cent of the world average and just over the EU15 average. (Italy's GDP per head is recorded on the horizontal axis in Figure 4.1. The population of each country is recorded on the vertical axis. As a result the volume of national GDP is represented by the area of each horizontal bar. Note that the population figures are cumulative, although a gap of 40,000, or in one case 80,000, separates each group of countries).[2] Grouped with the EU15 and the demographically small, high income European Free Trade Association (EFTA) countries, contemporary Italy finds itself at the southern tip of one of the largest markets in the world, with, in 2001, 330 million inhabitants (just over 5 per cent of the world's population), and just over 17 per cent of world GDP, measured at PPS.

Italy is also a Mediterranean country. It differs, however, from the other large Mediterranean/southern EU Member States (Greece, Spain and Portugal) in several ways. First, it is a longer-standing member of the EU with a longer continuous recent tradition of parliamentary democracy (dating from the defeat of Italian fascism in WWII rather than from the fall of right-wing dictatorships in the 1970s as in the cases of Greece, Spain and Portugal). Second, it has a significantly higher GDP per head than the three southern Cohesion Countries (Greece, Spain and Portugal) and also than the two small Mediterranean islands that joined the EU in 2004 (Cyprus and Malta). Central and northern Italy were of course amongst the most economically advanced parts of early modern Europe. After 1450, however, the Italian economy went into decline. Only in the years before Italian political unification in 1861 did growth resume. After 1880 and especially in 1896–1914 industrial growth accelerated in line with the upward swing in the world economy, creating in northern Italy a great industrial complex and modern urban civilization. Amongst the nations of southern Europe and the Mediterranean, Italy was however almost alone in making this qualitative developmental leap, putting itself in a class of its own and establishing an economic lead that it has yet to relinquish.

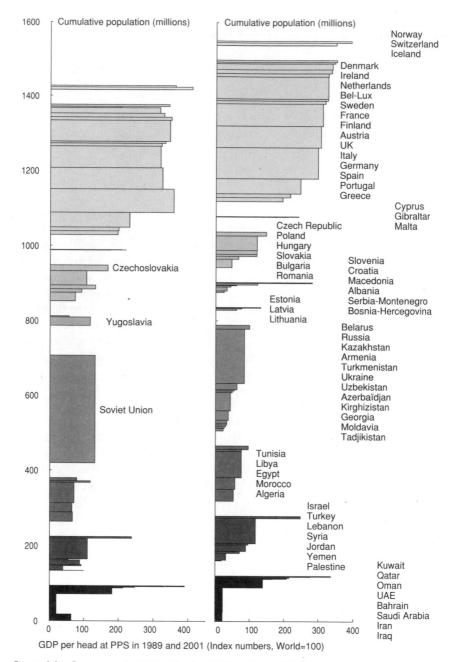

Figure 4.1 Economic potential in Italy, the EU15 and their neighbours.
Source: elaborated from CEPII, 2002.

Today Italy lies close to several peripheries. To the north and east lie 27 countries that have emerged as a result of the political fragmentation, sometimes into very small countries, of 7 formerly Communist states. At one end of the spectrum lie a group of new EU Member States (excluding Slovenia) and Candidate Countries with a GDP per head of 50 to 150 per cent of the world average. Included are the Visegrad Four (Poland, the Czech Republic, Slovakia, and Hungary),[3] three small Baltic States, that were formerly part of the Soviet Union, and two economically less developed Accession Countries (Bulgaria and Romania). Slovenia, which was the first of the republics of the former Yugoslavia to break away, is an exception in this group of new Member States in that its 1.93 million inhabitants have a per capita GDP that is close to that of the high income EU15 Member States.

At the other end of the spectrum there are two sets of countries: to the northeast the countries that emerged out of the break-up of the former Soviet Union, and to the east the adjacent Balkan countries ravaged by the civil wars of the 1990s. All of these countries are economically weak, due in part to the dramatic collapse of economic output that accompanied the transition to capitalism (compare the surface area of the horizontal bars for these countries for 1989 and 2002 in Figure 4.1).

A number of these countries are closely integrated into interregional and international divisions of labour and flows of goods, people and capital with Italy: to give just three examples Romania is emerging as an important focus for Italian investment, while Albania is an important source of often illegal migrants, and Montenegro is a centre from which criminal activities with a major impact on Italy (such as illegal trafficking in cigarettes) is organized. At the same time the fact that Communism in these countries has fallen, that all have embarked on a series of transitions to capitalism and that a substantial number will be integrated as a low-income periphery into the EU is shifting Europe's centre of gravity eastwards. As a result the economic marginalization of Europe's southern peripheries may well increase.

Across the Mediterranean lie a series of other countries with low average incomes per head. Included are the countries of North Africa, including two OPEC Member States (Algeria and Libya) with important oil and gas reserves, and the non-OPEC Middle East from which Israel stands out in no small measure as a result of the overseas support it receives. Further afield lie the OPEC member States that border the Arabian Gulf. What is striking about these countries is that a few have extremely high incomes per head, ranging from just under 140 per cent (Saudi Arabia) to just under 340 per cent (Kuwait) of the world average. As the population figures for this group shows, a fundamental feature of this part of the world (and a fundamental legacy of colonialism) is the existence of tiny countries with

Table 4.1 Trends in GDP per head at PPS, 1960–2001 (World = 100)

	1960	1975	1985	1991	1996	2001
EU15 plus GDR	248.4	283.6	299.0	313.5	302.2	301.2
EU High Income[1]	270.5	298.7	319.2	331.1	318.3	313.9
Cohesion Countries[2]	133.1	206.4	201.7	228.8	223.8	239.7
Italy	223.3	274.8	309.7	333.0	318.4	306.4
EFTA[3]	372.6	376.6	397.4	394.8	376.5	365.5
Mediterranean Small States[4]	88.2	100.9	171.4	208.9	228.3	232.0
CEECs[5]	73.8	107.9	112.9	94.5	98.2	101.3
Czech Republic, Hungary, Poland and Slovakia	91.2	129.5	125.3	108.9	115.9	124.7
Bulgaria and Romania	38.5	65.2	87.8	65.4	61.2	51.1
Balkans[6]	74.6	109.9	119.9	83.1	67.6	67.6
Former USSR	117.8	134.0	133.5	121.0	66.3	65.7
CIS[7]				121.2	66.2	65.4
Baltic States[8]				116.7	72.1	78.7
North Africa including Libya[9]	58.3	66.8	76.9	72.0	67.9	68.0
Non-OPEC Middle East including Israel[10]	98.8	109.3	106.0	105.9	109.8	102.3
Gulf OPEC excluding Libya[11]	101.9	133.8	89.6	68.4	65.4	59.3
United States	452.2	425.0	436.4	429.2	436.0	456.4
Canada	317.6	354.9	374.4	357.8	345.4	349.9
Japan	144.4	279.9	325.0	378.4	370.1	331.8
China	21.1	17.4	28.2	37.3	57.2	70.9
Average annual growth of real world GDP per head, 1960–75, etc. (%)		2.6	1.5	1.4	1.9	2.6

Source: elaborated from CEPII, 2002.

Notes
[1] EU High Income = EU15 less the Cohesion Countries
[2] Cohesion Countries = Greece, Ireland, Portugal and Spain
[3] EFTA = Iceland, Norway and Switzerland
[4] Mediterranean Small States = Cyprus, Gibraltar and Malta
[5] CEECs = Bulgaria, Czech Republic, Hungary, Poland, Romania and Slovakia
[6] Balkans = Former Yugoslavia and Albania
[7] CIS = Armenia, Azerbaijan, Belarus, Georgia, Kazakhstan, Kyrgyzstan, Moldova, Russian Federation, Tajikistan, Turkmenistan, Ukraine and Uzbekistan
[8] Baltic States = Estonia, Latvia and Lithuania
[9] North Africa including Libya = Algeria, Egypt, Libya, Morocco, Tunisia
[10] Non-OPEC Middle East including Israel = Israel, Jordan, Lebanon, Palestinian Authority, Syria, Turkey, Yemen
[11] Gulf OPEC excluding Libya = Bahrein, Iran, Iraq, Kuwait, Oman, Qatar, Saudi Arabia, United Arab Emirates

vast oil reserves and high per capita incomes, a few relatively populous oil-rich countries (Iraq and Iran), and, outside the Gulf, populous countries with no oil. In relation to these countries Italy's strategic position has played a significant part in shaping its development strategy, as we shall see when we examine the evolution of an oil and chemical sector in Italy and in the Mezzogiorno.

Italy sits in other words at the intersection of a number of worlds. What matters, however, is not just the contemporary levels of development of these worlds but also their recent economic trajectories. These trajectories are summarized in Figure 4.1 and especially Table 4.1 which records the average annual rate of growth of real World GDP per head and the GDP per head relative to the world average of a series of territorial entities (including the groups identified in Figure 4.1).

As Table 4.1 shows, the EU15 and the high income EU countries saw their relative positions improve significantly until 1991, at which point they lost ground. A similar pattern prevailed for the EFTA countries except that their highpoint occurred in 1985. Compared with the EU group of which it is a member, Italy grew much more rapidly than average until 1991. In the 1990s, by contrast, its performance was significantly worse than average for several reasons. The first was the strong deflation associated with the reduction of public sector deficits. In 1992 there was a currency crisis. At that stage, public expenditure as a share of GDP was 4.1 per cent higher than the EU15 average, while the share of expenditure on goods and services was 2.4 per cent less than the EU15 average due to the weight of debt service. Given the requirement that Italy satisfy the convergence criteria for Economic and Monetary Union, Italian governments implemented an intense and rapid process of fiscal adjustment that significantly reduced growth rates, demonstrating the way in which macroeconomic demand management can affect comparative economic growth. The second reason was that the supply side the Italian economy suffered from its specialization in traditional industries with a low technological content (see Chapter 6), and from relatively low rates of labour force participation.

In eastern Europe the collapse of Communism opened up new markets and provided new investment opportunities for western European capital. The potential advantages were limited, however, by the dramatic, but differentiated, economic collapse that accompanied the transformation of the CEEC economies. Generally speaking, the Communist economies, which had grown rapidly after WWII, but had stagnated in the 1980s, suffered from sharp declines in output with transition. In the case of the former USSR, for example, output per head declined by nearly one-half (from 121 per cent in 1991 to 65.7 per cent in 2001). In the Visegrad Four the fall in output was less sharp, and 1989 levels of GDP per head were generally achieved again in the second half of the 1990s (with a per capita GDP of 124.7 in 2001 compared with 125.3 in 1985).

A similar picture prevailed amongst Italy's neighbours in North Africa, the east Mediterranean and the Middle East. After achieving faster than average rates of growth in the years up until 1975 or 1985, growth rates declined, and GDP per head fell sharply compared with the world average. As a result Italy's potential markets in these areas grew more slowly, although the negative effect of their deteriorating economic performance on Italy and Europe was offset by several factors. One was the reduction in oil prices which accounted for much of the decline in per capita income in oil producing states. Another was the way structural adjustment and trade liberalization further opened up their markets for European exports and investment. Yet another was the reconstruction programmes in war-affected economies which provided large contracts for individual western companies.

Official Statistics, Unrecorded Activities and the Measurement of Output

In using official GDP statistics to measure the size of an economy, it is important to recognize that GDP is an estimate of the new wealth created in a particular territory in a particular period, and that contemporary GDP estimates include estimates of the size of some economic activities that are unrecorded. The main unrecorded economic activities comprise household self-provisioning and informal, irregular and criminal activities. An important characteristic of the latter is that they involve usually conscious attempts to avoid official detection. One reason why is the criminal character of some of these activities. Another is a desire to avoid legal reporting requirements and the institutional rules of the game, usually to circumvent the costs, especially taxation, associated with them, while foregoing some of the benefits of the framework of legal and institutional regulation. A third reason why activities are unrecorded is a failure to complete official documentation.

In the case of Italy, ISTAT (Italian National Statistics Office) produces estimates of irregular work. Irregular work is defined in accordance with the OECD definition as:

> illegal production, defined as those productive activities that generate goods and services forbidden by law or that are unlawful when carried out by unauthorized producers; underground production, defined as those activities that are productive and legal but are deliberately concealed from the public authorities to avoid payment of taxes or complying with regulations; informal

sector production, defined as those productive activities conducted by enterprises in the informal sector; production of households for own use, defined
as those productive activities that result in goods or services consumed or
capitalized by the households that produced them. (OECD, 2002: 13–14;
ISTAT, 2004a: 2)

Compared with other developed economies, Italy has a particularly large
nonobserved economy partly as a consequence of the combination of a
relatively large share of domestic self-employment and a strong orientation
towards nondeclaration. This orientation towards nondeclaration is itself
due to several factors. One is the relatively low productivity of many small-
scale and underground activities. Another is the existence of relatively high
wage costs and a strong system of social protection of wage earners. A third
is the reproduction of labour supply conditions that permit the continuing
recruitment of suitable labour. The most recent example is the recruitment
of illegal immigrants to work in agriculture, construction, care of the
elderly, cleaning, catering, other personal services and street selling which
complements underground domestic manufacturing. A fourth factor is the
weak enforcement of legislation. A fifth is the existence of a moral code that
does not discourage free riders who benefit from public expenditure but do
not help pay for it (Reyneri, 1998).

Estimates of the size of Italy's shadow economy differ. Schneider and
Ernste (2000) reported estimates for 1989–93. Two principal methods
were used. A physical (electricity) input method suggested that Italy's
shadow economy accounted for some 19.6 per cent of Italian GDP. An
alternative currency demand method generated estimates of from 20.4 to
24 per cent. In 1997, the shadow economy was considered to have reached
27.3 per cent of GDP. More recent ISTAT estimates are smaller. Although
ISTAT estimates do not include illegal activities, the underlying methodology is far more reliable than the macroeconomic methods deployed by
Schneider and Ernste. According to ISTAT (2004a: 4), in 1992, the
shadow economy accounted for 12.9 to 15.8 per cent of GDP, while in
2002 it reached 15.1 to 16.2 per cent of GDP. Compared with other
developed market economies the estimates for Italy are large. Although
the shadow economy is a phenomenon that is growing in importance in all
wealthy countries, its high share of Italian output does make Italy unique in
the panorama of western economies.

To improve its GDP estimates, ISTAT also estimates the scale of irregular work (Table 4.2). These estimates suggest that in 2002 14.2 per cent of
work was irregular, with the highest shares in agriculture (33.7 per cent),
services (15.5 per cent) and construction (13.9 per cent). At the same time
there were marked regional contrasts, with irregular work accounting for
approximately 9.5 per cent of the total in the northwest, 10.3 per cent in the

northeast, 13.3 per cent in the centre and 23.1 per cent in the Mezzogiorno. In the case of the south 42.1 per cent of work in agriculture, 27.4 per cent in construction, 21.2 per cent in services and 16.5 per cent in manufacturing was irregular. Calabria is the region where irregular work is most widespread in all sectors of the economy.

These estimates are combined with estimates of the productivity of irregular work to derive adjusted GDP estimates. The figures we shall use to examine the size of Italy's regional economies therefore include estimates of the size of the nonobserved economy.

At the same time the Italian government has sought to eradicate irregular work. In 2001, Law 383/2001 offered a series of tax concessions for companies and workers that submit a request for regularization. This law also offered some small financial incentives and guaranteed that no administrative, legal or financial penalties would be imposed for past irregularities. So far the results have been extremely disappointing. One reason why is that the fiscal concessions were not sufficiently attractive, and are widely expected to increase in the years to come. Censis (2003) suggested that another reason is the universal nature of the measures: an effective regularization policy requires, Censis argued, a set of measures that allow for the territorial and sectoral nature of the phenomenon (as it is more widespread in the south, centre and to a lesser extent the northeast, and as it is more concentrated in sectors such as construction, personal services, agriculture and tourism where production units are less efficient).

The underestimation of GDP is not just a result of the existence of undeclared work. GDP is also underestimated because of the impact on estimates of the Gross Operating Surplus of widespread corporate tax avoidance. Major corporations and wealthy employers recruit accountants and lawyers to organize their affairs and their economic transactions so as to avoid national taxes. A common practice is to place novel interpretations on the concepts of residence, domicile, transaction and jurisdiction. Individuals can, for example, declare that they reside officially in another country to avoid paying taxes in countries in which they earn income. In this way, these individuals can either avoid paying taxes in any country, or pay them only in tax havens. Corporations can similarly reduce their tax liabilities by operating through offshore companies. Similarly, multinational companies can price goods and services involved in intragroup transfers to avoid taxes: inflating inventory purchase prices to raise costs and reducing inventory sales prices to reduce revenues will reduce declared profits in high tax countries, while the corresponding inflation of sales prices and reduction of purchase prices in areas with no or artificially low taxes will raise declared profits and reduce aggregate profits taxes. Some jurisdictions make provision in exchange for a fee for the establishment of special international companies and for tax-exempt companies which

Table 4.2   Share or irregular work in 2002 (full-time equivalent irregular jobs as a share of the total)

	Total	Agriculture	Industry	Manufacturing	Construction	Services
Piemonte	9.8	19.8	3.2	3.2	3.6	12.7
Valle d'Aosta	15.3	29.6	2.6	1.3	4.1	19.0
Lombardia	8.9	21.4	2.4	1.7	5.8	12.1
Trentino-Alto Adige	11.9	21.1	6.0	7.0	4.3	13.1
Veneto	9.8	28.9	2.3	1.7	5.0	13.0
Friuli-Venezia Giulia	13.0	30.5	3.7	2.9	6.6	15.9
Liguria	12.0	26.2	6.6	5.1	9.3	12.9
Emilia-Romagna	9.8	24.6	3.3	3.6	1.8	11.9
Toscana	10.8	18.4	4.8	4.4	6.3	13.1
Umbria	13.8	23.8	6.6	5.7	9.3	16.4
Marche	11.4	27.2	2.8	2.7	3.2	15.5
Lazio	15.5	39.0	14.8	8.5	24.6	14.7
Abruzzo	13.7	28.7	9.5	5.1	21.2	14.2
Molise	20.9	30.0	18.0	17.0	20.2	21.0
Campania	25.1	43.3	20.5	17.5	26.6	24.8
Puglia	21.1	41.6	18.1	13.8	26.9	18.6
Basilicata	21.5	34.5	26.4	28.3	23.3	17.1
Calabria	30.0	53.0	34.4	28.2	41.1	23.1
Sicilia	25.0	44.6	25.8	22.2	31.2	22.1
Sardegna	17.7	25.6	12.2	10.6	14.7	18.3
Italy	14.2	33.7	7.6	5.5	13.9	15.5
Northwest	9.5	21.5	2.9	2.2	5.5	12.4
Northeast	10.3	26.3	3.1	2.9	4.0	12.9
Centre	13.3	29.1	7.9	5.3	15.0	14.4
Mezzogiorno	23.1	42.1	20.4	16.5	27.4	21.2

Source: ISTAT, 2004a.

trade with no local parties yet are permitted to register profits in the jurisdiction concerned. A consequence is once again an artificial inflation of value added in tax havens and a reduction of measured value added in high tax areas.

The impact of these practices on measured output and income is substantial: a 1997 report by Deloitte and Touche for the European Commission is thought to have estimated that tax avoidance schemes run by companies and wealthy individuals cost EU15 Member States in the region of £85 billion to £100 billion per year (Sikka, 2003).

Crime is another factor that has impacts on the relationship between GDP and its measurement. The GDP of areas in which crime and in

Table 4.3   Incidence of violent and organized crime, 1988–2001

	Violent crimes per '000 residents per year		Organized crime per '000 residents per year
	1988–94	1995–2001	1995–2001
Campania	20.1	21.1	2.2
Puglia	14.4	10.7	4.1
Basilicata	8.2	7.7	2.0
Calabria	11.9	10.0	7.0
Sicilia	20.4	14.9	5.9
Sardegna	10.1	10.7	6.1
Italy	11.1	11.9	2.9

Source: elaborated from ISTAT, 2000c; 2003.

particular organized crime is most present is adversely affected by the the negative impact of criminality on the attraction of investment. In Italy, in 1988–2001, the areas most affected included Campania, in which the Camorra operates, Sicilia and Puglia, where the incidence of criminality rose due in no small measure to the fact that it lay just across the Adriatic from Albania and the disintegrating former Yugoslavia (Table 4.3). At the same time, crime can impose other costs on an economy which paradoxically are seen as making a positive contribution to GDP. Examples include the costs of increased security and many of the costs associated with dealing with the consequences of crime. Additionally, criminal activities generate income, employment and investment funds (Arlacchi, 1986).

In the case of the province of Brindisi in Puglia, for example, a recent report of the parliamentary Commissione Antimafia (1999) pointed out that a Mafia emerged out of the illegal smuggling, first, of cigarettes, and, later, of drugs, arms and illegal immigrants (considered to exceed 10,000 per year). In the north of the province these activities fell under the control of the Sacra Corona Unita, and in the south under Albanian Mafia control. Together these illegal activities have a major economic effect: Italian police estimates valued them at up to €5 million per day when the sea was calm.

Extortion is another major source of criminal revenue. The Sacra Corona Unita imposes a fee of €5 on every case of cigarettes brought in by foreign smugglers. Many local figures also help finance this illegal trafficking by placing bets on particular speedboats and sharing in the proceeds if the speedboat's cargo is successfully shipped across the Adriatic. Illegal profits are laundered with the collaboration of local enterprises and banks. The Ufficio Italiano dei Cambi (Italian Foreign Exchange

Office) identified 244 suspicious cases in Puglia in one year. Only two regions had more cases. At the same time corrupt bank officials and Swiss-based financiers facilitate the export of up to €515–€775,000 per week to Switzerland.

Some of the money earned from trafficking is reinvested in construction projects. Attempts are made to influence the *piani regolatori* (land use plans) drawn up by public authorities. In addition, some local officials have been suspended for suspected collusion with criminal organizations, while some public procurement tenders have been subject to judicial investigation. Other illegal profits are reinvested in local retailing and distribution where the mechanism of overinvoicing permits the laundering of criminally derived monies alongside clean money. As a result of their access to finance for investment, smugglers have gained more and more power and have purchased shops, travel agencies and wholesalers. Some criminal organizations employ people off the cards. Another course of action is to set up finance companies to lend money at usurious rates of interest in part as a means of laundering money. The number of finance companies in the province of Brindisi far exceeds the number one would expect given the size of its legal economy. Illegal trafficking has generated orders for Adriatic-coast shipyards in Pesaro, Salento and Albania. Often scafisti steal speedboats from ports in Puglia. As a consequence, insurance companies will no longer insure speedboats anywhere in the region.

As a result of their economic muscle, local criminal organizations establish deep social roots and can rely on the support of large sections of the local population. In a province in which 50,000 people (12 per cent of the population) were out of work, the smuggling of cigarettes provided financial support for thousands of families that, for this reason, did not regard it as a criminal activity. Instead, cigarette smuggling was often seen as offering jobs and incomes. Young people, for example, missed classes to help smugglers unload speedboats for €77 per day. In aggregate these activities generated an annual turnover of some €515 million in a province whose GDP stood at about €3615 million and provided work and income for 5,000 families. These incomes depended, however, on on the extent to which the traffickers could outwit the Italian authorities in their struggle to combat this trade. An important step forward was made in the late 1990s when the Italian authorities negotiated the right of hot pursuit with the Montenegran authorities. As a result, in March 1998, the police chief of the Montenegran port of Bar, in the former Yugoslavia, was arrested by the Italian police. Subsequently, he was prosecuted and imprisoned for covering up the illegal activities of Italian criminals in Montenegro. This success played an important part in checking illegal trafficking from the port of Bar and stemming the associated flows of income in Puglia.

GDP, Net Transfers and Regional Income

Although GDP is a measure of the new wealth created in a particular area, it is not a measure of the income of the local population. The reason why is twofold. First, there are nonlocal claims over the wealth produced in any area (nonresidents' claims on the profits of economic activities, and wages paid to nonresidents), while an area's inhabitants may have claims over wealth created elsewhere. Second, there are transfers of income that result from national and international systems of income redistribution. These net transfers may be positive or negative. Areas with positive net transfers can import more than they export: net transfers fund net imports.

In the case of Italy financial transfers, associated with the primary distribution of income to the owners of factor inputs, and with the secondary distribution of income arising from state redistribution, are captured in regional accounts. Table 4.4 records the net imports of each of Italy's regional economies in 2000. Available or disposable regional income is equal to value added at factor cost plus net imports. As disposable income is either spent on consumption or saved, disposable income is also equal to final consumption expenditure plus fixed capital formation plus changes in stocks (including acquisitions less disposals of valuables), while final consumption expenditure can be divided into household final consumption expenditure and consumption expenditures incurred by general government and by nonprofit institutions serving households (NPISHs).

Table 4.4 shows that in 2000 per capita disposable income inequalities were substantially smaller than inequalities in GDP per head: while per capita GDP in the Mezzogiorno stood at 67.5 per cent of the Italian average, per capita disposable income stood at 80.1 per cent. Fixed capital formation was relatively weak (69.9 per cent of the national average), while household (82.9 per cent) and especially government consumption (99.4 per cent) were much closer to the national average.

The reason why per capita disposable income disparities are smaller than per capita GDP disparities is that, overall, Italy's net transfers reduce inequality. This redistributional role is a result of the fact that net transfers are inversely correlated with GDP per head, as the downward sloping relationship in Figure 4.2 shows. Even setting aside the special cases of a number of frontier areas in the north (Valle d'Aosta and Trentino-Alto Adige) the correlation is not perfect however. As a result the rank orders of regional economies differ on the two indicators. In the case of the south, for example, Calabria moves ahead of Campania and Puglia, while the two island economies of Sardegna and Sicilia are also recipients of relatively high net transfers per capita.

Table 4.4 Net transfers, disposable income, consumption and investment in 2000

	Net imports per head in 1995 Euro LIT	GDP per head	Disposable income per head	Final consumption expenditure per head				Gross fixed capital formation per head	Changes in stocks and valuables per head
				of house-holds	Total of NPISHs		of govern ment		
				as percent of Italian average					
Year	2000	2000	2000	2000	2000	2000	2000	2000	2000
Piemonte	−1,241	116.5	110.2	104.4	107.9	101.6	91.8	131.9	73.9
Valle d'Aosta	1,566	137.5	147.4	156.1	146.6	89.6	192.5	144.7	3,508.1
Lombardia	−3,264	129.5	111.6	108.4	113.1	112.3	91.6	122.7	15.4
Trentino-Alto Adige	2,296	132.9	147.0	134.3	130.7	173.5	146.1	197.6	361.5
Veneto	−1,338	118.2	111.3	108.4	112.6	117.6	93.0	121.8	58.3
Friuli-Venezia Giulia	−690	113.4	110.2	108.4	108.5	97.4	108.4	118.5	293.6
Liguria	−2	106.7	107.4	115.1	115.7	105.8	113.3	78.1	128.0
Emilia Romagna	−1,307	127.6	120.9	117.7	122.0	147.0	101.3	132.1	1.4
Toscana	−1,309	111.8	105.0	108.5	111.3	103.2	98.6	91.0	19.3
Umbria	−53	98.7	99.1	98.6	97.2	83.7	103.9	103.6	383.0
Marche	−203	103.2	102.7	103.3	103.1	112.6	103.8	101.7	245.7
Lazio	−21	106.5	107.0	108.7	107.7	142.0	111.4	100.0	37.5
Abruzzo	697	85.4	90.0	90.0	88.0	79.7	97.7	91.4	272.9
Molise	1,458	79.9	88.8	87.7	82.4	149.2	105.2	102.8	1,162.2
Campania	1,873	63.6	74.7	77.8	74.1	41.5	92.3	62.7	47.9
Puglia	1,749	67.3	77.8	80.8	78.7	77.6	88.7	66.3	76.0
Basilicata	1,192	74.1	81.4	79.4	73.2	84.0	101.7	94.5	671.7
Calabria	2,821	61.9	78.5	82.4	78.3	48.7	98.1	64.3	165.4
Sicilia	2,725	66.0	82.1	86.2	79.4	74.1	111.3	66.3	73.9
Sardegna	2,577	74.9	90.2	90.6	83.6	92.2	115.6	89.8	209.0
Italy	−116	100.0	100.0	100.0	100.0	100.0	100.0	100.0	100.0
Northwest	−2,302	123.4	111.0	108.4	112.2	108.4	94.8	120.7	71.9
Northeast	−933	122.5	117.9	114.2	117.3	131.3	102.5	132.0	90.0
Central	−457	107.1	105.2	107.2	107.5	121.4	105.7	97.6	85.1
Northern and central	−1,351	118.2	111.3	109.7	112.2	118.9	100.3	117.0	81.1
Mezzogiorno	2,107	67.5	80.1	82.9	78.4	66.5	99.4	69.9	133.4

Source: elaborated from ISTAT, 2003b.

The existence of net transfers to less-developed areas is a normal consequence of the operation of the public expenditure system under which tax
contributions tend to increase as income increases, while expenditures are

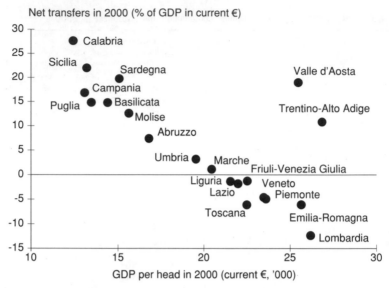

Figure 4.2 Net imports, net transfers and GDP per head in Italy in 2000.
Source: elaborated from ISTAT, 2003b.

related either to some concept of need (unemployment compensation and social security payments are examples) or to the distribution of the population (educational and health expenditures are cases in point). As a result, expenditures tend to exceed revenues in less-developed areas, and vice-versa in more-developed areas. In Italy, however, the existence of large net transfers to the south has emerged as a major political issue in some of the areas that are large net contributors. At the same time, as in other European countries, there are moves towards a reduction in the progressiveness of taxation which will reduce the degree of redistribution. Greater political decentralization will have a similar effect if a greater share of tax revenue is retained locally (see Chapter 5).

As Table 4.5 shows, there is already some evidence of a decline in the degree of redistribution in Italy. The figures, which relate to the period from 1980 to 2000, are derived from two different systems of national accounts. The figures for the real per capita value of net transfers to the Mezzogiorno rose from LIT 2,119 thousand in 1980 to 3,035 in 1990 but fell to 2,190 in 1995. Comparing the 1995 estimates, the new series produced higher estimates (17.0 per cent of GDP) than the old series (14.0 per cent). According to this new series, the real value of transfers to the south increased up to 2000. As a share of GDP, however, it remained beneath the 1990 peak of 19.6 per cent (which would conceivably had been

Table 4.5 Evolution of net imports and net transfers in Italy, 1980–2000

| Year | Net imports | | | | | Net imports as a percentage of GDP | | | | | |
| | Thousand 1990 LIT | | | 1995 EuroLIT | | Current prices | Constant prices | | | | |
	1980	1990	1995	1995	2000	1980	1980	1990	1995	1995	2000
Piemonte	−983	−1,794	−3,610	−2,010	−1,241	−3.8	−4.6	−6.8	−13.1	−10.6	−6.1
Valle d'Aosta	1,442	4,206	7,017	1,283	1,566	10.8	5.5	14.1	23.0	5.6	6.5
Lombardia	−2,465	−3,705	−4,646	−3,573	−3,264	−9.6	−10.6	−12.4	−15.0	−16.8	−14.4
Trentino-Alto Adige	2,053	2,323	2,214	1,426	2,296	16.7	9.0	8.4	7.4	6.6	9.9
Veneto	−34	−869	−3,209	−1,505	−1,338	3.0	−0.2	−3.3	−11.0	−7.9	−6.5
Friuli-Venezia Giulia	723	−615	−2,368	−1,548	−690	8.9	3.4	−2.3	−8.0	−8.3	−3.5
Liguria	−477	−437	−1,194	−371	−2	3.9	−2.2	−1.7	−4.4	−2.2	−0.0
Emilia Romagna	−1,568	−2,173	−3,582	−1,747	−1,307	−4.5	−6.5	−7.6	−11.4	−8.5	−5.8
Toscana	−579	−708	−931	−811	−1,309	1.5	−2.8	−2.9	−3.6	−4.6	−6.7
Umbria	621	1,249	806	117	−53	7.6	3.4	5.8	3.5	0.7	−0.3
Marche	302	282	40	−342	−203	4.6	1.5	1.2	0.2	−2.1	−1.1
Lazio	−1,804	−731	−953	−1,008	−21	0.2	−9.0	−2.9	−3.6	−5.7	−0.1
Abruzzo	1,477	2,015	505	384	697	19.1	9.0	10.0	2.4	2.8	4.0
Molise	3,920	2,684	2,521	1,075	1,458	38.0	28.5	15.9	14.2	8.6	10.4
Campania	1,643	2,339	1,761	1,726	1,873	21.2	12.8	15.3	11.9	17.0	16.8
Puglia	1,404	1,953	993	1,508	1,749	16.1	10.3	12.2	6.1	14.4	14.8
Basilicata	3,072	3,810	2,667	1,778	1,192	29.6	24.8	27.3	17.3	16.3	9.2
Calabria	3,292	5,485	4,402	2,663	2,821	38.3	29.9	43.1	32.3	27.4	26.0
Sicilia	2,673	3,636	2,766	2,194	2,725	29.1	21.0	24.3	18.5	20.9	23.5
Sardegna	2,004	3,642	3,066	1,959	2,577	24.3	14.1	21.7	17.7	16.3	19.6
Italy	28	123	−828	−452	−116	4.4	0.1	0.5	−3.5	−2.8	−0.7
Northwest	−1,765	−2,721	−3,872	−2,732	−2,302	−6.4	−7.8	−9.6	−13.1	−13.6	−10.6
Northeast	−353	−1,059	−2,781	−1,345	−933	1.7	−1.6	−3.9	−9.2	−6.8	−4.3
Central	−940	−443	−684	−773	−457	1.8	−4.7	−1.8	−2.7	−4.4	−2.4
NCE	−1,119	−1,562	−2,598	−1,743	−1,351	−1.9	−5.2	−5.8	−9.1	−9.1	−6.5
Mezzogiorno	2,119	3,035	2,190	1,818	2,107	24.2	16.2	19.6	14.0	17.0	17.8

Source: elaborated from ISTAT, 1998 and 2003b.

yet higher if the new accounting system had been used). At that stage, net transfers added nearly one-fifth to southern disposable income. Note also that the current currency estimate for 1980 stood at 24.2 per cent, although in constant 1990 prices the share was much smaller, standing at 16.2 per cent.

Figure 4.3 provides a graphical representation of the relationships set out in Table 4.5 for four regional economies on which we shall later concentrate. The income that accrues to a regional economy's inhabitants is equal to its GDP plus/minus net transfers. Called Gross Regional Product (GRP)

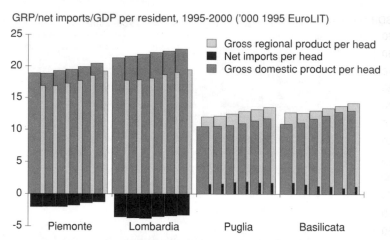

Figure 4.3   GDP, Gross Regional Product and net transfers, 1995–2000.
Source: elaborated from ISTAT, 2003b.

its per capita value is equal to the height of the grey column starting from
the zero line. In Piemonte and Lombardia, GDP exceeded GRP as there
were negative net imports, represented by the columns beneath the horizontal axis, while in the Puglia and Basilicata imports exceeded exports and
GRP exceeded GDP.

Territorial Inequality in Italy at the Turn of the Millennium

How large were the differences in development between Italian regions and
provinces at the end of the second millennium? To answer this question,
and bearing in mind the qualifications concerning the size of nonobserved
economic activities, Figure 4.4 plots as circles 2000 estimates of value
added net of an adjustment for financial services[4] for each of 102 Italian
provinces. The provincial estimates are grouped by region and expressed as
a percentage of the EU15 average, while the regional averages are plotted as
diamonds. More detailed data for the largest, richest and poorest provinces
are provided in Table 4.6.

As Figure 4.4 and Table 4.6 show, in 2000 Italy's leading province was
Milan (159 per cent of EU15 average). Next came Bolzano (147 per cent),
Modena (143 per cent), Bologna (142 per cent) and Parma (132 per cent).
At that date, Milan accounted for 10.2 per cent of Italian value added, net
of imputed financial services, or nearly one-half of the share of Lombardia

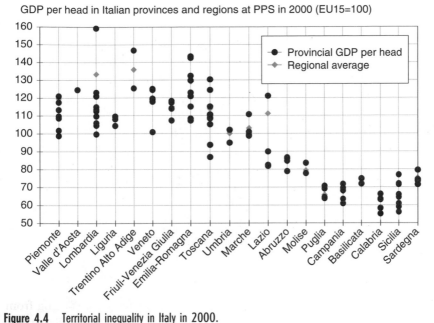

GDP per head in Italian provinces and regions at PPS in 2000 (EU15=100)

Figure 4.4 Territorial inequality in Italy in 2000.
Source: elaborated from ISTAT, 2003c.

as a whole (20.6 per cent). Bolzano (1.2 per cent of Italian value added), Modena (1.5 per cent), Bologna (2.2 per cent) and Parma (0.9 per cent) were substantially smaller economies (Table 4.6). Rome which accounted for 7.9 per cent of value added, Turin 4.7, Naples 3.4 and Brescia 2.3 were the next most important regional economies after Milan (Table 4.6).

At the other end of the spectrum lay a series of southern provinces in Calabria, Sicilia, Puglia and Campania with the lowest per capita value added figures for Crotone (54.2 per cent), Agrigento (55.5 per cent), Vibo Valentia (57.3 per cent), Enna (58.0 per cent) and Caltanissetta and Lecce (59.8 per cent). Of these economies Lecce accounted for 0.8 per cent of national value added. The others were in the range 0.2–0.4 per cent (Table 4.6).

The second feature of Figure 4.4 is the sharp north-south contrast it reveals (Figure 4.5 plots the same data cartographically). Of the 46 provinces in the regions from Emilia-Romagna northwards, just two lay (narrowly) beneath the EU15 average. By contrast, the highest score for the 36 provinces in the regions entirely in the Mezzogiorno was 86.6 per cent of the EU15 average for the coastal province of Pescara in Abruzzo, ranked 65th nationally. Twenty-eight of these southern provinces lay beneath

Table 4.6 Italy's provincial economies: the largest, the richest, the poorest, the fastest growing and the slowest growing

Province	Region	GDP in 2000 (% of Italy)	GDP per head in 2000 (EU15 = 100)	GDP per head in 2000 rank	Average annual growth 1995–2000 rank	Average annual industrial growth rank	Average annual construction growth rank	Average annual service growth rank	Change in GDP per head rank, 1995–2000
The largest									
Milan	Lombardia	10.2	159	1	39	65	51	7	0
Rome	Lazio	7.9	121	16	56	55	43	87	4
Turin	Piemonte	4.7	126	8	37	59	45	20	9
Naples	Campania	3.4	64	93	13	81	13	23	2
Brescia	Lombardia	2.3	121	18	83	70	68	47	–4
Bologna	Emilia-Romagna	2.2	142	4	48	69	18	46	0
Florence	Toscana	2.1	130	6	20	27	6	38	6
Bergamo	Lombardia	2.0	121	19	74	72	41	62	–1
Bari	Puglia	1.9	72	81	16	12	25	35	3
Padua	Veneto	1.7	118	24	26	15	63	34	10
The richest									
Milan	Lombardia	10.2	159	1	39	65	51	7	0
Bolzano-Bozen	Trentino-Alto Adige	1.2	147	2	23	45	11	36	1
Modena	Emilia-Romagna	1.5	143	3	57	44	47	54	–1
Bologna	Emilia-Romagna	2.2	142	4	48	69	18	46	0
Parma	Emilia-Romagna	0.9	132	5	67	43	58	90	2
Florence	Toscana	2.1	130	6	20	27	6	38	6
Reggio Emilia	Emilia-Romagna	1.0	130	7	91	50	40	97	–1
Turin	Piemonte	4.7	126	8	37	59	45	20	9

Table 4.6 Italy's provincial economies: the largest, the richest, the poorest, the fastest growing and the slowest growing

		1.0	125	9	66	80	52	71	2
Trento	Trentino–Alto Adige								
Belluno	Veneto	0.4	125	10	34	19	75	37	9
The poorest									
Crotone	Calabria	0.2	55	103	1	3	1	5	0
Agrigento	Sicilia	0.4	56	102	77	14	66	89	−2
Vibo Valentia	Calabria	0.2	58	101	28	23	93	42	0
Enna	Sicilia	0.2	59	100	5	2	74	8	2
Caltanissetta	Sicilia	0.3	61	99	59	99	76	4	−3
Lecce	Puglia	0.8	61	98	25	42	39	30	1
Reggio Calabria	Calabria	0.6	63	97	55	61	84	51	−4
Foggia	Puglia	0.7	63	96	11	28	42	17	1
Cosenza	Calabria	0.8	63	95	7	13	54	3	3
Benevento	Campania	0.3	64	94	69	46	99	59	−6
The fastest growing									
Crotone	Calabria	0.2	55	103	1	3	1	5	0
Matera	Basilicata	0.3	72	80	2	1	65	2	10
Siena	Toscana	0.5	115	28	3	8	22	18	18
Grosseto	Toscana	0.3	94	62	4	24	28	9	4
Enna	Sicilia	0.2	59	100	5	2	74	8	2
Isernia	Molise	0.1	84	68	6	4	78	10	5
Cosenza	Calabria	0.8	63	95	7	13	54	3	3
Genoa	Liguria	1.7	108	46	8	18	27	33	6
Sassari	Sardegna	0.6	80	72	9	96	81	1	4
La Spezia	Liguria	0.4	109	41	10	20	8	27	10
Foggia	Puglia	0.7	63	96	11	28	42	17	1

(Contd.)

Table 4.6 Italy's provincial economies: the largest, the richest, the poorest, the fastest growing and the slowest growing

Province	Region	GDP in 2000 (% of Italy)	GDP per head in 2000 (EU15 = 100)	GDP per head in 2000 rank	Average annual growth 1995–2000 rank	Average annual industrial growth rank	Average annual construction growth rank	Average annual service growth rank	Change in GDP per head rank, 1995–2000
Potenza	Basilicata	0.5	75	76	12	7	97	29	2
Naples	Campania	3.4	64	93	13	81	13	23	2
Pistoia	Toscana	0.5	105	50	14	25	2	19	5
Nuoro	Sardegna	0.3	74	77	15	78	56	6	2
The slowest growing									
Aosta	Valle d'Aosta	0.3	125	11	103	103	102	98	−6
Viterbo	Lazio	0.4	82	69	102	53	103	102	−6
L'Aquila	Abruzzo	0.4	79	73	101	102	79	103	−6
Lecco	Lombardia	0.6	113	32	100	84	77	85	−11
Mantua	Lombardia	0.8	123	14	99	86	71	77	−6
Frosinone	Lazio	0.7	82	71	98	89	94	95	−6
Como	Lombardia	1.0	110	39	97	92	85	78	−10
Pavia	Lombardia	0.8	100	57	96	91	34	92	−8
Brindisi	Puglia	0.5	68	87	95	97	59	64	−10
Pordenone	Friuli-Venezia Giulia	0.6	118	23	94	76	82	60	−8

Source: ISTAT, 2003c.

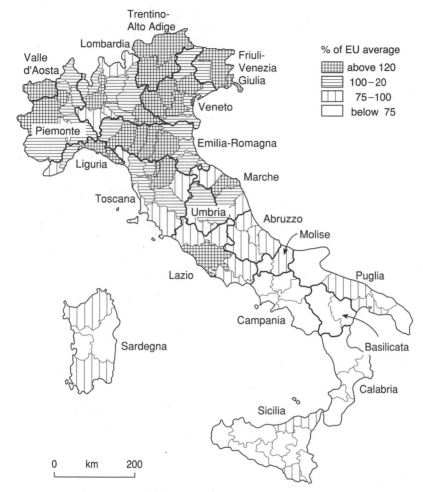

Figure 4.5 Mapping territorial inequality in Italy in 2000.
Source: elaborated from ISTAT, 2003c.

75 per cent of the EU15 average. Included were the whole of Campania, Basilicata, Puglia and Calabria.

An important way of measuring the sharpness of this north-south divide is by decomposing the Theil coefficient (Table 4.7 and Appendix 1). (The Theil coefficient is a measure of inequality. The coefficient can vary from 0 to infinity with 0 representing an equal distribution and higher values greater degrees of inequality). Table 4.7 shows that in 2000 the Theil coefficient (computed using natural logarithms and multiplied by 1000) for inequalities in value added per capita stood at 34.2 for the four

Table 4.7 Decomposition of the Theil coefficient, 1995–2000

	Theil index × 1,000						Share of total (%)
	1995	1996	1997	1998	1999	2000	2000
Between Areas	36.9	37.2	35.7	35.4	34.4	34.2	81.5
Within Areas	2.5	2.6	2.3	2.3	2.0	2.0	4.8
Between Regions	39.4	39.8	38.0	37.7	36.4	36.2	86.3
Within Regions	5.0	4.7	4.8	5.0	5.6	5.8	13.7
Between Provinces	44.5	44.5	42.8	42.7	42.0	41.9	100.0
Within Areas	2.5	2.6	2.3	2.3	2.0	2.0	4.8
Northwest	1.1	1.1	1.0	1.0	0.9	0.8	1.9
Northeast	0.2	0.3	0.2	0.3	0.3	0.3	0.7
Centre	0.2	0.2	0.2	0.3	0.2	0.2	0.4
South	1.0	1.0	0.9	0.8	0.7	0.7	1.7
Within Regions	5.0	4.7	4.8	5.0	5.6	5.8	13.7
Piemonte	0.4	0.2	0.2	0.2	0.2	0.2	0.5
Valle d'Aosta							
Lombardia	2.2	2.1	2.2	2.5	3.0	2.9	7.0
Liguria	0.0	0.0	0.0	0.0	0.0	0.0	0.0
Trentino-Alto Adige	0.0	0.0	0.0	0.1	0.0	0.1	0.2
Veneto	0.1	0.1	0.1	0.1	0.1	0.1	0.2
Friuli-Venezia Giulia	0.0	0.0	0.0	0.0	0.0	0.0	0.0
Emilia-Romagna	0.4	0.4	0.5	0.4	0.4	0.5	1.1
Toscana	0.5	0.5	0.4	0.4	0.4	0.4	1.0
Umbria	0.0	0.0	0.0	0.0	0.0	0.0	0.0
Marche	0.1	0.0	0.0	0.0	0.0	0.0	0.1
Lazio	0.8	0.8	0.8	0.9	1.0	1.1	2.7
Abruzzo	0.0	0.0	0.0	0.0	0.0	0.0	0.0
Molise	0.0	0.0	0.0	0.0	0.0	0.0	0.0
Campania	0.1	0.1	0.0	0.0	0.0	0.0	0.1
Puglia	0.1	0.1	0.1	0.1	0.1	0.1	0.2
Basilicata	0.0	0.0	0.0	0.0	0.0	0.0	0.0
Calabria	0.1	0.1	0.1	0.0	0.0	0.0	0.1
Sicilia	0.2	0.2	0.2	0.2	0.2	0.2	0.4
Sardegna	0.0	0.0	0.0	0.0	0.0	0.0	0.0

Source: elaborated from data in ISTAT, 2003c.

territorial divisions (northwest, northeast, centre and south), 36.2 for the 20 regions and 41.9 for the 103 provinces. Consider the case of interprovincial differences. A feature of the Theil index is that interprovincial differences can be disaggregated into differences between the territorial divisions, differences within the territorial divisions (regional differences), and differences within the regions (provincial differences within the regions). As

Table 4.7 shows, 81.5 per cent of the differences are due to differences between the 4 territorial divisions. Just 4.8 per cent are due to regional differences within these four divisions, with the largest contributions made by differences in the northwest (1.9 per cent) and south (1.7 per cent). Another 13.7 per cent were due to differences within the regions. The regions with the largest internal differences were Lombardia (7.0 per cent) and Lazio, containing Roma, (2.7 per cent). In Italy, therefore, the most fundamental aspect of territorial inequality was the gap between the four territorial divisions, though there were nontrivial contributions from inequalities within the northwest and south and within Lombardia and Lazio. Clearly, Italy's territorial inequalities are first and foremost a macro-territorial phenomenon.

Catching Up, Falling Behind, Surging Ahead and Losing Ground: Trends in Italian Regional Development

The overall trend in Italian territorial inequality since 1951 is depicted in Figure 4.6 (see also Terrasi, 1999). This figure plots two different indicators (see Appendix 2) of the degree of inequality in GDP per head between 19 regions (as Abruzzo and Molise are treated as a single region, since data were not published separately for them until 1963). Whichever indicator is chosen, the results are essentially the same.

After a sharp decline in 1951–53 territorial inequalities increased in the 1950s in spite of the putting in place in 1950 of an agrarian reform, the establishment also in 1950 of the Cassa per il Mezzogiorno to implement a preindustrial development programme, and the extraordinary wave of emigration from the south, with a net exodus of 1.8 million people in 1952–61. In 1960, the degree of inequality was close to its 1951 level.

In 1960–75 the situation changed dramatically. For 15 years there was strong catch-up as the less-developed parts of Italy closed the gap on the more developed. The Theil coefficient fell by almost 48 per cent from 57.3 in 1960 to 29.8 in 1975, while the weighted mean absolute deviation (WMAD) declined by 26.4 per cent from 28.4 to 20.9 (Figure 4.6).

After 1975, there was a reversal in the trend. Overall there was a clear increase in inequality, with the Theil coefficient reaching 37.8 and WMAD 24.6 in 1996. Within this phase there were, however, cyclical movements: a sharp increase in 1976 was followed by a renewal of convergence until 1983, after which the trend was decidedly upwards until 1996. A not insignificant part of the relative improvement that had occurred after 1960 was reversed in 1975 and after 1983. In the last few years, conversely,

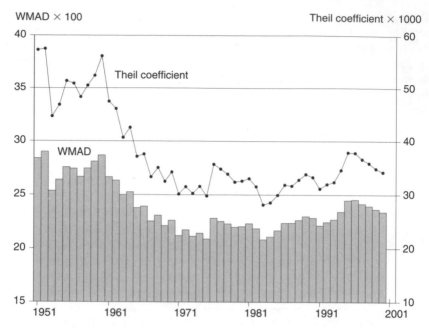

Figure 4.6   Trends in Italian regional inequality: disparities in regional GDP per resident, 1951–2000. *Source: elaborated from Istituto Tagliacarne and ISTAT data[5] and ISTAT, 1998; 2003b.*

there was a renewed phase of convergence, with the Theil coefficient declining to 34.2 in 2000.

Variations in GDP per head can be partitioned into (apparent) productivity and employment rate elements (Dunford, 1996). To examine the role of these two elements, Figures 4.6 and 4.7 plot the variation in productivity and employment rates, drawing on ISTAT data relating to the number of full-time equivalent jobs. As these data are only available from 1970, the graphs relate only to the end of the phase of convergence, the subsequent long phase of regional divergence and the very recent renewal of catch-up. The results are particularly striking.

In the first few years of the 1970s there was a rapid decline in productivity differentials. In 1976, productivity differences increased sharply. In 1976–86, there was once again a clear downward trend in differentials with some cyclical movements. In 1990, the degree of inequality descended beneath the level recorded in 1975, while in 1995–9 it declined quite rapidly.

The evolution of employment rate variations differed markedly from the evolution of productivity differentials. After a decline until 1972, employment rate differentials rose sharply, from a Theil coefficient of 8.6 in 1972

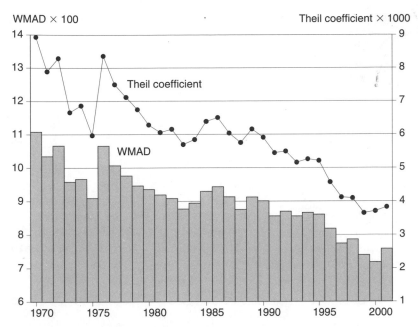

Figure 4.7   Territorial disparities in (apparent) productivity: disparities in regional GDP per full-time equivalent worker, 1970–2001.
Source: elaborated from ISTAT, 1998; 2003b.

to 13.1 in 1981. In 1981–8 and 1988—96, there were cyclical movements, with initial reductions in employment rate inequality followed by even stronger increases, and with the Theil coefficient reaching 18.6 in 1996. A renewed downturn started in 1996–2000.

Clearly, it is not productivity differentials that account for the recent overall increase in Italian regional inequality, as productivity differentials have tended to diminish, though productivity differentials remain a significant determinant of *differences* in the levels of regional development in contemporary Italy. At the root of the inequality increases up to 1996 were very sharp increases in employment rate variations. Similarly, the recent decline in disparities reflects the recent fall in employment rate differentials, although productivity convergence accelerated. A critical determinant of convergence and divergence tendencies seems, therefore, to be the extent to which the regional demand for labour is sufficient to absorb those seeking employment, as productivity increases. Growth in the demand for labour depends on the speed of accumulation and growth. Essentially, the speed of growth and of the demand for labour relative to the increase in the supply of labour was until recently weakest in the least

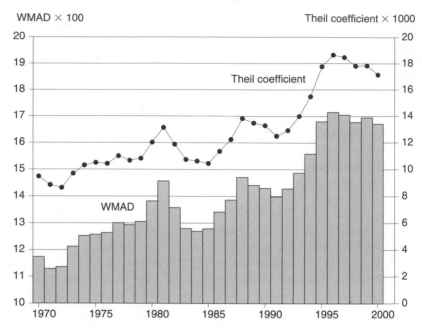

Figure 4.8 Territorial disparities in employment rates, 1970–2000.
Source: elaborated from ISTAT, 1998; 2003b.

developed parts of Italy. The recent relative improvement in the position of the south is in part a consequence of a relative decline in northern growth rates (along with important socially useful job creation schemes in the south).

Conclusions

In this chapter GDP per head was used as an indicator of the degree of uneven development in contemporary Italy. GDP per head is a measure of the new wealth created in each area in a particular year and makes partial allowance for unrecorded and criminal activities that are more important in less-developed areas. Analysis of this indicator shows that large development gaps essentially between the four main territorial units are a major feature of the geography of modern Italy. Although state redistribution serves to reduce these disparities, wide differences in average standards of living remain. In the final years of the post-war Golden Age these gaps diminished sharply, making a major contribution to the rapid rise in living standards that characterized this era. After the mid-1970s, however, divergence set in, lasting until 1996. In the latter part of this phase of development northern political movements contested state transfers to less-developed areas in the south. At the root of this divergence was a widening of employment rate differentials which themselves reflected the relative weakness of Italy's growth record.

Chapter Five

Institutional Dynamics and Regional Performance

Introduction

Although economic mechanisms play an important role in shaping the relative performance of regional economies, these mechanisms operate through the context-dependent actions of institutionally and spatially differentiated individuals and groups. At the same time, this context is shaped by these actions and their effects. This institutional and spatial context is examined in the literature on the social foundations/varieties of capitalism (Albert, 1993; Soskice, 1989) and on the social determinants of corporate and local development (such as the literature on the Third Italy). As we shall show, it includes a number of interacting tiers of governance, where the performance of one level depends on its interplay with the others, while the resulting multilevel framework defines a sociopolitical space in which actors and structures interact to determine industrial and employment trajectories. In other words, actors are differentiated across space, as are institutional endowments. Institutional interaction depends on these contexts (Magatti, 1991) and responds to context-dependent practices.

The aim of this chapter is to analyze the main characteristics of the Italian institutional system and to discuss its influence on the country's differentiated territorial development. What we shall show is that the Italian variety of capitalism involved remarkable state intervention in the economy, an extremely restricted number of large private companies and a vast number of SMEs. Underlying it was a stable political framework representing clearly defined socioeconomic interests. This institutional configuration played a crucial role in shaping individual and group economic strategies and in generating regional development paths and intraregional disparities.

In recent years, however, this institutional configuration has undergone dramatic change with in all probability important implications for future regional economic performance. Three major developments were involved. The first was a reform of the institutional architecture of the state. Included were a deconcentration and decentralization of central government responsibilities and resources in a number of fields, on the one hand, and the internalization of supranational rules and regulations, on the other. The second was a redefinition of the nature of state economic action, involving an ambitious process of privatization of publicly owned companies. The third involved a crisis and restructuring of the party system. These changes were associated with a reform of state regional policy and an important debate concerning the impact of institutional performance on regional development trajectories.

The Institutional Configuration and the Characteristics of Italian Capitalism

Until recently, Italy's institutional system was uniform and centralized. After the country's unification in 1861, reasons of territorial unity led to the adoption of a Napoleonic administrative system. After WWII, the Republic's fundamental law re-established elected communal and provincial authorities and provided in Article 117 for an additional tier of twenty regions. Four special statute regions had already been created in 1944–6 to accommodate ethnic, linguistic and cultural differences, and a fifth was added in 1964. Owing, however, to the desire of the ruling DC to limit the role of the PCI in local government, the Italian Parliament showed no initial willingness to implement Article 117 of the Constitution. Change had to wait until the 1970s. At that stage, however, the aim was not simply to create the multilevel system of governance envisaged in the Constitution. After the social conflicts of the late 1960s and early 1970s, the Italian state implemented a wide range of reforms that improved workers' rights, pensions and health care. With the subsequent growth slowdown, these reforms imposed a heavy burden on the state and its finances. Anxious to contain these costs. the state and governing parties also saw regional reform as a way of deflecting rising social demands from the central state to peripheral institutions. The performance of the newly created regions was however very mixed. The regions did not automatically acquire the recognition and legitimacy required to make them effective, while centrally determined legal provisions as well as tight financial constraints limited their autonomy (Piattoni, 1994:174–7, and 180–1). Institutional performance also varied due to differences in the availability of other critical resources such as the capabilities of their administrative and technical staff, the structure and dynamics of civil society, and the nature of political subcultures.

The political system was dominated by the DC, whose political precepts derived both from an anti-Communist and Catholic left wing tradition and a liberal tradition that had played a major role in the struggle against the Fascist regime. The DC also represented a specific set of socioeconomic interests. According to many scholars (Martinelli, 1998; Del Monte and Giannola, 1978), the strength of the DC lay in the institutional compromise it managed to forge, after 1945, between two major socioeconomic groups: the old landed aristocracy, the petty traders and the public administration concentrated mainly in the south; and a class of mainly northern entrepreneurs. Whereas the former were anxious to preserve their privileges or acquire new ones in a modernized institutional system, the latter wanted an environment conducive to the expansion of their economic activities. At the centre of this compromise was an expansion of the southern market as an outlet for northern products, although in the early years the south was also a large provider of cheap labour through a series of south-north migration waves. At the same time, economic inefficiency in the south was permitted and indeed sustained. One reason why was the systematic use of clientelistic practices in which jobs, subsidies and recommendations were exchanged for political support and social consensus. Another was the consequent emergence of middle and marginal groups capable of vetoing and obstructing reform.

The political strength of the DC and the socioeconomic interests it represented were important influences on, and reflections of, the configuration of Italian capitalism. Of particular significance were two features that distinguish the Italian model from the continental European and Anglo-American models.

The first was the scale of state intervention in the economy. After rejecting laissez-faire ideas, the DC opted for Keynesian economic management and substantial state interventionism in domestic economic life.[1] Two main reasons explain state interventionism. The first was the economic weakness of the private sector. The second was the twofold belief that (1) public ownership was the most appropriate way to provide public goods and to manage natural monopolies such as energy, water and telecommunications, and (2) the state should assume responsibility for the modernization of strategic industries. At the centre of this strategy were powerful state holding companies such as IRI and ENI and industrial credit agencies such as IMI.[2] Under the leadership of their influential managers (Oscar Sinigaglia and Enrico Mattei), these state owned companies drew on Marshall Plan resources and on their impressive earnings to provide infrastructures (including motorways and telecommunications) and to modernize the steel, energy, chemical and other sectors.[3]

As the years passed, political influence on the state owned companies increased. The roots of increased political influence lay in the emergence of

Fanfani as leader of the DC in 1954, and the putting in place of a new structure of political control of the state holding companies (comprising the Ministry for PP.SS established in 1956, CIPI and CIPE). This new structure allowed the state to do two things. The first was to manage financial resources directly and adopt an interventionist approach. The second was to sidestep the influence of Confindustria and its large, private sector interests (Castronovo, 1995). In this setting, general goals were defined politically, while respecting economic efficiency constraints. More specifically, the costs of politically-imposed social goals (the so called 'oneri impropri' or nonstatutory obligations) were expected to be small and to be covered by government transfers to the endowment funds of the various state holding companies. After 1957, the state holding companies were used as instruments of development and employment creation in the south (Barca, 1997; Dunford, 1988). In the 1960s and 1970s political parties and party factions exercized increasing influence, intervening in relation to location choices, subsidies, employment levels, diversification, the rescue of bankrupt private firms, the careers of managers and many other aspects of corporate life. As a result, political risks played an increasing role in shaping the perceptions and behaviour of managers.

A second important aspect of state interventionism dates from 1950 when the agrarian reform and the programme of EI in the Mezzogiorno were launched. Advocated by a range of actors, direct state intervention in the field of regional development was also justified by reference to several economic development theories. An example was Rosenstein-Rodan's (1943) theory of the 'big push' which suggested that, in underdeveloped areas, state intervention was required to create a range of external economies through the simultaneous growth of several interrelated sectors. To implement this type of policy, the DC set up a government ministry and an operating agency called the Cassa per il Mezzogiorno that managed the funds for EI. In the 1950s, the Cassa channelled these funds into infrastructure projects and investments in support of the land reform. As a result it also gave considerable support to the construction sector in southern Italy. This first wave of investments in water supplies, roads, telecommunications and electrification was followed after 1959 by a second phase of southern policy involving the creation of industrial zones, the provision of incentives, 'reserve measures' obliging state-controlled firms to locate in the south and infrastructure investment, including the development of a university system.

As Barca (1999) argues, the decision to establish extraordinary institutions was a result of a lack of confidence in the technical capacity of the ordinary administration to manage the reconstruction and the socioeconomic unification of the country. In spite of this negative judgement, no attempt was made to reform and improve the efficiency of the public administration, while the public utilities and other sectors were protected from competition.

The second important feature of the Italian model of capitalism was that the private sector consisted of a few large family dynasties (such as Agnelli, Pirelli and Falck) and a great number of SMEs and craft enterprises that flourished in the shadow of large companies. Italian capitalism is accordingly characterized as 'family capitalism' or 'granular/fragmented capitalism'.

After WWII some of these characteristics of the Italian model were also sustained politically. To assist craft enterprises, for example, a special administrative and financial regime was put in place (see Chapter 6). Confederations of artisans (Confartigianato and CNA), cooperatives, and SMEs (ConfAPI) were established and, through their connections with political parties, secured public support and political protection. SMEs were exempted from the application of the Statuto dei Lavoratori of 1970 that established a range of workers' rights, and from other measures such as the payment of welfare contributions for apprentices (Arrighetti and Seravalli, 1997). The consequences were twofold. First, Italian SMEs and craft enterprises enjoyed the most wide ranging and generous regime of incentives in Europe. Second, special legislation which helped prevent the dispersal of the skills present in these companies also reinforced the country's existing productive specialization by making change more difficult and less attractive (Arrighetti and Seravalli, 1997).

As far as large private corporations were concerned, what prevailed was an insider system of corporate ownership and control dominated by a small and closely knitted group of powerful families (family capitalism). These families dominated Confindustria and enjoyed direct access to decision-making centres. After WWII high profits permitted high rates of self-financing of investment. As the rate of profit declined, considerable reliance was placed on short-term bank finance. One reason why was that the special medium and long-term credit agencies that were alone permitted to take equity stakes in companies they financed were state owned and supported the state sector, although until 1992 most Italian banks and their credit creation activities were state controlled, stifling the private financial market (Castronovo, 1995). Italian banks and financial institutions did not, therefore, have significant shareholdings in firms, and did not play a significant role in corporate ownership and governance (in contrast to the past). Economic coordination by banks was, however, not entirely absent. Mediobanca, for example, was a merchant bank created in 1946 on the initiative of the Banca Commerciale Italiana (BCI, now Banca Intesa), Credito Italiano (now UniCredito Italiano) and the Banco di Roma (soon to be part of the group known first as Banca di Roma, then Capitalia). Mediobanca acquired stakes in all the large industrial and insurance companies, played a major role in ruling coalitions, intervened in relation to mergers, acquisitions and debt restructuring, played an important role in the transmission of power from one generation to another and even acquired the right to give control to external managers if the children of the ruling family were considered inappropriate candidates (see Chapter 8). A majority of the nonpublic share capital lay, however, in the hands of domestic households and nonfinancial enterprises. The importance of nonfinancial enterprises stemmed from the size of group holdings and cross holdings. An important feature of the resulting system was a very high degree of concentration of ownership. At the centre of this phenomenon was the dominance of pyramidal groups in which upstream firms have majority control of formally independent downstream firms, and direct shareholding in a firm gives rise to indirect shareholding in all of the firms in which the former owns shares. In this type of system just 7 per cent of the shares in a subsidiary at the fourth level of a pyramid are sufficient to secure direct control for shareholders who own 51 per cent of the shares in firms at each of the higher levels. A financial holding could, therefore, control directly or indirectly all the firms in the same group with a relatively small investment. Minority shareholding offered a means of raising finance without the risk of a loss of control, while minority shareholders were offered little legal protection. Extensive corporate cross-shareholding and interlocking directorships enabled leading families to establish mutual influence and diversify their investment portfolios. At the same time, there was no market for capital and corporate control: capital was not available for exchange on the market.

Institutional Context and Territorial Development Dynamics

The configuration of Italian capitalism into three worlds of production (state-owned enterprises, large family-owned companies and family-owned SMEs and craft industries) and the associated profit strategies had significant impacts on the macroterritorial differences in employment structures, employment rates, earnings and productivity identified in the last chapter. After WWII, for example, the expansion of Italy's large private sector groups was closely associated with the rapid growth of the metropolitan economies of the northwest. The catch-up of the Mezzogiorno in the 1960s and early 1970s was a result of a wave of industrialization in which first state-owned enterprises and subsequently large private groups played significant roles, while the rise of the districts and of the Third Italy stemmed from the dynamism of SMEs and craft enterprises. Italy's differentiated territorial development dynamics were in other words a result of the social characteristics of Italy's national model of capitalism.

Institutional and social factors also played a significant role in shaping the wider environment and mediating economic mechanisms at national, regional and local scales. The role of local social conditions was highlighted, for example, in the literature on the Third Italy which grew out of the early studies of Bagnasco (1977) and Becattini (1987). In this literature three institutional factors were seen as crucial. The first was the existence of a network of small and medium-sized urban areas that escaped Fordist industrialization, did not suffer from immigration and were the home of thriving commercial and craft activities. The second was the existence of small agricultural enterprises that were associated with mentalities conducive to the development of entrepreneurship and provided a flexible, cheap and skilled workforce (Paci, 1980; Ardigò and Donati, 1979). The third was the existence Catholic or Communist subcultures that helped engender trust and cooperation, harmonious industrial relations and efficient local public administration.

In the case of the south, dependency and centre-periphery economic relationships were institutionally mediated, while relative underdevelopment was also a consequence of internal institutional obstacles to modernization. As early as 1928, Gramsci had pointed out that the traditional Mezzogiorno was a zone of 'great social disintegration', and that southern society was a 'great agrarian bloc' which played an active role in national politics, yet constrained southern development. Agricultural workers and peasants were dominated economically and politically by urban-based landowners through the activities of a middle class made up of small owners and managers of land, doctors, lawyers and other professionals. This southern middle class dominated the local political and administrative scene, were completely lacking in any kind of entrepreneurship and, in the words of Rossi-Doria, 'were to social life what malaria was to physical life' (Dunford, 1988). In a subsequent study of a southern community, Banfield (1958) identified a situation called 'amoral familialism' in which people cared only for the short-term, narrow interests of their own families, opposed all figures of authority and made no attempt to improve collective resources such as schools or health care.

After WWII, clientelist politics played a crucial mediating role: southern modernization was profoundly shaped by the activities of powerful national political representatives who acted as central-local mediators. Most often these representatives were DC politicians. At the centre, they dealt with an inefficient, centralized and politicized public administration to secure resources for their localities. At the local level, these resources were used in order to confer benefits in exchange for consensus and political support. These clientelistic relationships were reflected in a range of indicators, that included the rate of use of preference votes in political elections, and the share of the population in receipt of invalidity pensions. This type of clientelism generally involved a conferment of benefits and resources that were individual rather than collective, and that were designed to satisfy particular local interests, often at the expense of principles of coherence and a concentration of resources on areas that had the highest potential for modernization and development. To take one example, in Basilicata in the 1980s, there was a very strong concentration of EI resources on the regional capital (Potenza). Other areas with perhaps greater relative economic potential, such as the Metaponto, Vulture and Maratea, in which majority support was for the PSI, secured little investment.

At the same time, the existence of centralized administrations for the management of special funds, the limited autonomy of peripheral offices and the absence of adequate financial and administrative monitoring and control of the use of financial resources fed a widespread culture of unaccountability. This system, in which vertical political relationships prevailed over horizontal ones, sustained the idea that political rewards lay in the capacity to attract money from the centre—a capacity that often depended on membership of the right political network rather than on good ideas or on effective use and good management of resources. As Pizzorno (1993) argued, at the local level politicians were interested in exploiting their position in order to build their future, upward, career. Often upward movement depended on the request that entrepreneurs benefiting from local public works contracts pay bribes. A part of this money was sent to party headquarters, while the rest would remain with the party's local organization.

An influential recent attempt to emphasize the role of institutional factors in shaping economic development is that of Putnam, Leonardi and Nanetti (1993), who argue that successful representative institutions and economic modernization are interrelated, and that institutional performance depends on the virtues of the underlying civic community. The study concentrated on the performance of the regional governments established in Italy after 1970. The central argument was that regional governments were successful in northern Italy, where there are strong civic traditions, while southern institutions performed poorly, due to the weakness of southern civic culture. Civicness itself was measured by the degree of (1) citizen interest and participation in politics, (2) political equality, (3) cooperation, solidarity, trust, and tolerance, and (4) development of secondary associations (which are a major source of social capital). According to Putnam, Leonardi and Nanetti southerners are used to autocratic rule, never developed the associations that make civil society lively and encourage participation, and have no tradition of trust or common action. Instead, their livelihoods depend on the will of powerful. Everyone expects administrators to cheat, and something is only done if it is solicited as a personal favour. To explain these north-south differences, Putnam, Leonardi and Nanetti go back some 1,000 years to the establishment of authoritarian, patron-client style governments in the south, and contrast it with the northern city-state tradition. Moreover, argue Putnam, Leonardi and Nanetti (1993: 157) 'economics does not predict civics, but civics does predict economics, better than economics itself.'

In our view Putnam, Leonardi and Nanetti go too far and not far enough. As far as the second point is concerned, a more complex set of indicators is required. In the south, for example, the indicators of institutional perform-ance suggest that Basilicata performed as well as Abruzzo and better than other southern regions in the implementation of government programmes. According to the indicators of political culture (electoral participation, newspaper readership and associationalism), however, Basilicata was weaker than Abruzzo. A number of political scientists (Piattoni, 1996; Mutti, 1994) have argued that only the inclusion of other variables enables these differences to be explained. A possible answer lies in the nature of the local political system, the stability of regional governments and the extent to which political power is concentrated in the hands of fewer political representatives. As in other parts of the south, Basilicata was an area of considerable electoral strength of the DC. It had a high rate of re-election of deputies, senators and regional councillors and displayed meaningful sta-bility in the relation between the regional political class and the national party system. Governments were also relatively stable: in 1970–89, there were fewer executives than in any other southern region, and the average duration of regional government coalitions increased. These factors are, however, far less deep-rooted in the evolution of the south and more amenable to change.

As far as the first point is concerned, while we agree that civics can affect institutional performance, we do not accept that civics is an independent variable, nor are we cultural determinists. To take the case of Abruzzo, it had advantages that enabled clientelism and dependence to produce a more positive outcome than in other parts of the Italian Mezzogiorno. Even if it had a greater degree of civic consciousness than other southern regions, it also suffered less from organized crime than any southern region other than Molise, had the relatively good fortune not to attract large capital-intensive investments in the crisis-prone industries of the 1970s and 1980s and, most importantly, occupies a strategic location near to Rome and on the north-ern frontier of the Mezzogiorno (see Dunford and Hudson, 1996).

At about the same time, Trigilia (1992) published a more systematic study of the impact of institutional variables on the divergent development paths of areas in the south. Dealing with the period from 1980 to 1988, Trigilia sought to identify institutional rather than the more traditional economic and structural reasons why southern Italy had failed to establish an autonomous development path in spite of the enormous public re-sources it had received. Three groups of variables were identified: socioeconomic traditions; social tensions; and state intervention (infrastructural aids, financial transfers, and electoral results).

Trigilia suggested that areas showing greater manufacturing dynamism, including all the provinces of Abruzzo and Molise, Bari in Puglia and Sassari in Sardegna, were those with a greater concentration of self-employed agricultural workers and commercial activities. Trigilia attributed this correlation to the idea that agricultural, craft and commercial traditions were preconditions for entrepreneurship (whose take-off also required favourable external economic conditions). The reason why is that these conditions are conducive to the creation of a flexible and cheap workforce, supportive family structures and social integration. The opposite applied to weakly industrialized areas with a large share of farm-labourers and insubstantial craft and commercial traditions. Some historians have argued that the presence of farm-labourers is negatively correlated with the development of urban structures and services. In addition, this type of agricultural work is associated with weaker extended family ties, a stronger development of nuclear families and partible inheritance.

Trigilia's examination of social tensions revealed a strong correlation between indicators of criminality, demographic growth and population density. An important cause of population concentration was the development of growth poles in the 1960s and 1970s. Highly urbanized growth poles with relatively high per capita incomes such as Taranto or Siracusa experienced a significant increase in social tensions in the 1980s as crises struck their key industries.

As far as state intervention was concerned, Trigila found an inverse relationship with growth: slow growing areas that comprised areas with peripheral service economies (Messina and Trapani in Sicilia, Cosenza and Catanzaro in Calabria, and Potenza in Basilicata), areas suffering from industrial restructuring (Naples and Palermo) and traditional industrial poles (Taranto in Puglia, Siracusa in Sicilia, Cagliari in Sardegna and Matera in Basilicata), were characterized by (1) a significant presence of state-owned enterprises, (2) high levels of employment in the service sector and, especially, in the public administration and (3) pension transfers that accounted for a high share of income. An analysis of election results showed that government parties were stronger in areas where reliance on public money was greater.

Trigilia also paid attention to two interconnected aspects of the southern political scene. The first was the low degree of legitimacy of its political class which he attributed to the fact that political alignment was founded on particularist interests requiring constant renewal rather than on general principles and relatively stable collective identities and values. The second was the unaccountability of the southern political class. Trigilia also argued that the southern political order encouraged the development of political rather than economic know-how and of a 'political' microentrepreneurship that relied on political connections rather than economic efficiency. Also it encouraged the diffusion of irregular work and the growth of organized crime. The development of the entrepreneurial capacities characteristic of a market society was conversely discouraged. This discouragement stemmed directly from a negative effect on the formation of appropriate responsibilities and values, and indirectly from unfair competition from politically protected or crime-related economic activities. Clientelism and corruption are of course different phenomena. Clientelism does not involve any exchange of money. In both cases, however, public resources are used to support private rather than collective interests (Signorelli, 1983).

Crime and Territorial Development

Collusion between politics and crime has played an important part in the history of southern Italy. The reason why is that, in some of its areas of responsibility, the institutional presence of the state has always been relatively weak, although it is also widely acknowledged that the Mafia is made up of criminal organizations that survive in the shadow of state activities rather than being 'anti-state'. At a national level, the Mafia mobilizes electoral support for political parties in exchange for the freedom to continue its illegal activities. At a local level, it presents itself as the protector of commercial and construction-sector SMEs who pay for protection from risks that the Mafia helps to create. Also, the Mafia lends money at extortionate rates of interest. In these ways it absorbs resources from the legal economy and alters the distribution of income. At the same time, it engages in a range of illegal activities (theft and drug trafficking), and in legal activities where it derives competitive advantage from criminality: intimidation of competitors; savings in wage costs; availability of financial resources; avoidance of the costs of corruption; and the creation of barriers to entry (Arlacchi, 1983). As a number of recent reports demonstrate (Censis and Fondazione BNC, 2003), the Mafia distorts the market for public contracts where the possession of substantial resources of illegal origin and a completely diverse rationality enable it to undercut enterprises without criminal connections. The Censis report suggested that every year the activities of criminal organizations reduce the value added of southern SMEs by € 7.5 billion (equal to 2.5 per cent of southern GDP).

Alongside the Sicilian Mafia, other criminal organizations have come into the spotlight in other southern regions. In Calabria the success of the 'Ndrangheta is widely acknowledged to be due to the construction of the Salerno-Reggio Calabria motorway. According to the Ministero dell'Interno (2002), in order to avoid conflict, 'Ndrangheta groups whose

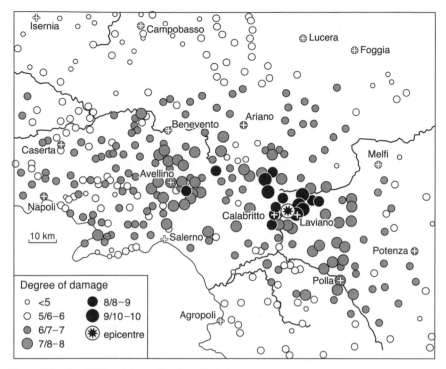

Figure 5.1 The 1980 Irpinia earthquake: affected areas.

areas were involved in recent construction work would collect money from construction companies and distribute a part of it to groups whose areas were affected by the work. Another cause of concern is the possibility of criminal involvement in the recently approved project for a bridge connecting Sicily and Calabria. The 'Ndrangheta which the DIA (2002) considers the best structured and interconnected organization is also fully integrated into international drug trafficking. Although formed only at the beginning of the 1980s, the Sacra Corona Unita grew rapidly in Puglia as a result of its involvement in the more traditional smuggling of cigarettes and immigrants and the more recent trafficking in heroin (Commissione Antimafia, 1999). Sacra Corona Unita is also unusual in the existence of a division of labour with the Albanian Mafia under which the latter brings drugs into Italy and the Sacra Corona Unita sells them on the Italian market. In Campania, the Neapolitan Camorra grew dramatically after the 1980 earthquake centred on Irpinia in the province of Avellino (see Figure 5.1 and Appendix 1), and subsequently as a result of involvement in a wide range of illegal activities (smuggling of arms and drugs, usury, gambling and fraud). In addition, it entered the key business of waste disposal exploiting slow and 'distracted' decision making with respect to the identification of waste disposal sites and the selection of contractors (Lega Ambiente, 2001; Cianciullo and Fontana, 1995). In its report to the Italian Parliament, the Ministero dell'Interno (2002) suggested that these four organizations exercize tight control over their areas.

Changes in the 1990s: The Political Scene

The 1990s witnessed an extraordinary restructuring of the Italian political scene, the system of governance, the economic order and territorial development strategies, altering the social profile of Italian capitalism, the political representation of economic interests, and the institutional and policy contexts in which economic practices are situated.

In the field of politics, the event that precipitated the upheaval was a judicial inquiry in 1992 into a bribery case involving a Milan hospice. The inquiry progressively uncovered systematic corruption (referred to as the Tangentopoli/Mani Pulite scandal). Corrupt practices involved all of Italy's main political parties, especially the ones comprising the ruling coalition, and were uncovered at all levels of government. At the centre of this system was an unlawful exchange in which political parties accepted bribes in return for the improper allocation of public works and public service contracts to particular companies. This unlawful exchange had negative effects on political and institutional life: politicians used the money they received to finance expensive election campaigns, to corrupt other members of the political establishment and to increase their influence. It also had negative economic effects. The reason why was that the rules of market competition were systematically broken by 'institutional' arrangements that had nothing to do with efficiency and competition. As Ginsborg (1998) indicated, successful development is strictly dependent on the transparency of administrative procedures as well as on control over the public actions of political actors.

In just three years from 1992 to 1994, the five ruling parties making up the centre-right coalition (the DC, PSI, PRI, PSDI and PLI) were swept from the political scene. Most remarkable was the collapse of the DC, as it had played a major part in every post-war government (Table 5.1) (Agnew, 2002). No less significant was the crisis and the subsequent extinction of the PSI. Under the leadership of Bettino Craxi, the PSI had rejected a political alliance with the PCI and had positioned itself at the centre of the political spectrum. As a result, the PSI gained the support of around 15 per cent of the electorate, making it the most important force in the formation of coalitions at the national level.

Table 5.1   General elections in Italy, 1972–92

	1972	1976	1979	1983	1987	1992
DC	38.7	38.7	38.3	32.9	34.3	29.7
PCI	29.0	34.4	30.4	29.9	26.6	16.1
PSI	9.6	9.6	9.8	11.4	14.3	13.6
PSDI	5.1	3.4	3.8	4.1	3.0	2.7
PRI	2.9	3.1	3.0	5.1	3.7	4.4
PLI	3.9	1.3	1.9	2.9	2.1	2.8
MSI	8.7	6.1	5.3	6.8	5.9	5.4
Lega Nord	—	—	—	—	—	8.7
Rifondazione Comunista	—	—	—	—	—	5.6
Others	2.1	3.4	7.4	7.0	10.2	10.0

Source: adapted from Piattoni, 1996.

In the midst of this political turmoil, another remarkable event sanctioned the definitive demize of the traditional party system. In a referendum held in April 1993, the majority of Italians decided to replace the existing system of proportional electoral representation with a 'first-past-the-post' system. A majority system was deemed most likely to achieve three objectives: a reduction in the number of political parties; the introduction of the principle of rotation between different coalitions; and, especially, government stability. After a complex political debate, the new national electoral Laws 176/1993 and 177/1993 established a mixed system according to which 75 per cent of seats are distributed on the basis of a 'first past the post' system and the remaining 25 per cent on the basis of a proportional system. The incomplete reform of the electoral system meant that it did not fully achieve the objectives originally expected of it. Government stability did increase, but the number of parties did not decline.

The demize of the ruling DC and the start of the so-called Italian Second Republic (Agnew, 2002) saw the emergence of two new parties. FI is without doubt the party that gathered the legacy of the former DC (Table 5.2). Silvio Berlusconi, a media magnate who turned to politics, quickly set it up before the 1994 General Election.

In contrast to the DC, however, FI was committed to a neoliberal economic agenda involving a drastic reduction in state economic intervention and state aid and increases in the flexibility of labour and service markets. The LN catalysed the interests of the northern petty bourgeoisie. This group was the unhappiest with the management of collective resources (Giordano, 2000). In many ways, the LN was the political force that most contributed to the breakup of the historical economic alliance between the north and the south of Italy. Although its political rhetoric was (and still is) extremist and often racist, the LN managed to put the 'Northern Ques-

Table 5.2   General elections in Italy, 1994–2001 (proportional system)

Parties	1994	1996	2001
Forza Italia	21.0	20.6	29.4
PDS-DS	20.4	21.1	16.6
AN	13.5	15.7	12.0
Margherita	—	—	14.5
PPI	11.1	—	—
LN	8.4	10.1	3.9
Rifondazione	6.0	8.6	5.0
Patto Segni	4.7	—	—
Lista Pannella-Bonino	3.5	1.9	2.2
Liste Verdi	2.8	2.5	—
Lista Dini	—	4.3	—
Partito Socialista	2.2	0.4	—
CCD-CDU	—	5.8	3.2
PPI-PVR-PRI	—	6.8	—
Fiamma	—	0.9	0.4
La rete	1.9	—	—
Alleanza democratica	1.2	—	—
SVP	0.6	—	—
Nuovo PSI	—	—	1.0
Biancofiore	—	—	
Girasole	—	—	2.2
Communisti Italiani	—	—	1.7
Lista Di Pietro	—	—	3.9
Democrazia Europea	—	—	2.4
Others	2.9	1.4	1.5

Source: Deputati DS, various years.

tion', namely, the sustainability of northern development, on the national political agenda alongside the more enduring 'Southern Question' (Diamanti, 1996). Although the political solutions it proposed did not secure widespread political support (after first proposing the separation of Padania from the rest of Italy, the LN argued for a strong devolution of power to regions removing from the central state much of its role), the LN did have the merit of starting a new discussion of the country's development priorities. The LN also assumed a regional dimension. The Liga Veneto, for example, emerged as the voice for the interests and complaints of the small and medium-sized entrepreneurs of an area in the northeast comprising a large part of the Third Italy.

Changes in the 1990s: The System of Governance

A series of other reforms, implemented with the aim of improving Italy's system of governance (Figures 5.2, 5.3 and 5.4), resulted in substantial institutional discontinuity. These reforms culminated in the approval of Constitutional Law 3/2001 which reformed Title V of the constitution and sanctioned a move in the direction of a federal system. Approved after a

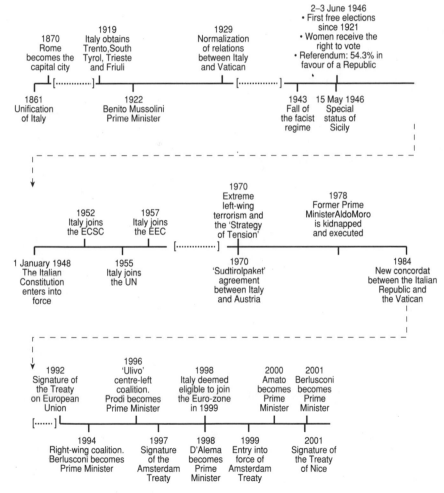

Figure 5.2   Major political events in Italy.

Source: adapted from http://www.aldobattista.it/sussidia/s005devolution.pdf.

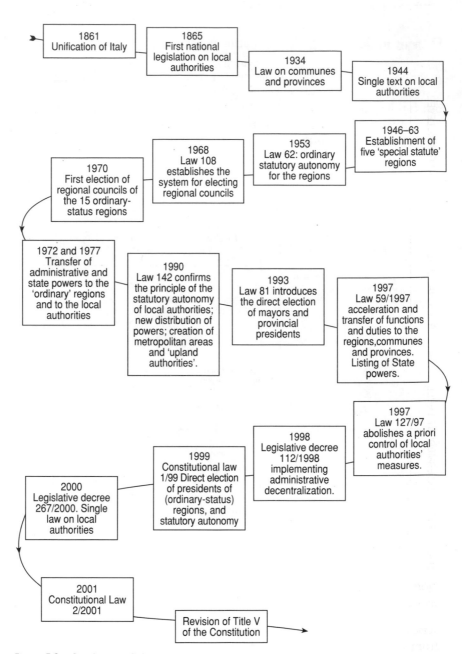

Figure 5.3 Devolution in Italy.
Source: adapted from http://www.aldobattista.it/sussidia/s005devolution.pdf.

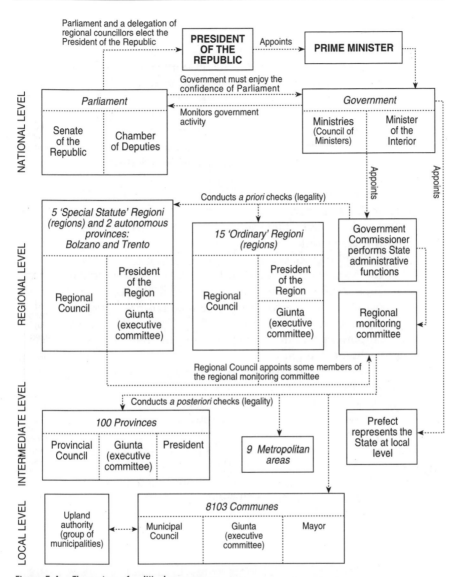

Figure 5.4 The system of political governance.
Source: adapted from http://www.aldobattista.it/sussidia/s005devolution.pdf.

series of steps that had transferred responsibilities, personnel and resources from central to peripheral administrations, this reform was considered the most profound institutional change since the approval of Italy's constitution in 1948 as it modified the traditional relationships between centre and periphery (ASTRID, 2002).

Law 142/1990 confirmed the principle of statutory self-governance of local authorities and proposed a new distribution of powers. Law 81/1993 provided for the direct election of mayors and provincial presidents. In 1997, the path-breaking Bassanini Law 59 implemented for the first time Article 5 of the Constitution: functions were transferred from deconcentrated state administrations to regions, provinces and municipalities. Law 133/1999 reformed regional finance. Other reforms affected the state bureaucracy. The Cassese Law 29/1993 gave senior civil servants more discretionary power but made them much more accountable for their actions. In addition, their jobs ceased to be tenured to avoid the establishment of clientelistic relationships and personal power. Laws 80/1998 and 145/2002 applied the spoil system to central administrations. Finally, three Constitutional Laws were passed (Constitutional Law 1/1999, 2/2001 and 3/2001) and a referendum was held to introduce a federal state system. This reform was revolutionary in three respects. First, it acknowledged the equal institutional dignity of all subnational authorities. Second, it ended the so-called 'parallelism' of state and regional legislative functions. As a consequence, the state and the regions were given different responsibilities: state responsibilities were substantially limited, and preventative state controls over regional legislation were annulled. Finally, Article 119 granted regions fiscal powers, within the framework of the national fiscal system.

Alongside the devolution of state power, the process of European integration has involved a transfer of some institutional power upwards to European institutions. The EU has a complex system of internal law which has direct effect on the legal systems of Member States. European law has three sources: the treaties; directives, regulations, recommendations and opinions; and the decisions of the European Courts. EU directives and regulations bind everyone. A directive establishes the objectives to be pursued by Member States, but leaves Member States free to choose the way in which they are transposed into national law. A regulation has a Community character and direct effect which means that they are in and of themselves a part of national law. A Member State can neither apply a regulation incompletely or selectively nor can it implement domestic legal provisions or practices that preclude the mandatory application of a regulation.

Arguably, these processes of institutional transformation have ushered in a new era in which the nation-state remains important but is embedded in a system of multilevel governance. In this system there are overlapping spheres of authority, and a complex mosaic of rights, responsibilities and relationships involving local, regional, national and supranational tiers of government and governance, and in which a principle of subsidiarity plays a significant role (Jeffrey, 2000; Hooghe, 1995).

Changes in the 1990s: Debt Reduction and Privatization

The 1990s also witnessed a decisive step in the redefinition of the relationship between politics and economics. To reduce government debt, to finance a more strictly controlled programme of public expenditure and to increase private sector involvement in economic life, a series of financial institutions, state-owned companies and public monopolies were privatized, altering corporate conduct and modes of coordination of economic life (Cella, 1997).

In the 1970s, inflation soared after the collapse of the Bretton Woods order and the first oil crisis. To pay for oil imports, Italy was forced to expand exports. Competing in traditional sectors implied devaluation, which reinforced the wage-price spiral, and encouraged capital flight. Government expenditures grew, in part to finance state holding company investment. Industrial decentralization impacted negatively on tax revenues. In the early 1980s, Italian government debt was well ahead of the EU average (Table 5.3), while the second oil shock had seen the rise of monetarist policies of demand compression and interest rate increases in the struggle against inflation (De Cecco, 2000). Government finances were further weakened by private sector rationalization, and the associated socialization of the costs of plant closures, early retirements and redundancies, on the one hand, and increasing debt service costs, on the other. At the same time, the political class pressed state controlled companies to provide employment. As compensation, the state increased their endowment funds, though not sufficiently to prevent a catastrophic increase in indebtedness. The return of growth in the late 1980s saw private sector growth, but no real resolution of Italy's underlying structural difficulties. Nonetheless, the Italian government decided to participate in the project of EU monetary and economic unification. In effect, the Italian authorities planned to use external EU discipline to restructure the public finances, while fixing the exchange rate to control inflation, attract foreign capital and increase domestic competitiveness. This strategy failed. In 1992, the Italian Lira left the exchange rate mechanism, and was devalued.

At that time, government debt stood at 107.7 per cent of GDP, debt service charges were onerous, and the government deficit exceeded 10 per cent of GDP. As a consequence, the Amato government adopted a policy of 'rigour' that subsequent governments also pursued. Attempts were made to reduce fiscal evasion, and especially to reduce public expenditure, primarily on health and other welfare expenses. In 1993, GDP declined by 0.9 per cent. In 1992, public expenditure was 56.6 per cent of GDP. In 2002, it stood at 47.2 per cent. Also in 1992 the short-lived Amato government

Table 5.3 Public finance in Italy and the EU15, 1980–2002

	1980	1981	1982	1983	1984	1985	1986	1987	1988	1989	1990	1991	1992	1993	1994	1995	1996	1997	1998	1999	2000	2001	2002
Gross public debt (as % of GDP)																							
Italy	57.7	59.9	64.7	70.8	76.0	82.3	86.3	90.4	92.6	95.4	97.2	100.6	107.7	118.1	124.3	123.7	122.7	120.2	116.3	114.9	110.6	109.5	106.7
EU	38.3	41.8	45.4	48.8	51.2	53.3	54.1	55.3	54.4	53.5	54.3	54.8	58.9	64.5	66.6	70.1	72.2	71.0	68.9	67.3	64.2	63.0	62.7
Total revenue (as % of GDP)																							
Italy	33.9	35.1	37.4	39.5	38.6	38.9	39.7	39.8	40.1	42.0	42.6	43.8	46.0	47.3	45.1	45.6	45.8	48.0	46.5	46.7	45.8	45.5	44.9
EU	42.8	43.6	44.4	44.8	44.8	45.3	44.9	44.9	44.5	45.9	45.7	46.1	46.8	47.2	46.8	46.1	46.8	46.8	46.6	46.9	46.6	46.2	45.5
Total expenditure (as % of GDP)																							
Italy	42.2	46.4	48.5	49.9	50.1	51.2	51.1	50.7	50.8	51.8	54.4	55.5	56.6	57.6	54.3	53.2	52.9	50.7	49.3	48.4	47.6	48.1	47.2
EU	46.2	48.2	49.2	49.6	49.5	49.7	49.1	48.8	47.8	48.0	49.1	50.2	51.9	53.3	52.3	51.3	51.0	49.3	48.2	47.6	46.9	47.1	47.4

Source: adapted from Banca d'Italia, 2004; 1999.

announced the decision to sacrifice the state owned sector to reduce government debt. The culture of economic unaccountability and political interference was considered irreversible, while the earlier model of economic dynamism and economic modernization was considered irrecoverable. At the same time, however, the Amato government chose to act in ways designed to secure international credibility and to please Anglo-American international financial institutions that were to earn vast amounts of commission from the consequent privatizations. The first step was the transformation of autonomous public institutions (such as IRI and ENI) into limited public companies. A similar process took place in the public banking sector (Banca Commerciale Italiana, IMI, Credito Italiano, Banca di Roma, Banca Nazionale del Lavoro). To permit the entry of competitors, electricity generation and distribution, telecommunications and other markets served by public utilities were liberalized. In 1994, an earlier decree dealing with the rules for the sale of the state holdings in corporations was converted into Law 474/1994. Amongst other provisions, it required the use of the resources raised to reduce public debt. In addition it provided for the creation of regulators for privatized/liberalized public utility markets and for Treasury retention of a 'golden share'.

These measures opened the way to a rapid withdrawal of the Italian state from manufacturing, public utilities, banking and insurance (Table 5.4). As a result of these sales, government debt was significantly reduced: in 1993–2003, these asset sales generated € 95,040 million (Camera dei Deputati, 2004). At the same time, stock market capitalization increased dramatically from 11.5 per cent of GDP in 1992 to 37.4 per cent in 2003, compared with a peak value of 70.3 per cent in 2002 (http://www.borsaitalia.it).

Changes in the 1990s: Territorial Development Policies

The 1990s also witnessed an end of existing territorial development policies. In 1980, Law 717/1965 that had extended the life of the Cassa expired. After several short-term emergency measures to refinance it, the Cassa was finally wound up. In 1986, Law 64/1986 put in place a new southern development system for the period up to 1993. New aims were defined. Included were the enhancement of entrepreneurship and support for training, human capital, producer services, scientific research and technological innovation. Industrial development and investment in large-scale infrastructures did remain however with provisions for large firms to agree to investment projects through Contratti di Programma (Table 5.5) and infrastructural projects through Accordi di Programma. In addition, subsidized social security arrangements were envisaged to protect employment. Subnational

Table 5.4 Privatizations in 1992–9

Year	Company	Share %	Seller	Buyer	Revenues (LIT billion)
1992	Acciaierie e Ferrerie di Piombino (1)	100.0	IRI	Lucchini	289
	Cementir	51.8	IRI	Caltagirone	480
	Tubi Ghisa	75.0	IRI	Pont-à-Mousson (F)	121
1993	Agip Argentina	100.0	ENI	YPF (AR)	110
	Cirio-Bertolli-De Rica	62.1	IRI	Fisvi	311
	Credito Italiano	54.8	IRI	Market	1,801
	EBPA	39.3	IRI	Market	271
	Italgel	62.1	IRI	Nestlé (CH)	431
	Pavesi (2)	51.0	IRI	Barilla	255
	SIV	100.0	EFIM	Pilkington (UK)	210
1994	AgipCoal (America)	100.0	ENI	Various foreign groups	171
	Alcantara	51.0	ENI	Toray (J)	195
	AST	100.0	IRI	Krupp (D)	624
	Comit	51.3	IRI	Market	2,891
	EVC International	34.2	ENI	Market	194
	IMI 1	27.4	Treasury	Market	1,794
	INA 1	47.2	Treasury	Market	4,530
	Liquipibigas Distribuzione	100.0	ENI	Primagaz (F)	221
	Nuovo Pignone	69.3	ENI	Genral Electric (USA)	699
	SME 1	32.0	IRI	Benetton & others	723
	STMicroeletronics 1	14.8	MEI	Market	165
1995	Carbones del Guasare	49.5	ENI	Shell (UK)/ Ruhrkolhle (D)	236
	ENI 1	15.0	Treasury	Market	6,300
	EniChem Augusta	85.0	ENI	RWE (D)	355
	ILP	100.0	IRI	Riva	2,332
	IMI 2	19.0	Treasury	Various	1,200
	INA2	18.4	Treasury	Various	1,687
	Inca International	80.0	ENI	Dow (USA)	262
	ISE	74.0	IRI	Edison	370
	Italtel	40.0	IRI	Siemens (D)	1,000
	Sidermar	1000.0	IRI	Coeclerici	129
	SME 2	14.9	IRI	Benetton & others	341
	STMicroeletronics 2	11.3	MEI	Market	277
1996	Alumix	100.0	EFIM	Alcoa (USA)	450
	AMGA	48.9	Commune of Genoa	Market	208
	Dalmine	84.1	IRI	Rocca	301

Continued

Table 5.4 Privatizations in 1992–9

Year	Company	Share %	Seller	Buyer	Revenues (LIT billion)
	ENI – fertilizzanti	100.0	ENI	Norsk Hydro (N)	166
	ENI 2	15.8	Treasury	Market	8,872
	IMI 3	6.9	Treasury	Marketed block trade	501
	INA 3	31.1	Treasury	Various	3,260
	MAC	50.0	IRI	Marconi (UK)	223
	Montefibre	66.4	ENI	Orlandi	187
	Nuova Tirrena	91.1	Consap	Toro	548
	SME 3	15.2	IRI	Benetton & others	238
1997	Aeroporti di Roma	45.0	IRI	Market	594
	Alfa Romeo Avio	100.0	IRI	Fiat Avio	238
	Banca di Roma	14.4	IRI	Market	666
	ENI – cavi in fibra ottica	100.0	ENI	Albacom	242
	ENI 3	17.6	Treasury	Market	13,230
	Fincantieri –ramo diesel		IRI	Wartsila (FL)	325
	Nuovo Pignone 2	9.3	ENI	General Electric (USA)	150
	San Paolo 1	23.2	CSP	Market	2,629
	SEAT	44.7	Treasury	Market	1,653
	Telecom Italia	28.0	Treasury	Market	22,880
1998	AEM	49.0	Commune of Milan	Market	1,473
	Alitalia	18.4	IRI	Professional investors	787
	BNL	62.2	Treasury	Market	6,706
	EBPA	62.0	IRI	ABB (CH)	1,300
	ENI – ramo alberghi	100.0	ENI	Closed funds	244
	ENI 4	14.2	Treasury	Market	12,994
	NST	100.0	ENI	Albacom	116
	SAIPEM	18.7	ENI	Market	742
	STMicroelectronics 3	13.1	MEI	Market	985
1999	ACEA	49.0	Commune of Rome	Market	1,808
	Autostrade 1	56.6	IRI	Market	8,105
	Banca MPS	27.0	MPS	Market	4,292
	ENEL 1	31.7	Treasury	Market	32,045
	Mediocredito Centrale	100.0	Treasury	Banca di Roma	3,944
	STMicroelectronics 4	10.5	MEI	Market	2,089
2000	Aeroporti di Roma 2	51.2	IRI	Gemina & others	2,572
	Autostrade 2	30.0	IRI	Benetton & others	4,911
	FinMeccanica	43.7	IRI	Market	12,359

Source: adapted from R&S, 2000.

Table 5.5 Planning and Enterprise Agreements approved in 1986–99: situation as of 31/12/2002 (€ million)

Agreement	Investment (€ million)	State commitment (€ million)	Envisaged employment	Of which new employment
Planning and Enterprise Agreements approved in 1986–95				
A.C.M.	287.4	153.9	3,780	2,241
Barilla (Foggia)	261.3	78.3	639	351
Bull HN	45.2	35.7	250	250
Compla. Sint	50.1	20.6	280	280
Eni (Val Basento)	637.1	191.6	3804	719
Fiat 1	1,032.2	519.5	30,209	1,265
Fiat 2	3,280.9	1,271.9	16,492	9,210
G.T.C. (Castrovillari)	96.4	40.3	390	390
IBM Semea (Bari)	27.7	20.9	180	151
IRI	747.3	421.9	6,073	1,860
Olivetti (Campania)	289.9	198.2	1,769	561
Piaggio	16.4	6.8	484	484
Saras I	247.0	88.4	892	277
Snia Btp (Pisticci)	153.8	61.4	624	624
Tari	43.8	25.1	383	348
Texas Instruments (Avezzano)	674.8	324.5	1,710	1,710
Total	7,891.4	3,459.1	67,959	20,721
Planning Agreements approved in 1996–9				
EDS	27.9	16.9	500	500
Getrag	210.5	104.3	800	800
IPM	74.2	51.2	780	340
Natuzzi	232.2	106.2	5,039	2,939
Nuova Concordia	255.9	137.9	1,574	1,574
Saras 2	258.7	139.0	274	274
SGS Thomson	297.3	172.2	315	315
Unica 1	44.3	27.8	693	432
Unica 2	44.7	29.9	365	303
Telit/Finmek	82.1	47.9	510	140
Total	1,527.8	833.3	10,850	7,617

Source: SVIMEZ, 2003: 439.

authorities were given greater initiative in the development of programmes and projects in the context of a strategic national development plan designed, monitored and evaluated by a government department (Dipartimento per il Mezzogiorno) and financed by a new agency (Agenzia per il Mezzogiorno). Agencies responsible for technical assistance, such as

FORMEZ, IASM and INSUD were made into corporations with their own budgets.

Although thoughtful in its aims, the new policy achieved little largely due to implementation failures (Martinelli, 1998). In 1992, to avoid a potentially divisive national referendum on the abolition of regional policy for the south, the policy was terminated and a new policy was established.

The new policy hinges upon Law 488/1992, Law 341/1995 that foresees automatic incentives to companies for the purchase of plant and equipment in depressed areas, and Law 662/1996. The new set of measures differs significantly from the previous ones in a number of aspects. First, it provides for incentives to support investment in depressed areas not just in the south but throughout Italy. These incentives were initially provided for investment in Objective 1, 2 and 5b regions and, after March 1995, in a series of areas that the European Commission considered eligible for national state aids under Article 92.3c of the Treaty. Article 92.3c permits derogation for problem areas of the general prohibition of state aids under EU competition policy. At that time, derogation was permitted for areas with a per-capita GDP lower than 85 per cent of the national average, and an unemployment rate 110 per cent higher than the national average. In Italy, the use of other indicators such as migration flows and productivity, and the use of the northern instead of the national average as the comparator made many northern areas that did not satisfy the first two criteria eligible for national financial assistance, although the grant ceiling decreased with increases in the degree of development of the area concerned. Second, the law required the transfer of responsibilities from 'extraordinary' institutions to the ordinary administration (Viesti and Prota, 2004).

An important factor permitting the extension of regional policy to all areas was the 1994 decision of the Italian government to succumb to EU pressure and to issue a decree providing for the elimination of the social security subsidies for companies operating in the south. The ending of these subsidies was to be instantaneous for the province of Frosinone in Lazio and for Abruzzo and Molise, and was to be completed in 1997 in the rest of the south. These subsidies were one of the main measures supporting employment creation and cost competitiveness in the south. According to Viesti (1996), the Berlusconi government took the side of northern and central Italian lobbies to secure a modification of the EU criteria for depressed areas in Italy in exchange for the end of the policy of social security subsidies in the south. As a result, in 1995 the cost of labour was estimated to increase by 10 per cent, with further increases in subsequent years (Del Monte, 1996).

As a result of these reforms, the geography of regional aid was changed. In particular, areas in the centre-north that were far more affluent than their southern counterparts were included in the list of depressed areas. Not only, however, were the resources provided by Laws 488/1992 and 341/ 1995 shared with the centre-north. Areas in the centre-north were made eligible for other aid programmes, under, for example, the law on youth entrepreneurship, and the Tremonti Law 549/1995, which provided investment aid for SMEs and firms in areas eligible for regional assistance. Analyzing 1996 data, Giannola (1999) showed that, although the value of the incentives requested in the centre-north was lower than in the Mezzogiorno, due to the lower ceilings for the percentage rate of grant, the number of requests for incentives in the centre-north was very high: 47 per cent of the total, worth some € 7,500 million. Table 5.6 records the

Table 5.6 Incentives under Law 488/92 per inhabitant by region in 1996–8

	Incentives per inhabitant								
	Million LIT			*Index:* *Mezzogiorno = 100*			*Rank*		
	1996	*1997*	*1998*	*1996*	*1997*	*1998*	*1996*	*1997*	*1998*
Piemonte	1,310.7	1,226.8	650.1	203.1	332.8	96.8	3	1	10
Valle d'Aosta	26.6	346.8	39.4	4.1	94.1	5.9	19	12	19
Lombardia	518.7	339.2	807.1	80.4	92.0	120.1	13	13	3
Trentino-Alto Adige	25.6			4.0	0.0	0.0	20	20	20
Veneto	549.6	358.6	467.0	85.2	97.3	69.5	11	11	17
Friuli-Venezia Giulia	77.1	79.9	356.0	11.9	21.7	53.0	18	19	18
Liguria	454.8	394.8	500.3	70.5	107.1	74.5	15	8	15
Emilia-Romagna	671.5	395.9	772.4	104.1	107.4	115.0	7	7	5
Toscana	833.4	543.7	640.2	129.2	147.5	95.3	6	4	11
Umbria	530.9	813.5	845.1	82.3	220.7	125.8	12	3	2
Marche	567.7	458.6	707.6	88.0	124.4	105.3	10	5	9
Lazio	842.2	1,042.4	486.5	130.5	282.8	72.4	5	2	16
Abruzzo	1,578.0	326.0	729.4	244.6	88.4	108.6	2	17	7
Molise	2,539.3	362.2	765.9	393.6	98.3	114.0	1	10	6
Campania	650.5	448.7	716.5	100.8	121.7	106.7	8	6	8
Puglia	571.7	334.5	610.6	88.6	90.7	90.9	9	15	12
Basilicata	900.3	278.9	1,517.2	139.5	75.7	225.8	4	18	1
Calabria	421.4	382.6	799.8	65.3	103.8	119.1	17	9	4
Sicilia	464.4	336.0	547.7	72.0	91.2	81.5	14	14	13
Sardegna	450.6	326.4	506.2	69.8	88.6	75.3	16	16	14
Mezzogiorno	645.2	368.6	671.8	100.0	100.0	100.0			
Centre-north	736.2	659.3	578.4	114.1	178.9	86.1			
Italy	677.8	472.9	638.3	105.1	128.3	95.0			

Source: SVIMEZ, 2000: 509.

regional per capita distribution of subsidies under Law 488/1992 in 1996–98 and shows that the greatest beneficiaries included Piemonte and Emilia Romagna. A paradoxical aspect of this law is that in the centre-north it provides grants for highly industrialized areas. Areas in the centre-north eligible for incentives had 112 manufacturing employees per 1,000 inhabitants, whereas Objective 1 areas in the south had just 40.5.

Alongside these incentives to companies, a complementary law (Law 662/1996) established the idea of 'Programmazione negoziata' (negotiated planning), which is a territorial approach to development. A number of instruments were envisaged for its implementation: Intesa Istituzionale di Programma, Patto Territoriale, Contratto d'Area, and Contratto di Programma (Table 5.7).

As in the case of incentives, responsibilities for negotiated planning were assigned to the ordinary administration. In this case, subnational institutions play an important role in the elaboration and implementation of plans, often in partnership with other social actors. Increasingly, the use of national funds is shaped by regional priorities. As a result, subnational

Table 5.7 The measures of negotiated planning

Intesa Istituzionale di Programma	An agreement between the central government and a regional or provincial government, needed for projects that require a joint financial effort and implemented through an 'Accordo di programma'
Patto Territoriale	An expression of social partnership and planning between local institutions, employers and workers organisations, and private firms, aimed at local development, and first introduced by the Law 341/95.
Contratto d'Area	Aims to encourage the location of new industrial activities and creation of new employment in areas affected by de-industrialisation and employment crises. The contract contains an agreement among social actors to increase labour flexibility, an agreement among local authorities to simplify and accelerate administrative procedures and an agreement with the local Prefettura (Prefect's office) to improve security.
Contratto di Programma	Agreement between the Treasury Ministry and either (1) a large industrial firm, (2) a consortium of SMEs or (3) the institutions representing an industrial district, tourism or agriculture and fishing. The aim is to carry out integrated projects involving industrial investments, research activity (centres and projects) and training (for new and old personnel).

institutions also assume greater responsibility for the outcomes. Conversely, central institutions play a planning, coordination, monitoring and evaluation role. At the centre of many of these developments is the idea that the pathway to competitiveness involves increasing the quality of the local context, making long term commitments to a development programme, evaluating outcomes and, as a result of decentralization, ensuring cooperation between the various tiers of government.

Justification for this radical shift in the country's regional policy was also derived from the empirical identification of a variety of development paths in the Mezzogiorno (De Vivo, 2001; Viesti, 2000; Meldolesi, 1999). Areas of economic crisis and lower growth, often associated with earlier investments by state-owned companies, coexisted with dynamic industrial districts, specialized primarily in export-oriented manufactures (Trigilia, 1992). This evidence gave support to the idea of putting in place local development strategies for specific commodity chains in particular territories (Barca, 1999). At the centre of this new development agenda was the valorization and mobilization of local resources, the search for a virtuous combination of endogenous and exogenous development factors and the differentiation of development approaches, all of which accorded with the existence of a variety of 'Mezzogiorni'. Underlying these arguments was a reconceptualization of the determinants of development. Although strongly contested in some quarters (Lovering, 1999), institutional and cultural conditions were seen as the most important cause of economic development. What determines development, it was argued, is the quality of governance structures and their human and material resources, the quality of social networks, the degree of civic participation and the associated capacity of actors to deliver adequate policies and mobilize local resources/assets.

These changes in incentive systems and from top-down to bottom-up development policies were taking place as EU funding was becoming increasingly important for Italian regional policy. Table 5.8 indicates the scale of Structural Fund support for Italy and its Objective 1 areas which are plotted in Figure 5.5. The commitment and expenditure of European money required, however, the application of a number of principles. Included were the principles of programming (planning), cofinancing and additionality, and partnership as well as the general principle of subsidiarity. At present, the general strategy for Objective 1 areas is set out in a CSF prepared by the Treasury's DPS in conjunction with the regions. At the core of the CSF are the major priorities. For the period 2000–6, the Italian government has identified six priorities: valorization of natural and environmental resources; valorization of cultural and historical resources; valorization of human resources; strengthening of local economic systems;

Table 5.8 Structural Fund support

	1989–93			1994–99			2000–06		
	Objective 1	Total (All objectives)	Outcome	Objectives 1 and 6	Total	Situation at the end of 1999	Objective 1 (including phasing out)	Total	Situation at the end of 2001
	ECU¹ million, 1993 prices		Payments/forecasts (%)	€ million, prices of year of adoption of programme		Payments/grant (%)	€ million, 1999 prices		Payments/grant (%)
Belgium	—	694	73.3	761	1,889	70.7	625	1,829	70.7
Denmark	—	291	78.0	—	754	80.2	—	745	80.2
Germany	3,144	5,907	77.6	14,238	20,397	78.1	19,958	28,156	78.1
Greece	7,497	7,654	83.4	14,531	14,531	72.8	20,961	20,961	72.8
Spain	10,999	14,162	84.2	27,510	33,186	80.8	38,096	43,087	80.8
France	1,021	5,310	81.0	2,277	13,859	73.0	3,805	14,620	73.0
Ireland	4,162	4,253	92.6	5,826	5,826	86.5	3,088	3,088	86.5
Italy	8,431	10,803	70.3	15,407	20,856	63.0	22,122	28,484	63.0
Luxembourg	—	31	58.2	—	86	73.3	—	78	73.3
Holland	714	714	76.4	159	2,249	61.1	123	2,635	61.1
Portugal	7,964	8,141	89.5	175	1,566	77.4	19,029	19,029	77.4
United Kingdom	871	6,285	80.3	14,590	14,590	88.8	6,251	15,635	88.8
Austria	—	—	—	521	1,638	63.8	261	1,473	63.8
Finland	—	—	—	310	1,341	69.9	913	1,836	69.9
Sweden	—	—	—	2,459	11,829	66.2	722	1,908	66.2
Total	44,089	64,247	81.5	98,763	144,598	75.4	135,954	183,564	75.4

Note:
¹ European Currency Unit
Source: elaborated from SVIMEZ 1995; 2001 and 2003.

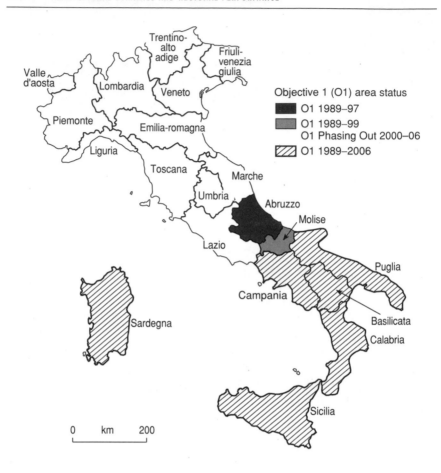

improvement of the quality of urban areas, local institutions and associational life; and strengthening of service centres and networks.

In the 1990s, however, the effectiveness of Structural fund programmes was adversely affected by the reorganization of the Italian state system. In particular, difficulties associated with the transfer of functions from national to subnational actors, as required by Law 488/92, resulted in a very low rate of take-up of EU resources. As of 2001, just 67 per cent of 1994–9 Objective 1 payments had been made compared with 78 per cent in the EU15, 87 per cent in Ireland and 89 per cent in Portugal (Table 5.8). In the 2000–6 cycle, the situation improved: at the end of 2001, Italy had committed 27.4 per cent of EU resources and paid out 7.3 per cent. In Ireland and Portugal, however, the corresponding figures were 39.3 and 14.6 per cent and 31.8 and 14.3 per cent (SVIMEZ, 2003). A poor absorption record is not without negative consequences for any country concerned, as the EU recovers resources initially allocated to a country but not committed and spent in the programme period.

At a national scale, there were also delays in committing and spending resources and making payments, especially in the areas that were least developed. Once again, devolution along with the transition from a centralized development policy model to a much more fragmented set of interventions in favour of depressed areas moved ahead more quickly than the capacities of subnational administrations to manage their new tasks (Martinelli, 1998; Viesti, 1996).

In addition, the new national and EU model envisaged the involvement of a multiplicity of institutional actors, making policy implementation complex. To deal with some of these difficulties, in 1995 the Cabina di Regia Nazionale (a national coordinating cabinet) was established in the Treasury to coordinate and monitor initiatives. The Cabina's impact was limited, and in 1998 the DPS was created to plan and implement structural initiatives.

Another consequence of the reforms embodied in Law 488 and its successors stemmed from the fact that the procedures for obtaining funds were complicated. To apply applicants often need the help of consultants. As a result, the reforms are contributing to the creation of a class of professionals, specialized in attracting financial resources. The emergence of this class is reminiscent of the famous category of the 'mediatori' (Della Porta, 1980). Donatella Della Porta's mediators were professionals who gained enormous power as a result of their capacity to divert enormous sums of money from the centre to the periphery. These professionals served particularistic interests, and yet were not accountable for the inefficient use of public resources.

Alongside these institutional impacts, the reforms had a number of economic repercussions. Most striking was the decline of the construction sector. At the end of the 1980s, construction accounted for more than 8 per cent of southern added value, compared with under 5.5 per cent in the centre-north (Viesti, 1996). At that stage, major state investments in, for example, motorway construction were associated with the creation of companies, some of which were irregular and exploited the conditions of exceptionality, that often involved the derogation of normal administrative procedures. In allocating contracts, criteria of political and administrative affiliation were often more important than competitiveness and efficiency criteria, so that price and quality were often unrelated. In many southern areas, criminal organizations created construction companies that thrived on subcontracts earned precisely because of their links with corrupt policy makers and public administrators. In 2002, however, the construction sector was smaller, representing 5 per cent of the southern value added and 4.6 per cent in the centre-north (SVIMEZ, 2003).

Concluding Remarks: The Implications of Recent Trends

Any assessment of the implications for territorial economic performance of the reshaping of Italy's economic and political order runs the risk of generalizing from short-term trends. Some observations are however possible.

Independently of the outcomes of recent transformations, there is no doubt that regions are now seen as key economic actors. Two major reasons explain this development. The first is the demise of macroeconomic demand management, and the switch to supply-side economics, which have called for action at the level of the regional economy: as fiscal and monetary policies are increasingly constrained by the quest for price stability and fixed exchange rates, growth policies are confined to supply-side competitiveness, infrastructural and workforce training measures. The second is the growing influence of new strands of literature (see Chapter 2) suggesting that the most dynamic industries are organized at a regional scale and depend for their success on a range of local economic, political and cultural externalities and collective goods.

The move in the direction of a federal system is also seen as an important opportunity for Italy since its great internal differentiation might be better addressed from subnational tiers of governance (Brancati, 1995). Municipal, provincial and regional authorities are more aware of the strengths and weaknesses of their territories than their central counterparts, and are, therefore, more able to make adequate development decisions. After analyzing preliminary evidence, Trigilia (2002) argued that regional and subregional institutional and administrative reforms have had a number of positive effects. First, the stability of subregional governments increased. Second, accountability has increased as mayors have increased powers, choose their own collaborators and are held responsible for government performance and for the strategies and policies governments implement. Third, citizens consider that the reforms have had a positive effect on city management and on cleaning and refuse collection, but not on social policies and urban planning, in spite of moves in the direction of 'strategic planning'.

Nonetheless, crucial political disagreements remain. In relation to questions as to which responsibilities are better dealt with at the regional level, there are two stances. The first is associated with a radical vision of federalism, that some political parties consider as the solution to all problems. According to this perspective, the devolution of responsibilities (and of the power to raise taxes) from the centre to the periphery is always appropriate and is the best ultimate option. A consequence is the acceptance of differentiated rights of citizenship. The second is associated with a cooperative alternative to this model of competitive federalism (Brancati, 1995). At the centre of this alternative model is the view that it is inappropriate and unacceptable for citizens of the same nation-state to have different rights from one region to another. In this model central government retains a vital role and can establish common and inviolable rights of citizenship (to which are added a coordinating role in growth and development policies).

As far as the radical shift in Italian regional policy is concerned, some short-term results sit alongside a number of long-term expectations. These results include greater cooperation and coordination between different tiers of governance, serious multiannual planning, and a careful spending of public financial resources in the light of potential development impacts. More importantly, the expectations include an improvement in the quality of the territorial context in which they are applied (Viesti and Prota, 2004). To date, however, the new policy measures have not set in motion virtuous processes of economic development. Nor do they seem sufficient by themselves to deal with the structural problems that are at the root of the persistent gap between Italy's regions, notwithstanding the recent renewal of convergence (see Chapter 4). Some scholars argue that the new policies involve a certain degree of continuity with previous ones, especially with respect to capital incentives (Giannola, 2000), and that automatic incentives tend to reproduce existing patterns of industrial specialization (La Spina, 2003). Others have emphasized the administrative obstacles to implementation (De Vivo, 2000; Trigilia, 1999) and the difficulties associated with the mediation of interests, especially in the area of negotiated planning: in some cases, cooperation among local institutions emerged when the area applied for funding, only to dissolve once funds were received (Pichierri, 2002). These interest-related issues are connected with issues of development quality. In some cases, development in less-favoured areas has been achieved at great costs for those involved. Especially where the district model was imitated, development saw the growth of 'pre-Fordist forms of workforce exploitation' (Rossi, 2003) including wage rates that fell beneath national standards, long working hours, restrictions on trade-union activity and unhealthy working conditions. In these cases, the need for development has affected its quality, and territorial competitiveness has been pursued at the expense of workers' rights and social inequalities. On the basis of policy experience and statistical data, La Spina (2003) suggests, finally, that there is a risk that the benefits of development policies will be geographically concentrated if strong and binding institutional and financial rules are absent.

The coexistence of continuity and significant change is also a feature of economic reforms. Alongside Italy's massive wave of privatizations (that raised more than € 850 billion in 1993–2001, and significantly increased the capitalization of the Italian stock market to € 714 billion or 65 per cent of GDP in 2000) rules governing the conduct of private enterprises were revised. New laws reformed banking and finance, an Antitrust Authority was created, and the electricity, telecommunications, natural gas and retail sectors were liberalized. State intervention has declined, yet there are few signs of a managerial revolution in the field of corporate governance where, if anything, family capitalism has emerged stronger and, with the eclipse of Mediobanca, more direct than ever (Pagano and Trento, 2002). De Cecco (2000) argues that privatization was implemented too quickly and without a clear strategic plan. As a result, it neither encouraged greater industrial competition nor strengthened those private groups capable of giving Italy a central role in strategic sectors of its economy. Effectively, the decision to privatize was taken to please financial markets and international observers at the expense of a clear strategy for the reorganization and modernization of the production system.

Even when macroeconomic conditions improved, rapid privatization was preferred to mixed economy alternatives. For some actors, however, privatization did not go far enough, while for yet others it occurred without sufficient liberalization to create a competitive framework. An example of the former is Confindustria which maintains that privatization was partial, that the state retained a significant degree of direct and indirect control over privatized enterprises and that there was insufficient international involvement, especially in the case of the utilities (De Nardis, 2000). An example of the second is the central Autorità Garante della Concorrenza e del Mercato (2004) which argued that energy and natural gas liberalization have not created sufficient competition.

Change did occur, however, in the corporate organization of the privatized industries. A study commissioned by the Camera dei Deputati (2000) indicates that in sectors such as steel the previous managers were retained because of their expertise, whereas in others new managers were appointed. The new companies have smaller boards, and more decision-making autonomy. The reason why is that former public companies were subject to a double level of control: industrial plans and strategic operations had to be assessed first by the sector holding company, and then by IRI or ENI. Companies that merged or were taken over have benefited from the influence of foreign colleagues. Employment relations were also affected by privatization. In state owned companies, a company-level social pact existed: trade unions managed human resources, managers were responsible for operational strategy while the holding company decided on long-term strategies. This social pact became untenable in the privatized companies. Industrial relations were individualized, and more flexible employment contracts were introduced, in order primarily to increase functional flexibility. In no case, however, were trade unions prevented from representing workers and discussing company policies. Arguably, these changes have modified the relationships between politics, economics, external markets and entrepreneurial culture. Also the criteria used in the market for public works have become more transparent, while the decrease in expenditure has squeezed out some of the more inefficient companies.

To conclude, recent reforms have introduced significant discontinuities in state-economy-politics relationships. A transition to supply-side policies coupled with a reduction in public expenditure and the emergence of more neoliberal political parties has already had an impact on the relationships between entrepreneurial cultures, politics and the public administration. In Italy, however, these changes have unfolded more slowly than in Anglo-American countries. Elements of continuity remain as a result in part of the resilience of the old hegemonic bloc and the interests it represented, and in part of the absence of a clear political project along with the influence of populist forces in the right-wing governments headed by FI. Indeed, the 1990s were still years of social bargaining, as the early 1990s' crisis was tackled by the Amato and Ciampi governments together with trades unions and employers, and the programme of debt reduction was pursued by the centre-left.

Chapter Six

Italian Regional Evolutions

Introduction

Aggregate trends in relative development are ultimately the result of the adding up of millions of differentiated, place-specific, microeconomic changes or constancies. More specifically, they are the result of underlying changes/inertia in the social and territorial division of labour that itself is a result of millions of actions of varying size and impact. At the same time, the speed and direction of change of the latter are affected by the system-wide results of the adding up of the parts. Connecting these microevolutions with wider trends, in order to place the former in context, and to reveal the variety the latter conceals, is a difficult task. There are, however, ways of making progress. The part that the evolution of an industry or a locality plays in creating the aggregate wealth of a country can be identified by disaggregating aggregate trends. Once a component of output, a sector or a regional economy that is expanding or contracting in certain ways has been identified, its contribution to the wider trajectories of which it is a part can be assessed. Additionally, detailed analysis of these mesoeconomic components can reveal variations and similarities in what is happening to their constituent elements. An explanation of these microeconomic changes and, in particular, of changes that work in the same direction or have a large impact on the whole contributes to an understanding of the latter.

The aim of this chapter is to start this process of disaggregation. As a first step, we shall identify the performance of individual regional economies. Greater attention to place-specific knowledge is one of the core contributions of geography. In this chapter, it will permit the identification of similarities and differences in the economic position and records of the Three Italies and their constituent regional economies, and will enablè us to ascertain whether there is significant intra-area differentiation and whether certain areas stand out in terms of their relative growth from the wider areas of which they are a part.

Attention to place-specific detail implies a reduction in scope. Accordingly, particular attention will be paid to four regions: Lombardia, Piemonte, Puglia and Basilicata and their constituent provinces (Figures 6.1 and 6.2), although we shall also deal with other parts of Italy and of Europe in which Italian capital is present or where capital invested in Italy is domiciled. Choosing these areas enables attention to be given to some of the most- and least-developed parts of Italy. This choice intersects with the subsequent concentration on certain industries (motor vehicles and chemicals and plastics) and their impact on the trajectories of regional economies.

In a second step we shall set in motion a process of disaggregation of the performance of individual regional economies. As a first step, we shall identify the respective roles of demography, investment and exports in shaping regional development trajectories. The analysis of export performance will permit the identification of trajectories centred on the district

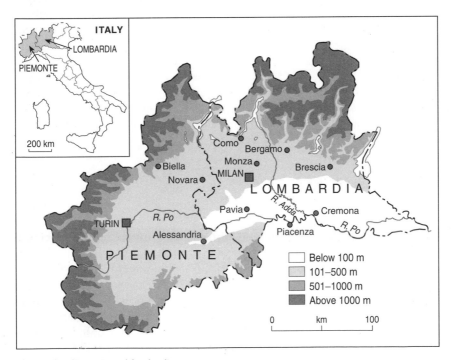

Figure 6.1   Piemonte and Lombardia.

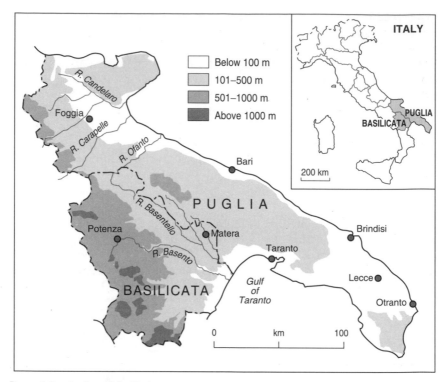

Figure 6.2 Puglia and Basilicata.

model, will reveal some of its limits and will point to a number of other models of economic and spatial organization that will be considered in more detail in subsequent chapters.

Italian Regional Evolutions

An important question examined in the recent academic and policy literature concerns the relative development records of individual regional economies. In this section, two methods of identifying relative regional performance will be used. The first employs an exponential growth model to compare cumulative regional rates of growth of a range of economic and demographic indicators with the EU15 average. The aim is to see which regional economies grew more rapidly/less rapidly than others and which increased/decreased their shares of EU15 output and population (see Appendix 1). The second examines the changing absolute position of regional economies. These methods permit an assessment of the claims frequently made in the literature about the dynamism of, for example, the Third Italy or the Adriatic Way.

Figure 6.3 records the cumulative growth of GDP relative to the EU15 average, G_{OT}^i. From the chart, it is clear that the relative performance of most regional economies has varied. Over the period 1951–2000, the three economies with the weakest growth records were Valle d'Aosta (−0.31 relative to the cumulative EU15 growth rate, or 73 per cent of the EU15 increase, as $e^{-0.31} = 0.73$), Liguria (−0.22 or 80 per cent) and Piemonte (−0.10). All were in the northwest. Next lowest was Friuli-Venezia Giulia (−0.06) although its trajectory was rather different in that strong relative decline was concentrated in the period up to 1957, with a subsequent upward cyclical trend interrupted in the early 1970s, early 1980s and 1990s. Of other areas with a high initial GDP per head, Lombardia grew slightly more than the EU15 average (0.06).

At the other end of the spectrum lay Lazio (0.46), with particularly strong relative growth in the 1950s and 1980s, the Veneto (0.41), whose growth relative to the EU15 dates from the second half of the 1950s and Marche (0.32). Marche declined relatively until the mid 1950s. At that point, rapid growth set in and lasted until the late 1970s. The growth of Emilia-Romagna (0.26) was more irregular, with strong relative growth from 1956 until the early 1960s, and strong growth in the 1970s wave of productive decentralization (Garofoli, 1991). In the cases of the more southern regional economies, growth dates from the end of the 1950s, when earlier tendencies to lose ground were reversed. Overall, the greatest southern GDP increases relative to the EU15 were recorded in Puglia (0.32), Abruzzo and Molise (0.24) and Sicilia (0.23). Of the areas in the Mezzogiorno, in the period from 1975 or 1980 until the mid-1990s, only Abruzzo and Molise, whose growth accelerated from the late 1960s, saw continued relative GDP growth. Indeed, in 1975–2000 as a whole, Basilicata (−0.09), Calabria (−0.04), Sardegna (−0.02) and Sicilia (0.01) were amongst the worst performing regional economies, although Basilicata saw strong relative growth after the early 1990s. These figures are a clear indication of the weakness of the performance of the Mezzogiorno after the mid-1970s. Also noticeable is that, in the Third Italy and the south, areas in the east bordering the Adriatic often grew more rapidly than areas in the west.

Figures 6.4 and 6.5 record GDP per head and population growth respectively. The GDP per head chart shows that in general cumulative GDP per capita growth was relatively fast in Italy compared with the EU15. The slowest relative rates were recorded in the more advanced regions such as Valle d'Aosta (−0.31), Piemonte (−0.06), Liguria (−0.03) and Lombardia (−0.02). Much of the relative decline occurred in the 1950s in Valle d'Aosta, and in the early 1970s in the other areas, most strikingly in

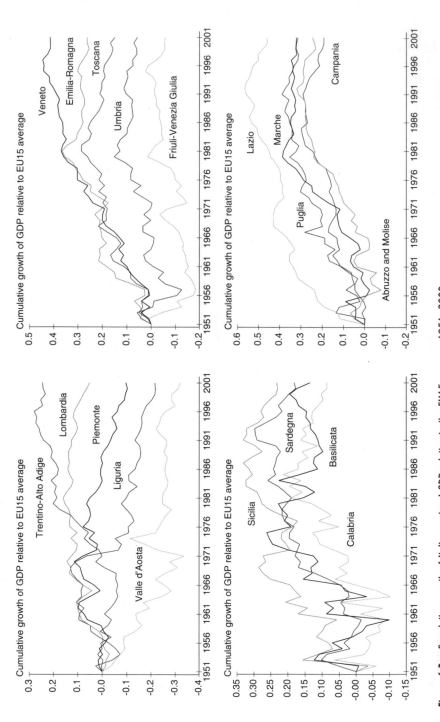

Figure 6.3 Cumulative growth of Italian regional GDP relative to the EU15 average, 1951–2000.
Source: elaborated from Istituto Tagliacarne and ISTAT data and ISTAT, 1998; 2003b.

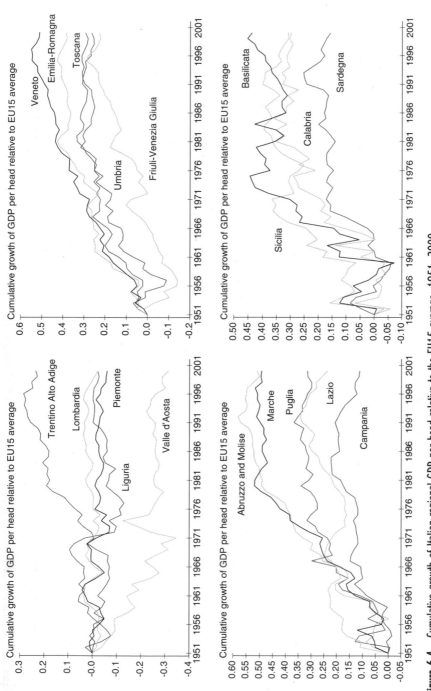

Figure 6.4 Cumulative growth of Italian regional GDP per head relative to the EU15 average, 1951–2000.
Source: elaborated from Istituto Tagliacarne and ISTAT data and ISTAT, 1998; 2003b.

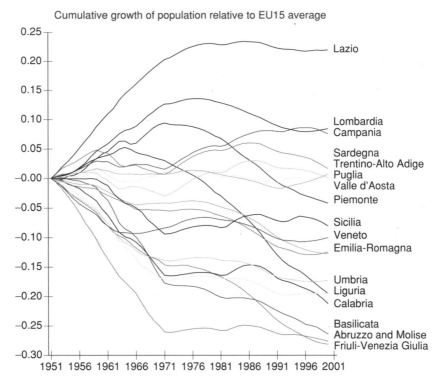

Cumulative growth of population relative to EU15 average

Figure 6.5 Cumulative growth of Italian regional population relative to the EU15 average, 1951–2000.
Source: elaborated from Istituto Tagliacarne and ISTAT data and ISTAT, 1998; 2003b.

Liguria, although most Italian regional economies also performed poorly in the 1990s. Areas in much of the Mezzogiorno also experienced slow relative growth. Included were Campania (0.11) and Sardegna (0.16). As with GDP, however, the south saw significant relative growth in GDP per head in the 1960s and early 1970s. The greatest relative growth occurred, on the other hand, in what were intermediate northern and central regions such as Veneto (0.51), Marche (0.49), Emilia-Romagna (0.39) and in certain small regions in the south such as Abruzzo and Molise (0.51) and Basilicata (0.44). In the case of Basilicata, for example, there was a strong upward trend with the wave of 1960s and early 1970s industrialization, centred around a cathedral in the desert model, and another weaker upward trend in the 1990s linked to major new industrial investments made in that region particularly with the creation of FIAT's Melfi complex.

More striking, however, are the remarkable shifts in population that occurred. In 1951–2000 southern regions such as Abruzzo and Molise and Basilicata saw their populations decline by 4.5 and 3.6 per cent respectively, while Friuli-Venezia Giulia declined by 3.2 per cent. Conversely, the population of Lazio grew by 58.0 per cent overall. Other areas that experienced large relative and absolute gains were Lombardy (38.3 per cent), Campania (32.9 per cent), Sardegna (29.1 per cent), Trentino-Alto Adige (28.8 per cent), Valle d'Aosta (27.9 per cent), Puglia (26.7 per cent) and Piemonte (21.8 per cent). Much of this redistribution occurred, however, in the years up to the early 1970s. Some of the areas whose populations grew sometimes quite strongly in this period experienced negative population growth in subsequent years. The examples are Liguria (17.9 per cent in 1951–75 and −12.2 per cent in 1975–2000) and Piemonte (27.5 per cent in 1951–75 and −4.5 per cent in 1975–2000).

Employment data are confined to the period since 1970 and are plotted in Figure 6.6. As this figure shows, there were strongly divergent regional evolutions in that period. In Italy as a whole, from 1970–2000, the number of full-time equivalent jobs increased by 17.8 per cent, with an increase of 21.2 per cent in the centre-north and 9.8 per cent in the south. Generally speaking, areas in the south other than Sicilia, Sardegna, Abruzzo and Campania, areas centred on metropolitan city regions in the northwest (Genoa, Turin and Milan) and Friuli-Venezia Giulia saw employment expand at rates that fell short of the EU15 average, while strong relative growth occurred in Trentino-Alto Adige (a cumulative increase of 0.21 compared with the EU15) and the Veneto (0.18) in the northeast, Umbria (0.14), Emilia-Romagna (0.05) and Marche (0.04) in the central part of Italy, in the capital city region (0.14) and in Campania (0.05). In Calabria (−0.17), Basilicata (−0.11), Molise (−0.11) and Puglia (−0.10) employment declined relative to the EU15, as it did in Liguria (−0.20), Piemonte (−0.10), Friuli-Venezia Giulia (−0.09), Valle d'Aosta (−0.02) and Lombardia (−0.02).

There were, however, sharp contrasts between the years from 1970 until 1984 and those from 1984 until 2000. In the first period, Italian full-time equivalent employment grew more rapidly than in the EU15 in every area except Liguria (−0.01) and Piemonte (−0.01), whereas in the second period it grew more slowly. In the second period, cumulative employment growth was less than in the EU15 in every region except the Veneto (0.00).

Figure 6.7 plots the productivity records of Italy's regional economies from 1970 until 2000. Again the regional trajectories are strongly differentiated. The greatest cumulative increases were recorded in the south. In rank order, the scores were: Molise (0.25), Abruzzo and Molise (0.16), Puglia (0.14), Calabria (0.14), Friuli-Venezia Giulia (0.13), Basilicata (0.10), Marche (0.08), Sicilia (0.02) and Emilia-Romagna (0.00). In all other areas, increases were smaller than in the EU15.

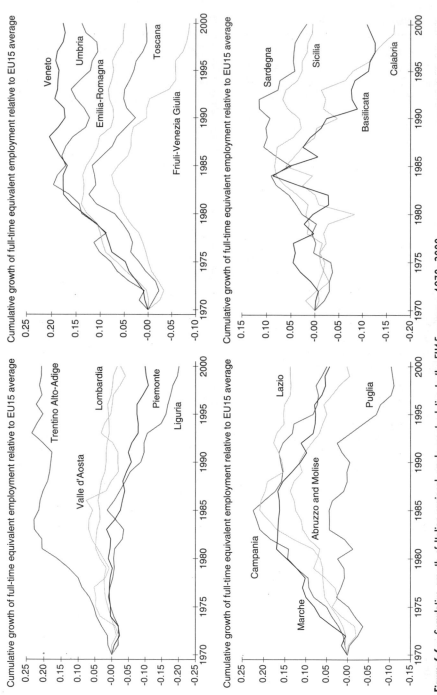

Figure 6.6 Cumulative growth of Italian regional employment relative to the EU15 average, 1970–2000.
Source: elaborated from ISTAT data and ISTAT, 1998; 2003b.

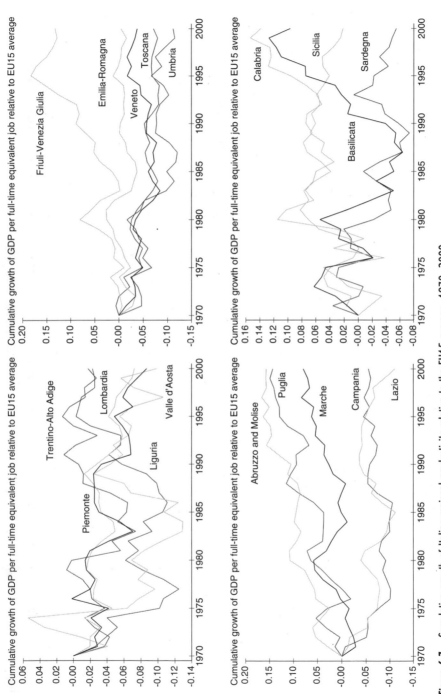

Figure 6.7 Cumulative growth of Italian regional productivity relative to the EU15 average, 1970–2000.
Source: elaborated from ISTAT data and ISTAT, 1998; 2003b.

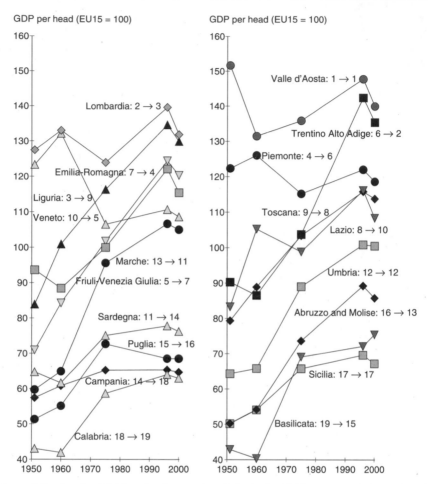

Figure 6.8 Growth of Italy's regional economies relative to the EU15 average and change in rank order in Italy, 1951–2000 (where 2-3 means that Lombardia's rank changed from 2nd in 1951 to 3rd in 2000).
Source: elaborated from Istituto Tagliacarne and ISTAT data and ISTAT, 1998; 2003b.

A second way of portraying the evolution of individual regional economies is to plot their position relative to an average at a series of points in time. In Figure 6.8, GDP per head is recorded as a percentage of the EU15 average in the years 1951 1960, 1975, 1996 and 2000.

Several features of this diagram warrant attention. The first is that the distribution of regional economies was less dispersed in 2000 than in 1951, largely as a result of the rise in the relative scores of the weakest regional economies, and the sharp relative decline of the Valle d'Aosta. The second characteristic of the chart is the way it confirms the strong convergence of the years from 1960 until 1975 and the relative weakening of the position of Italy's regional economies in the late 1990s. A third striking feature of the chart is the strong degree of inertia it reveals. The top regional economies in 1951 remained for the most part at the top end of the distribution in 2000, just as the regional economies at the bottom end tended to remain there. The chart records the changing rank order of regional economies. While these ranks do indicate a certain degree of inertia at the extremes, there is also evidence of some striking changes in relative position. The relative rise of a number of regional economies in the Third Italy is clear, with some striking ascents. Trentino-Alto Adige moves from 6th to 2nd, the Veneto from 10th to 5th and Emilia-Romagna from 7th to 4th. In the south, Abruzzo and Molise move from 16th to 13th and Basilicata from 19th to 15th. It is important to note, however, that a number of these regions are demographically small. In addition, it is clear that there are striking contrasts in the timing and degree of relative growth of areas comprising the Third Italy.

Figure 6.9 presents continuous (chained) time series for the four regional economies on which we shall concentrate. These data give different initial ranks for some regional economies and different initial GDP per head scores: Piemonte's initial rank is, for example, 3 rather than 4, while Basilicata's is 17 rather than 19. What this chart reveals is the existence of cyclical movements along with medium-term and secular trends in the relative positions of the four regional economies. Overall the two metropolitan economies in the north largely preserved their dominant positions. Lombardia saw its relative position strengthen up to the mid-1990s. At that point, however, it was particularly adversely affected by Italy's relative slowdown of the late 1990s. Piemonte also suffered in the late 1990s having held its ground relative to the EU15 until that date. In common with other areas in the Mezzogiorno, Puglia made significant relative progress in the 1950s and 1960s and Basilicata in the 1960s and early 1970s. After this phase of rapid catch-up, Puglia stood more or less still in relative terms, while Basilicata lost ground until the mid-1980s, after which point it saw its relative fortunes improve.

As is clear from Figures 6.8 and 6.9, there is a wide range of individual regional trajectories. In the northwest there is a certain degree of parallelism in the development paths of the metropolitan economies centred on Milan and Turin, with the former remaining ahead of the latter, and gaining some ground. The southern economies are quite differentiated: all started and ended as the least developed, although catch-up in the Golden Age meant that the size of the gaps diminished, while a few saw their relative fortunes improve in the late 1990s. These trajectories differ significantly from those

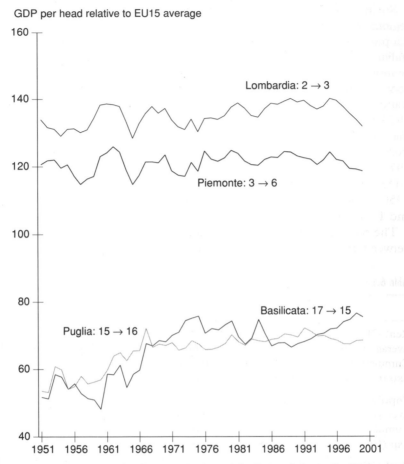

GDP per head relative to EU15 average

Figure 6.9   Growth of Lombardia, Piemonte, Puglia and Basilicata relative to the EU15 average, 1951–2000.
Source: elaborated from chained Istituto Tagliacarne and ISTAT data and ISTAT, 1998; 2003b.

of the Third Italy, within which there was very significant intra-area variation.

Comparative Regional Development

As indicated above, each regional economy has followed a distinctive trajectory. At the same time, however, there were certain broad similarities in the performance of groups of regional economies and some sharp differences between subperiods. In this section, attention will be paid to some of the temporal contrasts, with the help of essentially graphical techniques.

So far most emphasis has been placed on the performance of Italy's regional economies relative to the EU15 as a whole. A similar approach is adopted in this section. To contextualize the results, Table 6.1 reports Italian growth rates. As it shows, Italy achieved remarkably rapid rates of economic growth in 1951–75 (6.7 per cent per year in 1950–60 and 4.6 per cent in 1960–75). In the subsequent 21 years, growth rates were slower standing at an average of 2.4 per cent per year in 1975–96 (1975, 1983, 1993 and 1996 were cyclical troughs) and 1.9 per cent in 1996–2001. With the growth slowdown went a remarkable slowdown in population growth from 79 and 68 per thousand per year in the earlier periods to 15 and 12 in 1975–2001. Despite reduced population growth, the rate of growth of GDP per head also declined from 5.9 and 3.9 per cent per year in 1950–60 and 1960–75 respectively to 2.3 and 1.8 per cent in 1975–96 and 1996–2001.

The rate of growth of GDP per head is identically equal to the difference between the rates of GDP and population growth, while the rate of

Table 6.1   GDP and population growth in Italy, 1950–2001

	1950–60	1960–75	1975–96	1996–2001
Real GDP				
Average annual growth rate	0.0670	0.0460	0.0242	0.0192
Cumulative growth rate	0.6031	0.6903	0.5085	0.0962
Exp (Cumulative growth rate)	1.8278	1.9943	1.6628	1.1010
Population				
Average annual growth rate	0.0079	0.0068	0.0015	0.0012
Cumulative growth rate	0.0708	0.1018	0.0313	0.0060
Exp (Cumulative growth rate)	1.0734	1.1072	1.0318	1.0060
Real GDP per head				
Average annual growth rate	0.0591	0.0392	0.0227	0.0180
Cumulative growth rate	0.5323	0.5885	0.4772	0.0902
Exp (Cumulative growth rate)	1.7028	1.8013	1.6116	1.0944
Employment				
Average annual growth rate			0.0035	0.0036
Cumulative growth rate			0.0733	0.0143
Exp (Cumulative growth rate)			1.0761	1.0144
Real GDP per person employed				
Average annual growth rate			0.0207	0.0138
Cumulative growth rate			0.4352	0.0552
Exp (Cumulative growth rate)			1.5453	1.0568

Source: elaborated from data from ISTAT, 1990 (population 1950–9), CEPII, 2002 (real GDP and population 1960–2001), EUROSTAT, 2000 (employment 1970–2000) and Heston, Summers and Aten, 2002 (real GDP per capita 1950–9).

productivity growth is equal to the difference between the rates of GDP and employment growth. Algebraically, if *GDP* represents Gross Domestic Product, *POP* the population, *EMPL* employment and *G* (*x*) the rate of growth of *x*,

$$G\left(\frac{GDP}{POP}\right) = G(GDP) - G(POP) \tag{1}$$

and

$$G\left(\frac{GDP}{EMPL}\right) = G(GDP) - G(EMPL) \tag{2}$$

Figure 6.10 depicts the respective roles of GDP and population growth by region in 1951–2000, and in three subperiods: 1960–75, in which there was strong convergence, 1975–96, in which divergence prevailed, and 1996–2000 when there was renewed convergence.

In 1951–2000 as a whole, the areas whose GDP relative to the EU15 declined were Valle d'Aosta (−0.31), Piemonte (−0.10), Liguria (−0.22), and Friuli-Venezia Giulia (−0.06) in the north. In the first three of these regional economies GDP per head also grew less than in the EU15, as it also did in Lombardia (−0.02). The relative decline in Lombardia was proximately due to demographic growth exceeding GDP growth. Only four other regions experienced faster population growth than the EU15: Lazio, Campania, Sardegna and Trentino-Alto Adige, although Puglia and the Valle d'Aosta were close to the EU15 average. Generally speaking, in the south strong natural increase was offset by stronger net out-migration, except in areas with rapidly growing cities such as Naples and Bari. Conversely, in the Third Italy natural increase was comparatively weak (Table 6.2). The improved position in EU15 league tables of these two sets of areas resulted from a combination of this weaker than average population growth and stronger than average GDP growth.

In 1960–75, just five regions saw their GDP per head rise at less than the average EU15 rate. Three were in the northwest (Liguria, Valle d'Aosta and Piemonte), one was in the north (Lombardia) and the other was Lazio. From the height of the columns in Figure 6.10, it is clear that in Liguria and Valle d'Aosta the core factor was relatively slow output growth. In Piemonte above-average population growth and less than average GDP growth were equally important, whereas, in Lombardia and Lazio, demographic growth accounted for the decline in relative GDP per head. Conversely, the strong relative improvement in the position of the Mezzogiorno (a 22 per cent cumulative growth differential) depended primarily on a

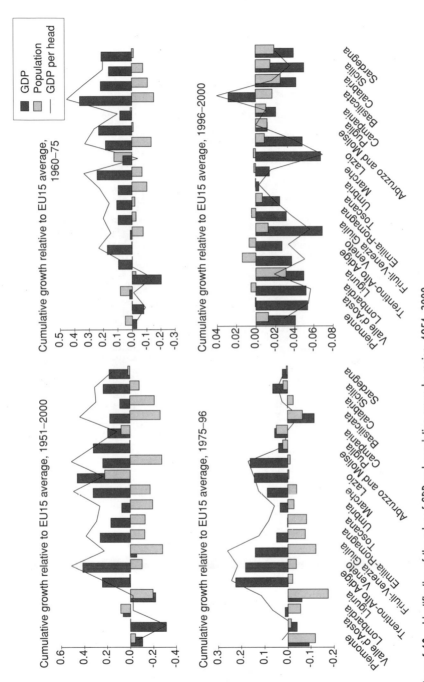

Figure 6.10 Identification of the roles of GDP and population grown by region, 1951–2000.
Source: elaborated from Istituto Tagliacarne and ISTAT data and ISTAT, 1998; 2003b.

Table 6.2 Average annual rates of population growth, natural increase and net migration per thousand, 1951–2001

	1951–71			1971–91			1991–2001		
	Population change	Natural increase	Net migration	Population change	Natural increase	Net migration	Population change	Natural increase	Net migration
Piemonte	11.6	1.5	10.4	-1.5	-2.0	0.5	-0.2	-3.6	3.3
Valle d'Aosta	7.4	3.2	4.5	3.0	-1.0	3.9	4.3	-1.9	6.1
Liguria	8.4	0.9	7.7	-5.1	-5.3	0.1	-3.5	-6.8	3.1
Lombardia	13.2	6.3	7.7	1.8	0.9	0.9	3.5	-0.7	4.1
Trentino-Alto Adige	7.3	8.9	-1.9	2.8	2.5	0.3	6.4	2.3	4.2
Veneto	2.6	8.0	-6.4	3.0	1.5	1.6	4.2	-0.4	4.5
Friuli-Venezia Giulia	-0.5	2.1	-2.7	-0.7	-3.9	3.0	-0.3	-4.9	4.4
Emilia-Romagna	4.1	4.1	0.0	0.8	-2.4	3.0	3.2	-3.9	6.9
Toscana	4.8	3.2	1.6	0.8	-2.1	2.8	0.8	-4.2	4.8
Umbria	-1.8	4.9	-7.4	2.3	-0.4	2.6	3.9	-3.3	7.0
Marche	-0.2	6.0	-7.0	2.5	0.6	1.9	3.3	-2.2	5.4
Lazio	17.1	11.4	7.0	4.6	3.6	1.1	3.5	0.2	3.4
Abruzzo-Molise	-6.2	7.1	-15.6	3.1	2.3	0.8	1.9	-1.3	3.1
Abruzzo	-4.5	7.2	-13.7	3.4	2.3	1.1	2.7	-1.1	3.8
Molise	-12.0	6.8	-22.0	1.7	2.0	-0.3	-1.4	-1.8	0.4
Campania	7.6	14.6	-9.4	5.3	8.7	-4.0	2.7	4.6	-1.9
Puglia	5.3	14.1	-11.8	5.9	8.2	-2.7	1.3	3.2	-1.9
Basilicata	-2.0	12.9	-19.9	0.6	5.7	-5.7	-1.1	1.0	-2.1
Calabria	-1.4	13.6	-20.4	2.0	7.2	-6.0	-1.8	2.1	-3.9
Sicilia	2.1	11.8	-12.5	3.0	6.4	-3.9	1.9	2.3	-0.4
Sardegna	7.2	14.4	-9.7	5.6	6.2	-0.7	-0.2	0.3	-0.4
Italy	6.5	8.4	-2.2	2.4	2.5	-0.2	2.1	-0.3	2.4
Northwest	12.1	4.2	8.5	0.0	-0.7	0.7	1.7	-2.2	3.8
Northeast	3.1	5.9	-3.1	1.7	-0.5	2.2	3.5	-2.0	5.4
Centre	8.7	7.1	1.8	2.9	1.0	1.9	2.7	-1.8	4.4
South	3.3	13.0	-12.8	4.2	7.1	-3.3	1.4	2.6	-1.3

Source: elaborated from ISTAT (1990) and EUROSTAT (2000).

17 per cent productivity catch-up. The relatively slow demographic growth of the Mezzogiorno, attributable to very high rates of net emigration in a context of high rates of natural increase (Table 6.2), accounted proximately for a 5 per cent cumulative growth differential. As the chart shows, however, the respective roles of demographic and output change varied from one southern region to another, and as Table 6.2 shows the role of natural increase and net migration differed quite markedly with, for example, relatively low rates of net emigration from Sardegna and from Puglia and Campania which contained the rapidly growing cites of Bari and Naples.

The situation from 1975–96 was rather different. In the northwest, only in the Valle d'Aosta was there a relative decline of GDP per head: while GDP growth in the northwest was just beneath the EU15 average, relative population growth stood well beneath the average, contributing to relative increases in GDP per head. As Table 6.2 shows, in these and a number of other regions in the centre-north, deaths exceeded births, as fertility rates fell. In the Mezzogiorno, only in Abruzzo and Molise did output per head grow significantly more rapidly than in the EU15. Only Basilicata and Calabria had significantly less than average demographic growth, although Abruzzo and Molise were just beneath the average. In all areas, relative rates of demographic growth were much reduced compared with 1960–75. As Table 6.2 shows, rates of natural increase in the south were lower than in the past, though so too were rates of net emigration, with the latter declining relative to the former. Overall, the Mezzogiorno had a cumulative GDP growth differential of 0.03 compared with the EU15, a population differential of 0.02 and a very small GDP per head differential of 0.02.

In 1996–2000, Italy's record was very weak. Only in Basilicata (0.04) did GDP per head growth exceed the EU15 average, while in Umbria and Puglia it lay just beneath the average. Areas in the north and Lazio fared worse than areas in the south, explaining the convergence of these years. In the south and in Liguria, Piemonte, Friuli-Venezia Giulia and Toscana demographic growth was weaker than in the EU15, whereas elsewhere it was close to the EU15 average. The major cause of the disappointing performance of Italy's regional economies was, however, relatively weak output growth, which itself stemmed from the deflationary impact of the reduction of public sector deficits and indebtedness on the one hand and the decreased competitiveness of Italy's traditional industries on the other.

To identify the role of employment growth, the differences between regional and EU15 cumulative growth are plotted in Figure 6.11 (see also Equation 2). This figure relates only to the period since 1975. In 1975–96, there were sharp regional contrasts. Employment and GDP growth were well beneath the EU15 average in Piemonte, Valle d'Aosta and Liguria in the northwest, and in Basilicata in the south. Significantly above average GDP and employment growth rates were recorded, conversely, in

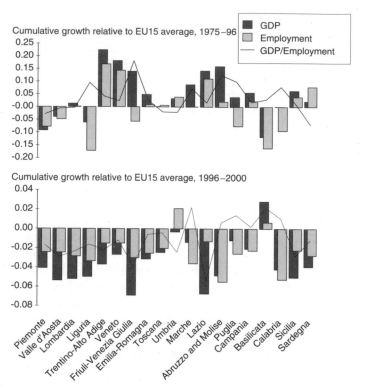

Figure 6.11   GDP, productivity and employment growth by region, 1975–2000.
Source: elaborated from Istituto Tagliacarne and ISTAT data and ISTAT, 1998; 2003b.

Trentino-Alto Adige and Veneto in the northeast and Lazio. As in all areas except Valle d'Aosta in the northwest and Puglia, Campania, Basilicata and Calabria relative cumulative employment growth exceeded relative cumulative population growth, relative employment rates generally increased. In absolute terms the result was rising employment rates in the centre-north and declining employment rates in the Mezzogiorno suggesting that the growth of employment was not sufficient to absorb those seeking work and helping explain the divergence that characterized these years: in less-developed areas, the supply of new jobs was not sufficient to match demand, made up of those who lost their jobs as a result of structural change, and net increases in the size of the potential workforce/population.

In 1996–2000, the situation differed radically. GDP, productivity and employment grew less rapidly in Italy than in the EU15, and, as areas in the centre-north fared worse than areas in the south, convergence resulted.

What, in short, this initial disaggregation indicates is that temporal variations in the relative performance of regional economies and the resulting convergence and divergence trends were a result of historically changing and regionally differentiated relationships between output, employment and demographic growth. An aim of later chapters is to identify and explain some of the components of these changes in output and employment.

Comparative Provincial Development

At a provincial level a more complex set of trends is observable. In 1995–2000, 11 of the 15 fastest growing provinces were in the Mezzogiorno (Table 6.3). Crotone's achievement of the fastest GDP growth, involving the fastest growth in construction value added, the third fastest in industry including energy and the fifth fastest in services, was not sufficient to lift it off the bottom of the provincial GDP per capita league table. Matera's growth conversely raised it 10 positions from 90th to 80th.

Just four of the slowest growing provinces were in the Mezzogiorno: L'Aquila, which lost it EU Objective I status at the end on 1997; Brindisi; Siracusa; and Cagliari. The coexistence of some of the fastest and slowest growing areas in the south is indicative of the emergence of strong infraregional contrasts. A similar phenomenon is noticeable in the centre-north with areas that surround large metropolises such as Milan and Rome strongly represented amongst the slowest growing areas.

Strong intraregional contrasts are also revealed in Table 6.4, which records the same indicators as Table 6.3, but for the four regional economies on which we shall concentrate. In Piemonte and Lombardia, the metropolitan economies of Milan and Turin preserved or enhanced their relative position. Milan retained top spot, while Turin rose 9 places. As far as growth of the components of GDP was concerned, value added in services was ranked highest. Conversely, the ranks for industrial value added growth were well beneath the ranks for overall GDP growth. In these regions, there were, however, significant falls in the relative position of a number of strongly industrialized provinces: in Piemonte, Biella fell from 27th to 33rd, while in Lombardia Como fell from 29th to 39th, Lecco from 21st to 32nd, and Brescia from 14th to 18th. Of these areas, Biella and Como are famous textile and clothing districts, while Lecco and Brescia are the home of districts specialized in metals. In Puglia and Basilicata, most provinces achieved sufficiently fast growth to raise their positions in the GDP per head league table by a few places, though Matera jumped 10 positions and Brindisi dropped 10 for reasons that we shall consider in Chapters 8 and 9.

Table 6.3 Trends in provincial growth: the fastest and slowest growing provincial economies, 1995–2000

Province	Region	GDP, 2000, share of Italy	GDP per head, 2000 (EU15 = 100)	GDP per head, 2000 rank	Average annual growth 1995–2000, rank				Change in GDP per head rank, 1995–2000
					GDP	Industry	Construction	Services	
The fastest growing									
Crotone	Calabria	0.2	55	103	1	3	1	5	0
Matera	Basilicata	0.3	72	80	2	1	65	2	10
Siena	Toscana	0.5	115	28	3	8	22	18	18
Grosseto	Toscana	0.3	94	62	4	24	28	9	4
Enna	Sicilia	0.2	59	100	5	2	74	8	2
Isernia	Molise	0.1	84	68	6	4	78	10	5
Cosenza	Calabria	0.8	63	95	7	13	54	3	3
Genoa	Liguria	1.7	108	46	8	18	27	33	6
Sassari	Sardegna	0.6	80	72	9	96	81	1	4
La Spezia	Liguria	0.4	109	41	10	20	8	27	10
Foggia	Puglia	0.7	63	96	11	28	42	17	1
Potenza	Basilicata	0.5	75	76	12	7	97	29	2
Naples	Campania	3.4	64	93	13	81	13	23	2
Pistoia	Toscana	0.5	105	50	14	25	2	19	5
Nuoro	Sardegna	0.3	74	77	15	78	56	6	2
The slowest growing									
Aosta	Valle d'Aosta	0.3	125	11	103	103	102	98	−6
Viterbo	Lazio	0.4	82	69	102	53	103	102	−6
L'Aquila	Abruzzo	0.4	79	73	101	102	79	103	−6
Lecco	Lombardia	0.6	113	32	100	84	77	85	−11
Mantua	Lombardia	0.8	123	14	99	86	71	77	−6
Frosinone	Lazio	0.7	82	71	98	89	94	95	−6
Como	Lombardia	1.0	110	39	97	92	85	78	−10
Pavia	Lombardia	0.8	100	57	96	91	34	92	−8
Brindisi	Puglia	0.5	68	87	95	97	59	64	−10
Pordenone	Friuli-Venezia Giulia	0.6	118	23	94	76	82	60	−8
Siracusa	Sicilia	0.5	77	75	93	95	73	74	−3
Biella	Piemonte	0.4	113	33	92	87	35	39	−6
Reggio Emilia	Emilia-Romagna	1.0	130	7	91	50	40	97	−1
Cagliari	Sardegna	1.0	74	78	90	90	89	82	−4
Latina	Lazio	0.8	90	63	89	68	33	96	−1
Rieti	Lazio	0.2	82	70	88	98	86	84	0

Source: elaborated from ISTAT, 2002.

Table 6.4   Trends in provincial growth: Lombardia, Piemonte, Basilicata and Puglia, 1995–2000

| Province | Region | GDP, 2000, % of Italy | GDP per head, 2000 (EU15 = 100) | GDP per head, 2000 rank | Average annual growth 1995–2000, rank | | | | Change in GDP per head rank, 1995–2000 |
					GDP	Industry	Construction	Services	
Turin	Piemonte	4.7	126	8	37	59	45	20	9
Vercelli	Piemonte	0.3	110	38	84	77	92	44	−3
Biella	Piemonte	0.4	113	33	92	87	35	39	−6
Verbano-Cusio-Ossola	Piemonte	0.3	99	59	47	40	60	50	−1
Novara	Piemonte	0.7	117	26	65	37	64	68	0
Cuneo	Piemonte	1.1	121	17	87	39	95	83	−4
Asti	Piemonte	0.4	102	54	41	35	23	55	2
Alessandria	Piemonte	0.8	109	43	61	30	36	79	−2
Varese	Lombardia	1.6	115	30	60	85	31	11	0
Como	Lombardia	1.0	110	39	97	92	85	78	−10
Lecco	Lombardia	0.6	113	32	100	84	77	85	−11
Sondrio	Lombardia	0.3	104	51	64	75	72	49	−1
Milan	Lombardia	10.2	159	1	39	65	51	7	0
Bergamo	Lombardia	2.0	121	19	74	72	41	62	−1
Brescia	Lombardia	2.3	121	18	83	70	68	47	−4
Pavia	Lombardia	0.8	100	57	96	91	34	92	−8
Lodi	Lombardia	0.4	106	49	75	82	12	94	−7
Cremona	Lombardia	0.6	113	34	72	58	21	86	−2
Mantua	Lombardia	0.8	123	14	99	86	71	77	−6
Foggia	Puglia	0.7	63	96	11	28	42	17	1
Bari	Puglia	1.9	72	81	16	12	25	35	3
Taranto	Puglia	0.7	70	85	18	57	3	16	1
Brindisi	Puglia	0.5	68	87	95	97	59	64	−10
Lecce	Puglia	0.8	61	98	25	42	39	30	1
Potenza	Basilicata	0.5	75	76	12	7	97	29	2
Matera	Basilicata	0.3	72	80	2	1	65	2	10

Source: elaborated from ISTAT, 2002.

Employment, Productivity and Investment

To help understand the regional contrasts identified in this chapter, Table 6.5 records a number of ratios derived from constituent parts of regional GDP. In a different way, it shows that there is a sharp productivity divide between the north and the south in particular, with more subtle differences between regions in these two broad territorial divisions. In the Mezzogiorno, productivity, measured by dividing gross value added net of an

Table 6.5 Structural differences between Italian regions, 1980–2000

Year[6]	Productivity[1]				Wage share[2]				Employee share[3]				Employment rate[4]				Employee wage relative to the national average[5]			
	1980	1995	1995	2000	1980	1995	1995	2000	1980	1995	1995	2000	1980	1995	1995	2000	1980	1995	1995	2000
Piemonte	100.0	102.1	107.1	105.4	52.7	47.0	45.9	44.9	69.8	69.2	68.0	69.3	44.1	43.3	43.2	45.1	108.8	104.1	105.8	104.7
Valle d'Aosta	107.3	100.2	111.8	105.4	45.3	41.1	43.9	44.9	64.8	62.3	64.3	66.6	50.8	48.5	47.9	49.0	96.1	101.0	111.6	104.8
Lombardia	111.6	113.8	113.2	111.8	52.0	45.8	47.3	45.7	76.1	72.3	71.7	71.7	43.7	43.4	46.1	47.1	109.6	108.5	109.1	110.2
Trentino-Alto Adige	97.7	95.9	103.7	103.6	46.9	43.2	47.6	45.9	66.1	66.7	68.0	68.7	48.4	49.7	50.2	52.2	92.7	94.1	106.2	106.3
Veneto	99.4	104.0	100.7	100.6	47.4	42.1	45.6	44.5	68.2	69.0	68.7	69.0	42.5	44.9	46.1	47.8	96.5	95.0	97.7	98.1
Friuli-Venezia Giulia	103.4	114.4	104.3	102.3	50.3	40.6	48.1	46.1	70.8	69.5	70.7	70.8	43.6	42.0	43.7	45.5	99.9	101.0	103.7	100.3
Liguria	114.9	113.7	104.6	105.9	50.4	41.7	46.7	43.5	70.3	67.3	68.0	68.5	39.8	37.8	38.8	40.8	109.9	109.3	105.0	102.5
Emilia-Romagna	108.8	110.2	103.6	104.4	43.9	41.8	45.0	43.9	65.9	64.8	65.8	67.3	46.5	45.5	48.7	49.9	101.3	105.7	103.6	103.0
Toscana	100.6	96.7	97.4	98.7	49.0	44.8	46.4	43.3	67.3	65.2	67.0	67.6	42.3	42.3	44.0	45.7	101.1	100.4	98.8	97.5
Umbria	95.0	93.6	96.8	94.6	49.6	45.9	47.0	45.3	70.5	69.7	67.8	69.0	40.9	40.0	39.9	42.6	94.9	91.6	98.2	94.2
Marche	92.7	95.4	90.9	94.3	45.9	41.6	46.0	44.9	63.5	62.3	65.8	68.4	45.2	42.4	43.4	44.1	92.5	93.9	93.0	93.6
Lazio	113.8	106.3	108.6	104.8	57.2	52.7	51.5	48.9	74.2	72.4	74.4	74.5	37.1	39.6	40.4	41.8	114.0	116.4	109.9	106.7
Abruzzo	92.1	92.7	90.5	90.6	42.5	42.8	45.4	45.4	58.3	64.4	65.0	67.8	37.4	36.4	37.3	37.5	89.6	92.6	92.6	91.7
Molise	77.4	86.5	89.7	92.6	43.7	44.0	45.3	44.5	51.8	60.0	61.8	64.5	38.1	34.1	33.2	34.9	89.9	92.8	96.2	95.8
Campania	83.8	83.3	85.5	87.7	55.8	48.3	50.2	48.4	67.3	67.3	70.8	71.0	32.5	28.6	29.0	29.8	91.0	90.2	88.7	91.7
Puglia	85.8	88.8	80.9	84.3	51.4	45.3	48.3	46.3	72.9	71.0	69.3	69.4	34.0	29.8	31.4	32.3	81.4	85.2	82.6	85.1
Basilicata	76.5	82.2	88.3	89.8	45.6	46.2	48.2	45.8	59.8	63.3	67.0	69.2	34.6	31.2	30.3	32.7	82.9	88.4	93.1	91.2
Calabria	80.2	76.1	77.5	82.7	51.5	47.3	50.0	46.3	69.4	67.0	70.7	70.3	29.0	28.8	30.4	30.2	80.3	80.7	80.2	83.6
Sicilia	87.7	81.8	90.5	90.6	52.3	50.0	47.7	46.2	69.4	70.9	68.8	69.6	30.4	29.2	28.0	29.3	87.6	87.1	91.8	91.4
Sardegna	93.1	86.1	87.9	86.7	51.6	48.4	47.8	48.3	65.5	66.7	66.6	68.9	32.3	32.0	32.9	34.5	95.0	94.7	92.3	93.2
Italy	100.0	100.0	100.0	100.0	50.7	45.8	47.4	45.7	69.8	68.8	69.3	70.0	39.1	38.1	39.3	40.7	100.0	100.0	100.0	100.0
North-west	108.5	110.3	110.6	109.4	52.0	45.7	46.8	45.3	73.5	70.8	70.2	70.7	43.4	42.8	44.5	45.9	109.3	107.3	107.9	107.9
North-east	103.4	106.7	102.5	102.5	46.2	41.9	45.8	44.6	67.4	67.2	67.7	68.5	44.6	45.2	47.1	48.7	98.4	99.5	101.3	101.0

| Centre | 104.6 | 100.7 | 101.6 | 100.6 | 52.3 | 48.3 | 48.9 | 46.3 | 69.9 | 68.4 | 70.2 | 70.9 | 40.2 | 40.9 | 41.9 | 43.4 | 105.3 | 106.7 | 103.4 | 101.2 |
| Mezzogiorno | 85.7 | 84.2 | 85.6 | 87.5 | 51.8 | 47.4 | 48.5 | 46.9 | 67.8 | 68.3 | 68.9 | 69.7 | 32.3 | 29.9 | 30.3 | 31.2 | 87.1 | 88.0 | 88.1 | 89.7 |

Source: elaborated from ISTAT, 1998; 2003b.

Notes

[1] Gross value added net of imputed financial services at factor cost in 1990 LIT (1980, 1985), or at basic prices in 1995 EuroLIT (1995, 2000) per full-time equivalent worker relative to Italian average (Italy = 100)

[2] Compensation of employees, in current LIT (1980, 1995), or in current € (1995, 2000) as a share of gross value added net of imputed banking services at factor cost in current LIT (1980 and 1995) or at basic prices in current € (1995, 2000)

[3] Employees as a share of full-time equivalent employment

[4] Full-time equivalent employment as a share of the mid-year resident population

[5] Compensation of employees in current LIT (1980, 1995), or in current € (1995, 2000), divided by the number of full-time equivalent employees relative to the Italian average (Italy = 100)

[6] Data for 1980 and first column for 1995 are from ISTAT, 1998; data for second column for 1995 and for 2000 are from ISTAT, 2003

adjustment for financial services by full-time equivalent employment, stood at just 87.5 per cent of the national average in 2000, while the employment rate in the Mezzogiorno stood at just 31.2 per cent compared with 40.7 per cent nationally, reflecting the south's more modest capacity to generate jobs and its relatively faster population growth. While its relative productivity had increased a little since 1980, its employment rate had fallen from 32.3 per cent.

At the same time, there have been important changes in the structure of employment and the distribution of income. In 1995–2000, the national share of employees increased slightly from 69.3 per cent to 70 per cent of full-time equivalent employment, yet the wage share continued to decline from 47.4 per cent to 45.7 per cent compared with 50.7 per cent in 1980 (see Dunford, 2002). The average compensation of employees differed widely from the north to the south and within the north. In the south, it stood, in 2000, at 89.7 per cent of the national average. In the centre-north, it exceeded the national average in all regions except Marche, Umbria, Toscana and the relatively dynamic Veneto. Also important is the evidence of relationships between some of these variables: the per capita compensation of employees is positively related to the apparent productivity of labour, and to the employment rate (Figure 6.12).

An important determinant of the structure of regional economies and their rates of growth is the rate of investment, which augments their resources and assets, and helps raise productivity. Trends in fixed capital formation are recorded in Figure 6.13. Figure 6.13 shows, first, that there are wide and enduring differences in real investment per resident and, more strikingly, in industrial investment per head. Consistently higher levels of investment characterize the northwest and the northeast, with the former constituting the main focus of reconstruction and modernization in the years following WWII, and the latter overtaking the former in terms of total investment, and catching up in terms of industrial investment, from the mid-1970s. The Mezzogiorno, conversely, had the lowest levels of per capita investment, except in 1963–88 in the case of total investment, and 1971–6 in the case of industrial investment.

Figure 6.13 also plots index numbers to record the growth of investment. It shows that the most striking increases occurred in the Mezzogiorno. Total investment increased more rapidly than in the other main territorial entities until the early 1990s recession, which saw the relative position of the south drop sharply in spite of the eligibility of virtually the whole of the Mezzogiorno for EU Structural Fund assistance. In the case of industrial investment, the Mezzogiorno saw the strongest increase, though what is also noticeable is the existence of several waves of relative growth, most strikingly in the early 1960s, with the first wave of southern industrialization, in the early 1970s, when there was a remarkable wave of industrial

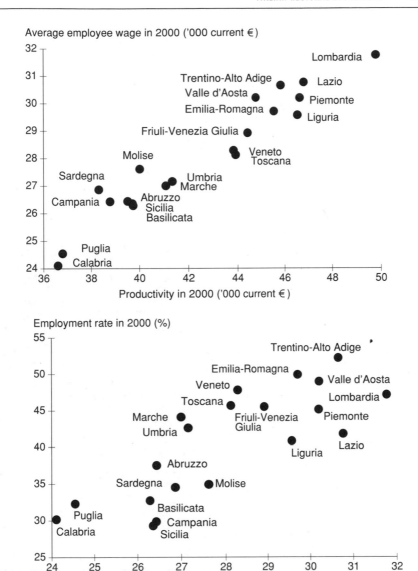

Figure 6.12 Compensation of employees, productivity and employment rates in 2000.
Source: elaborated from ISTAT, 2003b.

investment in the south, in the early 1990s, with the renewal of FIAT's investments in the south, and after 1996. These waves of investment are closely associated with phases of improvement in the relative position of the south.

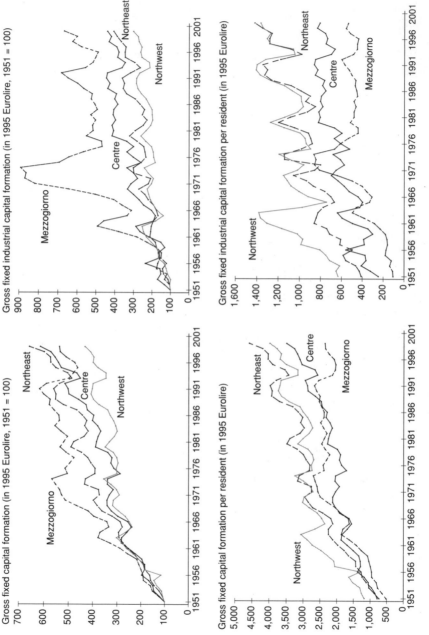

Figure 6.13 Gross fixed capital formation, 1951–2000.
Source: elaborated from ISTAT, 1975; 1983; 1998; 2003b.

Economic Specialization, Exports and International Integration

In 1991–2001, Italy's share of world exports declined from 4.8 per cent to 3.9 per cent. At the same time, the shares of Germany fell from 11.5 to 9.2 per cent, France from 6.2 to 5.2 per cent and the UK from 5.3 to 4.4 per cent (ICE, 2003). In the Italian case, this decline was in part a reflection of the sectoral and geographical profiles of Italian exports.

Sectorally, Italy was specialized in small and comparatively slow-growing sectors. As Figure 6.14 shows, in 1994–2001 Italian exports accounted on average for 4.7 per cent of world exports. Italy's highest shares were, however, for sectors that accounted for relatively small shares of overall exports, and which were characterized by relatively slow rates of growth: leather and leather products including footwear; nonmetallic mineral products including construction materials, glass and ceramics; agricultural and industrial machinery and equipment; furniture; textiles and textile products; and wearing apparel. Of the sectors in which Italy had above-average market shares, only rubber and plastic products and furniture had above-average growth in the value of world exports. Conversely, in a number of sectors that accounted for a large share of world exports, and which were characterized by strong relative growth, Italy had a weak specialization. The most striking case was

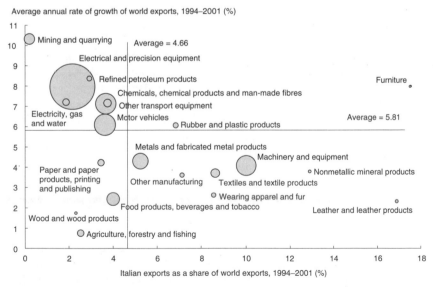

Figure 6.14 Italy's share of world exports by sector: export growth, sectoral exports and Italian specialization in 1994–2001.
Source: elaborated from data from ICE, 2003b.
Note: Circle size is proportional to the share of the sector in world exports in 1994–2001.

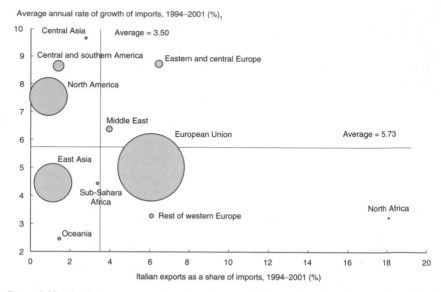

Average annual rate of growth of imports, 1994–2001 (%)₁

Figure 6.15   Italy's exports free on board as a share of imports at cost including insurance and freight by geographical area: import growth, market area size and Italian specialization in 1994–2001.
Source: elaborated from IMF, 2004.
Note: Circle size is proportional to the share of the geographical area in world imports at cost including freight and insurance in 1994–2001.

electrical and precision equipment. Within this sector, ICT products and precision equipment were particularly underrepresented, whereas there was a stronger Italian presence in electrical equipment and domestic appliances. Other examples of Italian weakness were chemicals and chemical products, including pharmaceuticals, and motor vehicles.

Geographically, Italy's presence was also strongest in market areas that grew more slowly. Figure 6.15 indicates that in 1994–2001 Italy accounted for 3.5 per cent of world imports. Its highest market shares were, however, in the relatively small and slow-growing North African market area, in the CEECs, and in the large but relatively slow-growing EU15. Conversely, Italy had small shares of the large and fast growing North American market and of the East Asian market, whose imports were adversely affected by the 1997–8 financial crisis.

A pattern of specialization that is negatively correlated with changes in the sectoral/market composition of world demand is a potential source of weakness. One reason why is that Italy could increase its share of each of the sectors/markets in which it competes and yet, as the growth of these sectors/markets is below average, see its share of world exports decline.

Another important feature of Italy's export structure is the major role played by large firms (ICE, 2002; 2003). In Italy there is a particularly large

Table 6.6   Italian exporters by export revenue, 2001

Export revenue ('000 €)	Number of exporters	Share of exporters (per cent)	Total export revenues	Share of total export revenues (per cent)
0–75	105,941	58.5	1,726	0.7
75–250	25,268	14.0	3,608	1.4
250–750	19,661	10.9	8,805	3.3
750–2,500	16,037	8.9	22,435	8.5
2,500–5,000	6,025	3.3	21,323	8.1
5,000–15,000	5,352	3.0	44,973	17.1
15,000–50,000	2,096	1.2	52,946	20.1
50,000 and over	676	0.4	107,186	40.8
Total	181,056	100.0	263,002	100.0

Source: ICE, 2003.

number of exporters. In 2001, for example, there were 181,056 exporters. Of Italian exporters in 2000, 43.9 per cent exported to just one overseas market. Also in 2000, 61.6 per cent of exporters had fewer than 10 employees, while 92.7 per cent employed fewer than 50. These groups accounted for 11 and 31 per cent of Italian exports respectively. If exporters are measured in terms of the value of their export sales, as in Table 6.6, however, it is clear that in 2001 just 676 exporters (0.4 per cent of the total) with export sales of at least € 50 million each accounted for 40.8 per cent of Italy's export revenues, and that 1.6 per cent of exporters accounted for 60.9 per cent of export revenues.

Trade is one aspect of internationalization. Foreign direct investment (FDI) is another. In the case of Italy, direct investment overseas and FDI in Italy are comparatively small. In 2001, Italy accounted for just 3.0 per cent of the stock of outward FDI compared with 3.3 per cent in 1990 (Table 6.7A). Comparable figures were 14.8 per cent for the UK, 7.1 per cent for France and 7.7 per cent for Germany. Italy's share of the stock of inward investment was also low standing at 1.8 per cent in 2001 and 3.1 per cent in 1990 (Table 6.7B). A number of sometimes overlapping factors explain this situation. One important reason for low inward FDI and for the lack of outward FDI is the fact that most mergers and acquisitions are in sectors other than the ones in which Italy specializes. In the case of inward FDI, the ownership structure of Italian firms is an obstacle to overseas acquisition. Outward FDI is restricted by the fact that SMEs are relatively numerous, and usually lack the organizational resources required to expand overseas.

Geographically, there are significant differences in the degree of international integration. FDI is largely located in the northwest. In 2000, there were 1,492 establishments that were, at least in part, foreign controlled in the four regions of the northwest. These establishments accounted for 50.5

Table 6.7A   Outward foreign direct investment, 1990–2001 (as % of world total)

Rank (1996–2001 flow)	Country of origin	Flows		Stock		
		1990–5 average	1996–2001 average	1980	1990	2001
1	United States	23.0	16.6	15.4	25.0	21.3
2	United Kingdom	10.1	15.5	15.4	13.3	14.8
3	France	9.4	10.7	4.7	7.0	7.1
4	Belgium-Luxembourg	2.8	10.3	1.2	2.4	6.3
5	Germany	9.3	8.4	8.3	8.6	7.7
6	Netherlands	5.7	5.8	8.1	6.2	5.1
7	Canada	2.7	3.7		4.9	3.7
8	Japan	9.9	3.6	3.8	11.7	4.6
9	Spain	1.4	3.5	0.4	0.9	2.7
10	Hong Kong	5.1	3.4	0.0	0.7	6.0
11	Switzerland	3.4	3.2	4.1	3.8	3.7
12	Sweden	2.7	2.4	0.7	2.9	2.0
13	Italy	2.5	1.6	1.4	3.3	3.0
14	Finland	0.6	1.4	0.1	0.7	0.9
15	Denmark	0.9	1.3	0.4	0.4	1.1
16	Singapore	0.9	0.8	0.7	0.5	0.9
17	Australia	1.0	0.7	0.4	1.8	1.1
18	Taiwan	1.2	0.6	0.0	0.7	0.8
19	Portugal	0.2	0.6	0.1	0.1	0.3
20	Ireland	0.1	0.4		0.2	0.3
	Total for 20 countries	59.8	62.4	65.2	95.1	93.5
	World ($ million)	253,302	765,884	521,486	1,721,462	6,707,141

Source: ICE, 2004.

per cent of the national total. Of this total, one-third were in Lombardia and 13.8 per cent in Piemonte. Next came Emilia-Romagna with 9.9 per cent, Veneto with 8.8 per cent and Lazio with 5.7 per cent. In the south, there were just 348, accounting for 11.8 per cent of the national total. In the case of exports, there were also sharp regional contrasts. In 2000, the northwest accounted for 41.3 per cent of Italian exports, although, as a result of tertiarization, this share was in decline. The northeast accounted for 30.8 per cent, the centre for 16.6 per cent and the south for just 11.1 per cent. Expressed as a share of GDP, exports vary from 35.1 per cent in the Veneto to 1.2 per cent in Calabria (Table 6.8) demonstrating a profound north-south divide in the degree of internationalization.

Also striking, finally, are the strong regional contrasts in the sectoral profiles of exports (Table 6.9). In the south, for example, in early 2001,

Table 6.7B Inward foreign direct investment, 1990–2001 (as % of world total)

Rank (1996–2001 flow)	Country of destination	Flows		Stock		
		1990–5 average	1996–2001 average	1980	1990	2001
1	United States	18.1	22.0	13.1	21.1	19.4
2	Belgium-Luxembourg	4.3	9.8	1.1	3.1	6.9
3	United Kingdom	7.8	8.0	9.9	10.9	7.0
4	Germany	1.9	6.7	5.8	6.4	7.2
5	China	8.6	5.3	1.0	1.3	5.6
6	France	7.2	4.5	8.8	5.3	4.1
7	Canada	2.8	3.3	8.5	6.0	3.2
8	Hong Kong	2.2	3.0	19.6	7.9	6.9
9	Brasil	0.9	2.9	2.7	2.0	3.1
10	Sweden	2.4	2.7	0.5	0.7	1.3
11	Spain	4.8	2.1	0.8	3.5	2.3
12	Mexico	3.6	1.8	1.3	1.2	1.6
13	Denmark	1.1	1.4	0.7	0.5	1.0
14	Ireland	0.5	1.3	0.3	0.2	1.0
15	Singapore	2.6	1.1	1.0	1.5	1.5
16	Italy	1.7	0.9	1.4	3.1	1.8
17	Bermuda	0.8	0.9	0.8	0.7	0.9
18	Poland	0.6	0.8		0.0	0.5
19	Japan	0.5	0.7	0.5	0.5	0.8
20	Portugal	0.8	0.4	0.6	0.6	0.4
	Total for 20 countries	73.0	79.6	78.3	76.6	76.6
	World ($ million)	225,321	812,337	635,543	1871,594	6993,409

Source: ICE, 2004.

refined petroleum products, motor vehicles and other transport equipment were relatively strongly represented. At a regional scale a whole series of specializations were identifiable. Of these specializations many were connected with the character of the area's main industrial districts: machinery and equipment in Emilia-Romagna; shoes in Marche and the Veneto; clothing in the Veneto and Toscana; furniture in Puglia and Basilicata; and so on.

After the Three Italies: The Origins and Limits of the District Model

In the last section, we pointed to Italy's international specialization in the 'Made in Italy' sectors, to the dominance of large firms and the northwest

Table 6.8 Exports as a share of regional GDP (%)

Regions and territorial units	1995	1996	1997	1998	1999	2000
Piemonte	33.2	30.9	30.1	29.0	27.2	29.5
Valle d'Aosta	14.7	10.3	8.4	9.8	9.5	13.0
Lombardia	31.3	29.6	29.3	28.8	27.8	31.0
Trentino-Alto Adige	18.9	16.2	16.4	16.2	16.4	17.2
Veneto	31.8	31.1	31.2	31.3	32.1	35.1
Friuli-Venezia Giulia	29.0	28.2	28.9	32.9	29.9	33.4
Liguria	11.2	10.7	10.7	8.8	8.5	10.0
Emilia-Romagna	26.9	26.3	27.0	27.5	27.1	29.2
Toscana	25.8	25.3	25.5	24.4	23.7	27.1
Umbria	14.1	13.2	13.2	13.0	12.6	14.2
Marche	24.0	23.3	25.1	25.2	22.5	25.2
Lazio	7.1	7.5	8.0	8.2	8.7	10.2
Abruzzo	19.9	18.8	20.3	21.4	19.1	23.8
Molise	9.3	9.5	10.3	10.2	10.1	9.6
Campania	8.5	8.2	8.5	9.4	9.1	10.2
Puglia	10.7	9.7	9.9	10.0	9.8	10.8
Basilicata	6.2	6.0	5.1	11.7	13.3	12.5
Calabria	0.9	1.1	0.9	1.0	0.9	1.2
Sicilia	5.2	5.0	5.7	5.7	5.4	8.1
Sardegna	7.0	6.6	7.7	6.4	6.5	9.8
Northwest	29.8	28.0	27.6	26.9	25.7	28.5
Northeast	28.4	27.5	27.9	28.5	28.4	30.9
Centre	15.7	15.6	16.2	15.8	15.5	17.8
Centre-north	25.5	24.5	24.6	24.3	23.7	26.3
Mezzogiorno	8.2	7.7	8.2	8.6	8.3	10.1
Italy	21.3	20.4	20.6	20.5	19.9	22.3

Source: elaborated from ISTAT, 2002b; 2003d.

(where 'Made in Italy' sectors and the district model are also found) and to the low degree of international economic integration of the south, alongside a not insignificant presence of exports of more modern sectors such as transport equipment. The different sectors of which an economic system is made up are interdependent parts of a division of labour. This division of labour has a national dimension with different regional economies specialising in different industries/functions with different modes of territorial organization (industrial districts, cathedrals in the desert, industrial metropolises, technopoles, logistic platforms, service centres, etc.). These specializations are not absolute: a range of sectors/functions and a range of modes of territorial organization can coexist in individual regional econ-

Table 6.9 Exports by sector and region, January–March 2001 (€ million)

€ million	1	2	3	4	5	6	7	8	9	10	11	12	13	14	15	16	17	18	19	20	21
Northwest	228	45	1,037	2,261	898	445	110	602	118	3,497	1,374	422	2,932	5,779	3,744	2,901	594	541	601	53	28,183
Piemonte	50	12	448	638	179	73	26	198	45	458	462	117	513	1,639	605	1,811	206	24	234	5	7,741
Valle d'Aosta	-	-	2	2	-	2	1	2	-	-	1	37	9	25	10	-	-	1	2	-	93
Lombardia	82	30	540	1,605	709	367	81	389	56	2,959	872	274	2,295	3,973	2,940	1,059	285	512	340	22	19,389
Liguria	96	2	47	15	8	4	3	13	17	80	39	32	88	158	175	23	104	4	25	26	960
Northeast	348	19	1,088	1,108	1,053	1,346	150	427	34	1,042	532	1,292	1,527	4,934	1,729	1,022	519	1,055	706	34	19,964
Trentino-Alto Adige	82	2	151	32	37	12	17	72	-	99	42	36	95	191	55	103	9	32	8	1	1,075
Veneto	110	10	379	662	589	1,127	61	207	25	401	237	331	738	1,881	876	237	322	461	618	17	9,286
Friuli-Venezia Giulia	34	2	91	49	6	13	35	78	5	67	64	46	236	493	285	37	24	433	10	7	2,014
Emilia-Romagna	122	5	467	366	421	194	36	70	4	476	189	878	459	2,368	513	646	164	129	71	9	7,588
Centre	139	33	358	956	503	1,633	59	270	87	1,270	227	313	619	1,423	937	349	428	322	536	48	10,510
Toscana	74	29	221	778	304	938	35	186	27	335	85	206	293	610	209	106	164	126	460	19	5,206
Umbria	11	-	41	62	54	21	9	7	-	30	10	24	129	108	22	7	10	11	3	-	558
Marche	14	-	19	42	100	617	10	23	24	57	68	21	116	561	91	10	8	156	34	1	1,972
Lazio	41	4	77	75	45	57	5	54	36	848	62	62	82	145	615	226	246	28	39	27	2,774
Mezzogiorno	239	23	499	158	277	464	33	85	1,017	663	226	157	437	460	682	1,174	350	362	50	46	7,403
Abruzzo	6	2	51	56	93	42	5	35	-	86	63	74	74	170	222	372	23	30	30	1	1,436
Molise	-	-	8	14	64	3	1	1	-	26	21	1	2	3	2	-	-	3	-	-	146
Campania	46	3	262	30	61	214	15	39	2	152	77	29	74	115	263	405	278	23	12	7	2,111
Puglia	83	6	57	37	51	202	1	4	14	89	31	22	206	140	30	139	19	251	2	5	1,388
Basilicata	6	-	6	4	1	1	-	2	-	13	19	1	1	4	3	188	9	53	2	-	309
Calabria	9	-	8	6	2	-	1	-	-	17	5	1	1	3	2	1	3	1	2	-	60
Sicilia	87	7	60	5	5	2	1	1	670	182	6	25	21	20	153	69	16	2	4	23	1,357
Sardegna	3	5	47	5	1	-	9	3	332	99	5	5	58	5	9	-	2	-	1	10	597
Italy	955	120	2,983	4,486	2,738	3,890	352	1,394	1,256	6,477	2,360	2,187	5,520	12,604	7,096	5,449	1,891	2,281	1,894	181	66,112

Source: elaborated from ICE, 2003.

Key: 1:Agriculture, forestry and fishing; 2:Mining and quarrying; 3:Food products, beverages and tobacco; 4:Textiles and textile products; 5:Wearing apparel and fur; 6:Leather and leather products; 7:Wood and wood products; 8:Paper and paper products, printing and publishing; 9:Refined petroleum products; 10:Chemicals, chemical products and man-made fibres; 11:Rubber and plastic products; 12:Nonmetallic mineral products; 13:Metals and fabricated metal products; 14:Machinery and equipment; 15:Electrical and precision equipment; 16:Motor vehicles; 17:Other transport equipment; 18:Furniture; 19:Other manufacturing; 20:Electricity, gas and water supply; 21:Total omies. In addition, this division of labour has an international dimension, with Italy occupying particular niches in an international division of labour.

As we argued in earlier chapters, economic geographies of Italy have largely concentrated on a subset of this interdependent division of labour: the 'Made in Italy' sectors, SMEs, the industrial district model of territorial organization and the regions in the Third Italy in which these parts of the division of labour and these modes of organization are of greatest relative importance. In subsequent chapters, we shall concentrate on other aspects of the division of labour and on other regional economies. As we shall argue in the conclusions, the evidence of a weakening of Italy's trade position presented in the last section itself suggests that greater attention should be paid to the role of these other sectors/functions. This evidence also implies, however, the development of a new strategy for the 'Made in Italy' sectors themselves, and a reconsideration of their roots. In this final part of this chapter, we wish, therefore, to consider briefly the extent to which economic geography after the Three Italies should review its understanding of the district model (which we outlined in Chapters 1, 2 and 3).

In the last section, we pointed to the recent weakening of Italy's trade position. This weakening was most marked in the fashion, household goods and other 'Made in Italy' sectors. Two factors were at work. First, there was a reduction in exports due in part to the delocalization of the production of goods for export. Owing to Italy's relatively high wage costs, a movement offshore, and a reduction in the scale, of certain wage-sensitive activities are probably conditions for the survival of the 'Made in Italy' districts. The creation of alternative job opportunities implies, therefore, a movement into different sectors, products and phases of production. Second, there was a decline in the competitiveness of Italian districts, due to an insufficiently rapid adaptation to the changing conditions of international competition. The reasons why were that small enterprises have more difficulties than large ones in internationalising their productive activities, establishing more direct links with overseas product markets and investing in distribution and after sales service to secure a greater degree of control over distribution channels. Not surprisingly, therefore, the districts that have done best in the recent past are often the ones in which there are a few dominant firms, able to promote the path of internationalization, or foreign controlled firms, with more direct links with overseas markets (ICE, 2002).

At a deeper level, the roots of Italy's competitive weakness, in a context of changing trends in world supply and demand, lie in those factors that led to its particular pattern of specialization, and that shape its capacity to renew its sources of competitive advantage. Essentially, for three decades Italy pursued a path of industrial development involving the growth of micro-enterprises, specialized in traditional sectors, and located in districts situated outside of the main industrial areas of the country, helping to explain the profound changes in Italy's industrial geography that generated some of the changing regional development trajectories identified earlier in this chapter. Three factors were particularly important. The first was the reaction to social conflict in large plants in established industrial areas. The second was the presence of small firms and, in particular, of craft industries. The third was the downward movement of Italy's exchange rate.

As far as the exchange rate is concerned, the starting point was the 1971–3 collapse of the Bretton Woods fixed exchange rate system. The breakdown of the Bretton Woods order opened the way to a system of fluctuating exchange rates and to a sharp differentiation in the trajectories of the EU15 economies. At one pole. the German Deutschmark emerged, alongside the Japanese Yen, as a strong currency, rising sharply in value against the US dollar (Figure 6.16). Connected initially to the world crisis of profitability, this relative decline in the exchange value of the $ was designed to improve US competitiveness, profitability, output and investment relative to Japan and Germany. The upward trend in these two strong currencies was reversed several times. The first reversal coincided with the

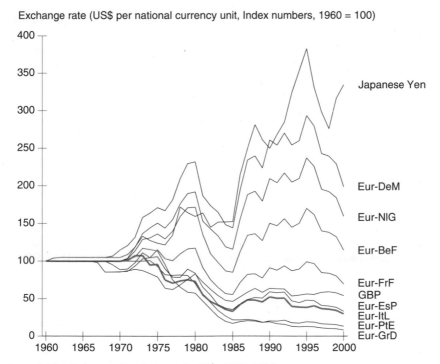

Exchange rate (US$ per national currency unit, Index numbers, 1960 = 100)

Figure 6.16   Exchange rates relative to the US$, 1960–2000.
Source: elaborated from data from EUROSTAT, 2000.

Reagan-Thatcher monetarist counter-revolution of 1979–80. The second dated from the Plaza Accord of 1985, which saw the resumption of dollar devaluation. The third was associated with the reverse Plaza Accord of 1995, designed to reverse the upward trend of the Yen, and lift Japanese manufacturing out of its protracted crisis, albeit at the expense of the East Asian economies, whose currencies were linked to the rising dollar (Brenner, 2002). At the other pole, a number of EU15 currencies, including that of Italy, followed a downward path. In the Italian case, outflows of money capital, inflation and devaluation interacted to lead to downward movements in the exchange rate until the mid-1980s. A further devaluation occurred in 1992, when Italy left the Exchange Rate Mechanism. Devaluations made Italian manufactures more competitive than those of the rest of world, while domestic demand weakened. These conditions gave strong encouragement to cost-reducing strategies in price-sensitive export-oriented sectors characterized by a high labour intensity, simple technologies and comparatively low productivity. Also involved was a movement of low-paid jobs towards the south. Manufacturers of goods requiring large investments also faced intensifying competition, as oil- and raw-material-importing countries sought to increase exports to pay for more costly imports. These companies, however, found adaptation more difficult, as we shall see.

The second factor was the existence of a large number of small and, in particular, of craft enterprises and the advantages they enjoyed. Craft enterprises are recognized and protected by the Constitution, have strong institutional representation and enjoy a host of other financial and legal advantages. Craft enterprises, for example, pay less tax, incur lower social charges, pay lower fees to Chambers of Commerce, have special rights in cases of bankruptcy and have fewer obligations in relation to written accounts than other enterprises.

The craft sector finds its roots in the preindustrial arts and crafts, and in the guilds that defended them. Essentially, crafts are entrepreneurial activities that are centred on the work of skilled individual-artisan-entrepreneurs, that involve the use of tools rather than machines and that do not involve the employment of external capital. These principles are embodied in legal definitions of craft enterprises: according to Laws 443/1985 and 133/1997 craft enterprises are ones in which a craft entrepreneur (1) participates directly in the work of the enterprise, (2) is qualified to exercize the profession, train apprentices and direct skilled workers, and (3) assumes direct personal responsibility for enterprise management. Craft enterprises provide goods and services including semi-finished goods, permitting them to act a subcontractors, have a maximum number of employees, which depends on the sector of activity (in clothing, for example, it is 40 excluding newly qualified apprentices, domestic workers, and the handicapped) and are legally defined as individual, family-run or limited firms.

A small share of Italy's craft enterprises warrant their artisanal status (makers of hand-made pottery, glassware, violins, etc.). The great majority are, however, construction and service enterprises and manufacturing firms that use machines rather than tools, apply line principles of production and often employ substantial numbers of people. Table 6.10 shows that, in 2001, according to Census data, craft enterprises accounted for 31 per cent of firms and 21 per cent of employment. More than one-half of craft sector jobs were in enterprises employing 3–15 people.

Table 6.10 also shows, however, that 55 per cent of craft enterprises had no employees. In the same year, Infocamere data revealed that 5.8 million firms were registered with Italy's Chambers of Commerce, and that 1.4 million were craft enterprises. Most were very small, indicating the extraordinary fragmentation of the Italian productive system. In the same year, in Lombardia, there were 896,000 firms, including 257,000 craft enterprises. Just 88,000 employed more than one person. The rest were one-person operations, or were part of the nonobserved economy, avoiding the need to reveal employment and income data. Some estimates suggest that declared income is only one quarter of what is earned, although, as we indicated earlier, these estimates are probably too high (Chapter 4). There are measures designed to bring these enterprises into the realm of legality. As matters stand, these enterprises pose a competitive threat to legal firms, are ineligible for support measures designed to improve competitiveness, are outside the system of social protection, and avoid taxes. The difficulty is that many of these producers have very small margins and consider that their survival depends on the avoidance of taxes and social security charges. 'Instead of "small is beautiful", I would say that "small is beautiful, if it is black", and if it does not change' (interview data, July 2000). In the medium-term the capacity of many of these enterprises to compete successfully and survive is in doubt. As far as the rise of the districts is concerned, however, the initial significance of craft status lay in the cost advantages it conferred, and the stimulus it gave to the growth of small firms.

The third factor was the related, two-fold decision, first, to strengthen workers' rights, mainly through the approval of the Statuto dei Lavoratori, and, second, to exclude firms with fewer than 15 employees. One consequence was a process of productive decentralization, involving the delocalization of a growing number of economic activities away from large firms in traditional industrial areas to microenterprises in areas of new industrial growth. Some companies, for example, re-established relations with redundant former employees, who reassumed their jobs, not as employees, but as self-employed craft entrepreneurs. As a result, Italy's industrial structure

Table 6.10 Size distributon of firms, 1981–2001

	All firms						Craft firms					
	Number of firms			Employment			Number of firms			Employment		
	1981	1991	2001	1981	1991	2001	1981	1991	2001	1981	1991	2001
1	1,368,430	1,582,487	2,382,398	1,368,430	1,582,487	2,382,398	655,169	532,745	690,069	655,169	532,745	690,069
2	674,945	717,952	713,790	1,349,890	1,435,904	1,427,580	222,019	219,544	226,855	444,038	439,088	453,710
3–5	485,029	615,549	576,576	1,767,372	2,259,089	2,105,502	202,546	208,888	208,080	748,351	772,349	765,796
6–9	150,992	186,182	190,848	1,078,225	1,336,265	1,362,906	71,115	74,372	74,527	504,707	534,855	532,470
10–15	81,125	92,755	99,425	971,239	1,115,790	1,198,111	26,377	33,744	35,610	302,646	403,008	425,774
16–19	22,054	28,861	29,854	380,776	499,705	515,526	2,283	8,874	9,538	38,899	153,351	164,722
20–49	43,387	51,112	52,832	1,281,163	1,480,367	1,555,079	1,099	6,062	8,479	28,723	143,723	218,267
50–99	11,799	11,707	13,660	808,715	800,209	934,672	84	5		5,522	335	0
100–199	5,538	5,137	5,822	760,990	701,237	796,232	16	1		2,069	103	0
200–249	926	920	1,072	205,740	203,882	238,190	1	0	0	206	0	0
250–499	1,809	1,703	1,931	621,602	578,708	662,048	1	0	0	305	0	0
500–999	765	728	790	529,390	492,736	537,443	0	0	0	0	0	0
1,000 and over	514	455	548	1,877,655	2,072,355	1,971,674	0	0	0	0	0	0
Total (Census)	2,847,313	3,295,548	4,069,546	13,001,187	14,558,734	15,687,361	1,180,710	1,084,235	1,253,158	2,730,635	2,979,557	3,250,808
Craft as % of total							41	33	31	21	20	21
Total registered		5,792,598					1,410,552					

Source: elaborated from ISTAT, 2004b; 2004c and Infocamere, 2001.

was transformed. In contrast to other advanced economies, this transformation did not involve an increase in industrial concentration, and a movement into sectors and methods with a higher technological content. Also transformed was Italy's industrial landscape, as industrial development diffused into nonmetropolitan areas in the northwest and the Third Italy, even if some of the areas in which industries grew had established industrial and craft traditions.

Most growth was of industrial districts. These districts were made up of large numbers of small enterprises. These enterprises specialized in light industries, drew on a large workforce and used simple technologies, with the possibility of the development of machine tools to permit subsequent automation. In these districts, different phases in the production of industrial goods were carried out in different establishments located in comparatively small areas covering a few communes, while each phase was fragmented across a multiplicity of firms. As Marshall showed, the concentration of small, specialized enterprises in a small area offers some of the advantages of large-scale production. Specialization itself permits the realization of some economies of scale, while within the district external economies offered other advantages. The geographical concentration of people engaged in the same trade encouraged the accumulation and diffusion of knowledge, and created possibilities for shared use of large-scale infrastructures, and common services, that are beyond the means of individual firms. The existence of a large number of employers created a large and constant market for special skills, that encouraged their reproduction from one generation to another. A mixture of competition and collaboration—the exchange of ideas, the sharing of knowledge, and the pooled use of specialized resources—characterized the relationships within and between entrepreneurs and skilled workers. Cooperation and collaboration were centred on the existence of relations of trust: formal contracts which raise transaction costs were infrequent. For the theorists of industrial districts, trust was rooted in the existence of regular personal contacts, common political and cultural values, and shared norms and rules. Derived from an experience of repeated contracting, trust was the source of the confidence that contracts and agreements would be respected, and that transactions would proceed as planned. Also significant, however, was the general absence of professional managers, and the cost advantages associated with large-scale recourse to work done at home or in small workshops by formally independent and often undeclared workers. Some commentators also suggest that local political cultures, civic pride and the efficiency of the public administration played an important role in encouraging industrial development in parts of the Third Italy and the northwest, while others point to the earlier importance of family labour and sharecropping in creating the entrepreneurial capacities that made them successful. All of these arguments were pertinent and of value. Some of the underlying societal conditions were, however, overlooked.

Growth of industrial districts led to relatively rapid increases in income and employment in the Third Italy and in parts of the northwest and south. Often entrepreneurs put in place successful upgrading strategies, improving their products, and reducing their costs, so as to compete effectively with low wage countries. With the help of Italy's declining exchange rate, many small enterprises established large export sales. As recent events indicate, however, significant constraints stand in the way of the further development of this model. On the one hand, economic and monetary union closed off the possibility of further currency devaluations relative to Italy's near markets in the EU, while more distant markets are more difficult to serve, especially in the absence of adequate advisory services and commercial assistance. On the other, the survival of many firms is in question. After thirty years of development, there are, first, difficulties associated with generational change. Entrepreneurial capacities and aptitudes are not necessarily replaced from one generation to another, so that, as one cohort of owners reaches retirement age, there is no guarantee that the new generation will want to follow in its footsteps. At the same time, the nature of property rights and the functioning of capital markets limit concentration. Second, many of these firms face acute labour shortages, not least as a result of the reluctance of new generations of workers to follow in the footsteps of their parents. As a result of high levels of parental investment in education, an excess supply of young people seeking nonmanual jobs coexists with a shortage of manual workers, only partially resolved through recourse to immigrant labour, which is itself limited due to housing shortages in small towns. Third, vertical integration sometimes offers advantages compared with the independent, small-firm, relational model (see Chapter 3). As research on the Vigevano shoe district showed (see Dunford et al., 1993), if survival and growth depend on expensive investments in research, development and innovation, smallness can be tragic, as the volume of sales may be insufficient to amortize investments. In fast changing conditions, the immobilism of individual small enterprises may prevent coordinated upgrading along the chain. Greater vertical integration can also help improve quality, and, if there are high rates of turnover of small enterprises, it can jeopardise the security of supplies of parts and components required, for example, to service and maintain equipment goods.

At present, it is this set of factors and circumstances that explain the supply-side difficulties faced in restoring the profitability and competitiveness of the district model and ensuring a renewal and further development of Italy's current pattern of sectoral and geographical specialization. Combined with trends in world demand, the emergence of new competitors and the fixing of Italy's exchange rate relative to those of other members of the

Eurozone, they play a major part in explaining the weakening of Italy's economic performance in the last decade.

Conclusions

In this chapter, we have identified a wide range of individual regional trajectories. Distinctions were made between the metropolitan economies of the northwest whose leading role was reproduced, the relative growth followed by the relative decline of much of the south, and the relative rise of the economies of the Third Italy, that occupied the middle of the distribution, Yet, within each group, there were significant differences in the speed and timing of development. An initial set of disaggregations resulted in the identification of historically changing and regionally differentiated relationships between output, employment and demographic growth. Different regional economies also specialize in different mixes of industries/functions with different modes of territorial organization. In the final section attention was concentrated on the 'Made in Italy' sectors and their district mode of organization: as we emphasized, the loss of momentum of these sectors is reflected in a deterioration in Italy's relative economic performance and points to the importance of examining the trajectories of other, more rapidly-growing sectors/functions sectors and their modes of territorial organization.

Chapter Seven

Industrial Change and Regional Development: The Changing Sectoral Profile of Regional Development and the Evolving Regional Profile of Industrial Change

Introduction

As mentioned earlier, aggregate differences and trends in GDP and its components are the sum of a series of sectoral contributions. The aim of this chapter is to identify these components, and to outline the changing geographies of two sectors whose regional impacts are identified. The ultimate aim is to split regional structure and performance into industrial sector and firm/plant contributions. If individual sectors/firms/plants reveal similar modes of spatial organisation, one can identify regional models/paths, of which districts, vehicle assembly parks, industrial estates, cathedrals in the desert and their trajectories are examples. Our aim is not fundamentally to identify these models. Instead, it is to identify and explain the contributions of different sectors/firms to regional structure and performance, and the changing geography of the sectors on which we shall concentrate. In some regions, sectors/firms that are organized on district lines play an important part. In others, they do not. The sectors we shall examine fall into this second group. In Chapters 8 and 9 we shall apply a set of tools developed in Chapter 3 for explaining these geographies and regional impacts.

The Sectoral Profile of Regional Economies

As a first step in identifying the significance of sectoral structures, Figures 7.1 and 7.2 identify the structure of four regional economies to which

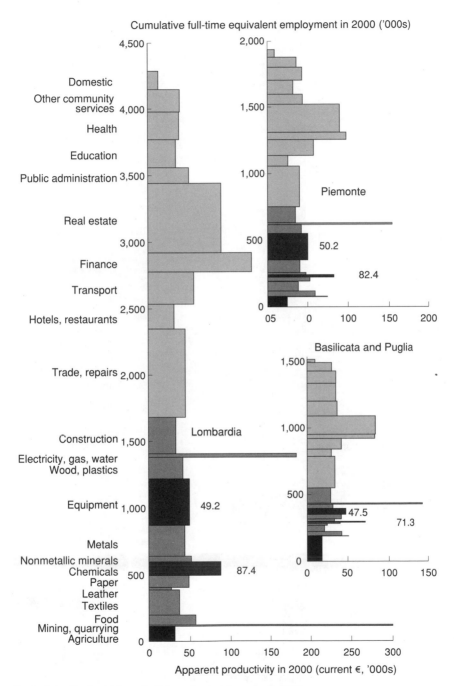

Figure 7.1 Employment, productivity and output by sector in 2000.
Source: elaborated from ISTAT, 2003b.

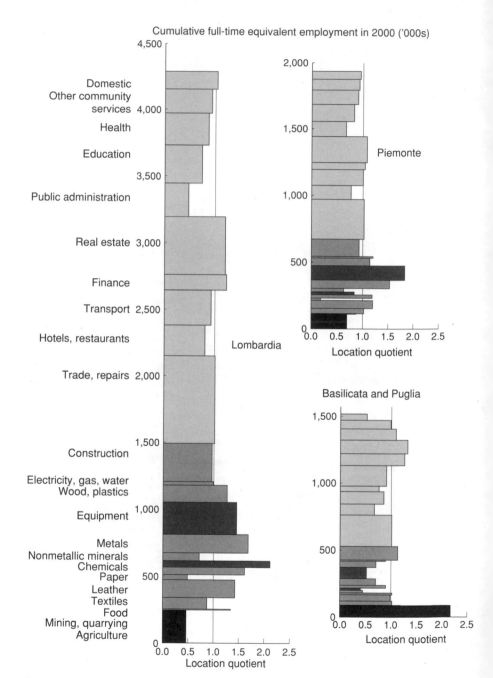

Figure 7.2 Specialization and employment by sector in 2000.
Source: elaborated from ISTAT, 2003b.

particular attention will be paid. Figure 7.1 records cumulative full-time equivalent employment on the vertical axis, and sectoral productivity on the horizontal axis. As a result, the area of each segment is proportional to aggregate value added. As is immediately clear, Lombardia is much larger in economic terms than Piemonte and than Basilicata and Puglia, which were grouped owing to the small size of Basilicata. Manufacturing industries are much more weakly developed in the south than in the two northern areas. In all of these economies, however, manufacturing and, in particular, the sectors we shall consider (vehicles, which is included in the machinery and equipment sector, and chemicals) account for a relatively small share of output. Also striking is the fact that productivity is roughly comparable in the two sectors in the two northern areas. The figures for the south are, however, somewhat smaller, standing at 95–7 per cent of the northern figure in machinery and equipment, and 82–7 per cent in chemicals.

Figure 7.2 records the location quotients for each sector (the share of employment in a particular sector in total regional employment divided by the equivalent national figure) on the horizontal axis,. The volume of employment is represented by the area of each segment. Clearly, Lombardia has a strong specialisation both industries. Chemicals is more than twice as important as in Italy (2.1), while the location quotient for machinery and equipment is 1.5. The Piemonte area has a stronger than average presence in machinery and equipment (1.8). Conversely, in the south manufacturing industries are weakly represented with location quotients of much less than 1. Just as Figure 7.1 showed that the two sectors account for small shares of regional output, Figure 7.2 shows that they account for small shares of employment.

Sectoral Structures and Uneven Development

These varying sectoral profiles and productivities of different regional economies help explain the overall differences in regional development examined earlier. To identify the magnitude of the productivity variations, Figure 7.3 portrays box and whisker plots of apparent productivity by sector and region in 2000. As is clear from the chart, average productivity was much higher in energy, with a median value of nearly € 150 thousand, and financial services (just over 91) than in textiles and clothing (just over 30), agriculture, hunting and forestry (nearly 25) and domestic services (somewhat over 9). At the same time, within these sectors there were varying and sometimes substantial interregional differences and some striking outliers. In the manufacture of machinery and mechanical, electrical, optical and transport equipment the range extended from € 30.9 thousand in the Valle d'Aosta to 63.1 in Basilicata. In textiles productivity varied by more than a factor of two

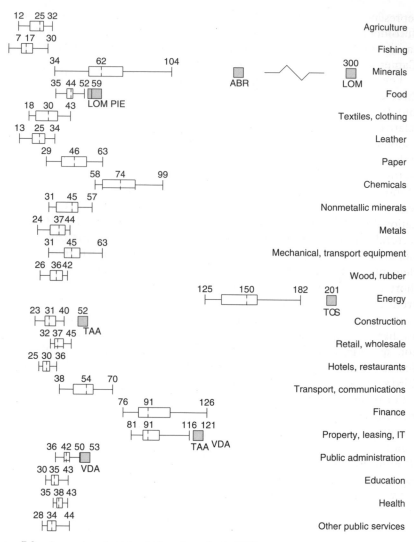

Figure 7.3 Apparent productivity of labour by sector in 2000.
Source: elaborated from ISTAT, 2003b.

from € 43.4 thousand in Trentino-Alto Adige, 42.1 in Friuli-Venezia Giulia, 38.3 in Piemonte and 37.1 in Lombardia to 21.2 in Puglia, 20.8 in Calabria, 20.7 in Sicilia, and 19.3 in Basilicata.

To examine more analytically this idea that variations in Italian regional development and regional performance are a consequence of intrasectoral productivity differentials on the one hand, and variations in the degree of representation of high and low-productivity sectors on the other, several decompositions of relative regional productivity growth can be used. In this case, the comparator is the Italian average rather than the EU15 (see also Cameron Muelbauer and Snicker, 2002; Dunford, 2002).

The first decomposition splits actual relative regional productivity into three elements, of which the first two are plotted for the periods 1980–95 and 1995–2000 in Figure 7.4. The first element is the amount by which relative regional productivity would differ from national productivity if the region concerned had regional productivity rates in each sector, but had the Italian average industrial structure. Called a 'relative productivity shift' factor, it compares regional and national productivity in each sector. Scores of more than zero indicate that on the whole the industries in a particular region achieve higher levels of productivity than the same industries at a national level. A score of 0.10, for example, would mean that differences in sectoral productivity would, controlling for industrial structure, cause regional productivity to lie 10 per cent above the Italian average. The second element is the amount by which relative regional productivity would fall short of or exceed the Italian average, if the region had Italian average productivity in each industry, but retained the regional industrial structure. Called a 'composition' factor, it indicates whether the region does or does not have an advantageous industrial structure, in the sense that it is, or is not, specialized in sectors with high productivity rates at a national level. Scores in excess of zero indicate an advantageous industrial structure. The third element is a covariance term.

Formally, if w denotes regional and W national productivity, i denotes the ith region, j represents the jth sector, where $j = 1, \ldots, n$, and e_j^i and E_j denote the share of employment in each sector regionally and nationally, then (see Appendix 1 for details)

$$\frac{(w^i - W)}{W} = \sum_{j=1}^{n} \left(\frac{w_j^i - W_j}{W} \right) E_j + \sum_{j=1}^{n} (e_j^i - E_j) \frac{W_j}{W} + \sum_{j=1}^{n} \left(\frac{w_j^i - W_j}{W} \right) (e_j^i - E_j)$$

(1)

Actual = Relative Productivity Shift + Composition + Covariance

Figure 7.4 plots this decomposition for all Italian regions. The underlying data are derived from two different sets of regional accounts, and involve the use of different industrial classifications, with finer distinctions within the service sectors, and cruder distinctions between manufacturing industries in the second series compared with the first. In a few cases, mainly of frontier areas, there are quite sharp breaks in the two series. In many cases,

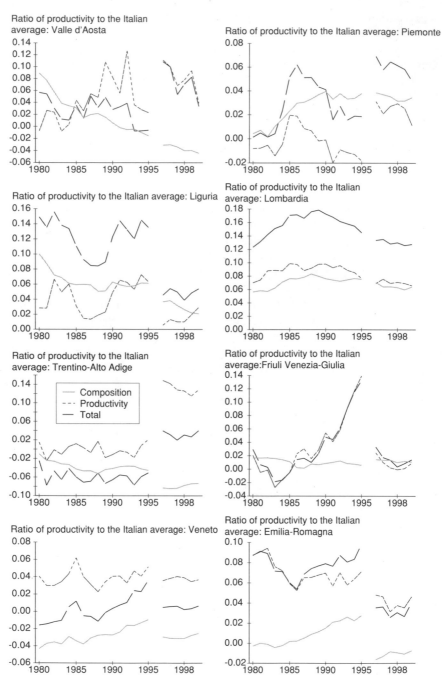

Figure 7.4 A decomposition of relative regional productivities, 1980–2000.
Source: elaborated from ISTAT, 1998; 2003b.

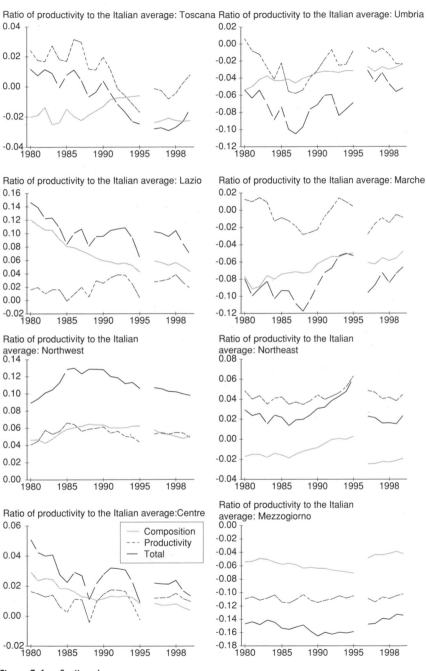

Figure 7.4 *Continued.*

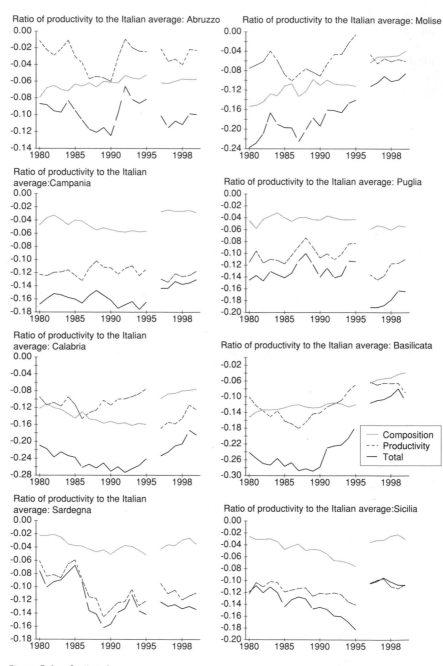

Figure 7.4 *Continued.*

however, there is a certain degree of consistency in the regional scores on the different indicators.

As the chart shows, the relative fortunes of the northwestern regions fluctuated and were mixed. Overall, an increase in total relative productivity in the early 1980s changed to decline in the 1990s. Consider two examples. The region of Lombardia followed and significantly determined the northwestern trend, though a significantly positive composition and relative productivity shift ensured that its total productivity remained well ahead of the national average. In the case of Piemonte, the situation was different. At the start of the period, Piemonte had a marginally advantageous 'composition', which meant that, if it had matched Italian average productivity in each sector, it would have an average productivity that exceeded that of Italy as a whole. At the same time, its 'relative productivity shift' score was negative, indicating that its industries did less well than the same industries in Italy. In the period 1980–95, there was a small rise in its relative productivity, due in the main to the upward trend in the composition component. An improved composition element reflected a move into sectors associated with higher than average productivities at a national level. On the intrasectoral relative productivity shift front, its record was less impressive: over the period as a whole, intrasectoral productivity grew more slowly than average. The 1995–2000 data suggest that relative total productivity was somewhat higher than the earlier series implied, although the direction of change remained downwards, as the size of the intrasectoral shift continued to decline.

In the northeast, there was strong relative growth. In the case of the Veneto, the dominant factor was its relative productivity shift. As its negative industrial composition scores indicates, it was specialized in sectors with productivities that were less than average nationally: if it had had Italian productivities in each sector, it would have had an overall productivity that was less than the Italian average. Over the years, this industrial profile became less adverse, yet the composition score remained negative. Indeed, according to the new regional accounts series for 1995 onwards, the size of this negative composition effect was yet larger than it was considered to have been in the past. Conversely, the productivity levels within these low productivity sectors were well in excess of national levels. An adverse or near-adverse industrial structure more than offset by strong positive relative productivity shifts also characterized Trentino-Alto Adige, Emilia-Romagna, Marche and, to some extent, Toscana.

Consider, finally, the south. What emerges is universally below-average relative productivities. In Campania, Puglia, Sicilia, Sardegna, and, after 1995, Calabria, the dominant factor was a poor relative intrasectoral productivity shift performance. In Abruzzo, Molise, Basilicata and, until 1995, Calabria industrial composition played an equally or more important role.

There were, however, some striking changes in relative fortunes. One of the most remarkable concerned the demographically small Basilicata region, whose relative total productivity increased sharply in 1988–99.

To go deeper, the productivities of different industrial sectors can be compared, and their role in explaining the change in relative regional productivity can be identified. To this end, the change in relative productivity can be split into a number of elements, using a decomposition that is similar to the one set out in Equation 1. In this case, the rate of growth of relative regional productivity is expressed as the sum of three elements. The first is an intrasectoral shift which is the change in productivity in each sector relative to the change in Italian productivity (where the comparison is of the change in productivity of each sector at the regional scale with the overall Italian average, indicated by the fact that W_t has no j subscript) holding employment shares constant. The second is an intersectoral shift, which is the change in relative productivity associated with the change in the employment share of each sector, holding relative sectoral productivities constant. The third element is a covariance term, not reported in the tables. Formally, as Appendix 1 indicates,

$$\frac{(w_t^i - w_{t-1}^i)}{(W_t - W_{t-1})} = \sum_{j=1}^{n} \left(\frac{w_{jt}^i - w_{jt-1}^i}{W_t - W_{t-1}} \right) e_{jt-1}^i + \sum_{j=1}^{n} (e_{jt}^i - e_{jt-1}^i) \frac{w_{jt-1}^i}{W_t - W_{t-1}}$$

$$+ \sum_{j=1}^{n} \left(\frac{w_{jt}^i - w_{jt-1}^i}{W_t - W_{t-1}} \right) (e_{jt}^i - e_{jt-1}^i) \qquad (2)$$

$$= \textit{Intrasectoral shift } + \textit{ Intersectoral shift } + \textit{ Covariance}$$

As is clear from an inspection of Equation 2, the first element identifies the contribution of the change in productivity within each sector relative to the overall national productivity change to the relative performance of the regional economy. An increase implies an upgrading in the position of the sector in the region concerned relative to the economy as a whole. An upgrading in this context means relative increases in value added per head through:

1 improvements in productive organization and technology, that reduce costs relative to prices and increase market share;
2 the introduction of new products, or improvements in the quality or design of existing products, which permit increases in the prices the products command and/or increases in output;
3 switches to new functions/higher value market segments within the sector (see Chapter 3 and the sectoral studies).

The second identifies the contribution of the changing shares of employment in different sectors to the performance of the regional economy, holding relative productivity constant. An increase in this case implies a movement of resources into sectors that are associated with higher value added per head regionally. The size of this term depends on the size of the employment change, the degree of the productivity advantage/disadvantage in the sectors that increase/decrease in importance and the extent to which national productivity increases. The sum of the three elements for any sector identifies the contribution of that sector to the overall change in regional productivity relative to the national change.

In 1980–96, in all regions the intrasectoral shift was more important than intersectoral shifts (Table 7.1), in part as negative and positive intersectoral shifts cancelled each other out, although the differences were often smaller in the south. In the shorter 1995–2000 period, conversely, in all of the four territorial entities the composition effect outweighed the intersectoral shift.

Consider, first, Piemonte (Table 7.1 and for sectoral detail Appendix 7.2). In 1980–95 overall relative productivity increased by 1.07, with positive contributions from all sectors except transport equipment (RPG = −0.01). The largest positive contributions were made by other market services (RPG = 0.26), trade (RPG = 0.19) and transport (RPG = 0.14). The intersectoral shift (IES) stood at 0.34, while the intrasectoral component (IAS) stood at 1.07. In the case of the intersectoral shift, however, the absolute value of the sectoral scores was equal to 1.07, indicating the coexistence of negative contributions, due to relative job losses in some sectors, and positive contributions, due to relative job growth in others. Of the negative effects, the most important were the ones associated with the metal products and machinery and transport equipment sectors, whose combined share of regional employment (Appendix 7.2) fell from 20 per cent (EM80 = 0.20) to 12 per cent (EM95 = 0.12). Contributing to this substantial decline were the crisis of the Olivetti group and the restructuring of the transport equipment sector (see Chapter 8). On the positive side there was a strong expansion of other market services (IES = 0.40), and trade, hotels and public stores (IES = 0.13). Of the intrasectoral changes, the relative increase in productivity in metal products and machinery (IAS = 0.27) and transport equipment (IAS = 0.16) made strong positive contributions, along with transport and communication (IAS = 0.09), several other manufacturing sectors including textiles (IAS = 0.11) and wood (IAS = 0.11), and agriculture (IAS = 0.09). Other (IAS = −0.08) and nonmarket services (IAS = −0.02) had a depressive effect on relative productivity growth.

In 1995–2000 Piemonte lost ground (RPG = 0.64) with intersectoral effects dominating (IES = 0.60). The sector that contributed most by

Table 7.1 An intrasectoral/intersectoral decomposition of regional productivity, 1980–2000

	1980–95				Employment share 1980 1995		Relative productivity 1980 1995		1995–00				Employment share 1995 2000		Relative productivity 1995 2000	
	Actual RPG	Composition IES	Within IAS	Covariance COV	EI80	EI95	RP80	RP95	Actual RPG	Composition IES	Within IAS	Covariance COV	EI95	EI00	RP95	RP00
Piemonte	1.07	0.34	1.07	−0.34	0.09	0.08	1.00	1.02	0.64	0.60	0.27	−0.24	0.08	0.08	1.07	1.05
Valle d'Aosta	0.81	0.14	1.12	−0.46	0.00	0.00	1.06	0.99	−0.47	1.07	−1.19	−0.34	0.00	0.00	1.11	1.03
Lombardia	1.21	0.34	1.24	−0.37	0.18	0.18	1.12	1.14	1.01	0.63	0.62	−0.24	0.18	0.18	1.13	1.13
Trentino–Alto Adige	0.87	0.26	0.74	−0.13	0.02	0.02	0.98	0.95	1.04	0.78	0.50	−0.23	0.02	0.02	1.04	1.04
Veneto	1.19	0.44	1.00	−0.24	0.08	0.09	0.98	1.04	1.04	0.67	0.59	−0.22	0.09	0.09	1.00	1.01
Friuli–Venezia Giulia	1.45	0.32	1.33	−0.21	0.02	0.02	1.03	1.13	0.64	0.46	0.33	−0.16	0.02	0.02	1.03	1.01
Liguria	1.09	0.37	1.03	−0.31	0.03	0.03	1.15	1.14	1.22	0.49	0.88	−0.14	0.03	0.03	1.05	1.05
Emilia-Romagna	1.13	0.34	0.95	−0.16	0.08	0.08	1.09	1.10	1.15	0.80	0.56	−0.21	0.08	0.08	1.04	1.04
Toscana	0.86	0.39	0.68	−0.21	0.07	0.07	1.01	0.98	1.25	0.66	0.80	−0.21	0.07	0.07	0.97	0.99
Umbria	0.88	0.37	0.74	−0.23	0.01	0.01	0.95	0.93	0.53	0.59	0.21	−0.26	0.01	0.02	0.97	0.95
Marche	1.03	0.38	0.79	−0.15	0.03	0.03	0.92	0.95	1.53	0.85	0.83	−0.15	0.03	0.03	0.90	0.93
Lazio	0.81	0.32	0.63	−0.14	0.08	0.09	1.15	1.06	0.38	0.51	0.09	−0.22	0.09	0.09	1.10	1.07
Abruzzo	0.94	0.40	0.62	−0.08	0.02	0.02	0.91	0.92	0.93	0.63	0.49	−0.19	0.02	0.02	0.90	0.90
Molise	1.15	0.42	0.68	0.05	0.01	0.01	0.76	0.86	1.47	1.34	0.65	−0.51	0.00	0.00	0.89	0.92
Campania	0.85	0.40	0.58	−0.13	0.08	0.08	0.83	0.84	1.13	0.63	0.67	−0.17	0.07	0.07	0.86	0.87
Puglia	0.98	0.31	0.91	−0.23	0.06	0.06	0.85	0.89	1.37	0.55	0.92	−0.09	0.06	0.06	0.81	0.84
Basilicata	1.02	0.46	0.57	0.00	0.01	0.01	0.76	0.82	1.01	1.05	0.39	−0.43	0.01	0.01	0.88	0.89
Calabria	0.66	0.30	0.44	−0.08	0.03	0.03	0.79	0.76	1.80	0.87	1.05	−0.12	0.03	0.03	0.77	0.82
Sicilia	0.62	0.32	0.46	−0.15	0.07	0.07	0.88	0.82	0.84	0.82	0.24	−0.23	0.06	0.06	0.89	0.89
Sardegna	0.66	0.35	0.43	−0.13	0.02	0.02	0.92	0.86	0.64	0.80	0.08	−0.24	0.02	0.02	0.88	0.87
Italy	1.00	0.35	0.85	−0.20	1.00	1.00	1.00	1.00	1.00	0.68	0.52	−0.20	1.00	1.00	1.00	1.00
Northwest	1.16	0.35	1.15	−0.35	0.30	0.29	1.09	1.11	0.91	0.61	0.53	−0.23	0.30	0.30	1.11	1.10
Northeast	1.16	0.36	0.98	−0.17	0.21	0.21	1.03	1.06	1.04	0.70	0.54	−0.20	0.22	0.22	1.02	1.02
Centre	0.88	0.38	0.67	−0.16	0.20	0.20	1.05	1.01	0.84	0.62	0.43	−0.20	0.20	0.21	1.02	1.01
Mezzogiorno	0.81	0.35	0.57	−0.11	0.29	0.29	0.85	0.84	1.13	0.71	0.59	−0.17	0.28	0.28	0.85	0.87

Source: elaborated from ISTAT, 1998; 2003b.

some way was real estate, renting and business services (RPG = 0.34) into which there was a significant shift of employment from 9 to 11 per cent of the regional total (Appendix 7.2), although the relative productivity of this sector dropped beneath the Italian average (RP00 = 0.98). Next came transport (RPG = 0.17), wholesale and retail trade (RPG = 0.13), financial intermediation (RPG = 0.09) and machinery (RPG = 0.09), which includes the transport equipment sector, and in which there was strong relative productivity growth, in part offset by a falling regional employment share.

In Lombardia, in 1980–95 (Appendix 7.2), relative productivity growth was much stronger (RPG = 1.21). Once again, service sectors played a leading role, though metal products and machinery (RPG = 0.10), chemicals (RPG = 0.08) and textiles (RPG = 0.08) made quite large positive contributions to relative productivity growth. The contribution of industrial sectors stemmed from strong relative productivity growth, which was invariably offset by declining shares of regional employment. In metal products and machinery, for example, there was an intrasectoral relative productivity shift (IAS) of 0.37 offset by an intersectoral shift (IES) of −0.14 and an associated fall in the share of regional employment from 15 to 10 per cent. In chemicals, there was an increase of 0.14 in the intrasectoral productivity shift. In textiles it increased by 0.13, indicating a strong upgrading of this sector, in a situation in which its contribution to regional employment declined from 9 to 8 per cent. A different pattern emerged in the service sectors. All increased their share of jobs and were associated with positive intersectoral growth. The most striking example was other market services with an intersectoral shift (IES) score of 0.51 and a doubling of its share of regional jobs from 8 to 16 per cent. In this sector, however, productivity growth was smaller than the national average, making intrasectoral growth (IAS) negative (−0.08).

In 1995–2000, the productivity performance of Lombardia (RPG = 1.01) relative to Italy was stronger than that of Piemonte, with intersectoral effects equalling the impact of intrasectoral effects. Financial intermediation (RPG = 0.42), real estate (RPG = 0.39), transport (RPG = 0.17), and machinery (RPG = 0.06) were the main contributors, while a number of industrial sectors including textiles (RPG = −0.09), mining (RPG = −0.08), leather (RPG = −0.03) and metals (RPG = −0.02) had negative effects.

In the case of the south, two important qualifying remarks must be made. The first is that levels of productivity were much lower than in the centre-north and played a significant role in explaining disparities in Italian development. The second is that, since the start of the 1980s, relative productivity differences have not changed much (Figure 7.4 and Table 6.5). Of course, this outcome is itself important, as standard catch-up theories suggest that less-developed areas will normally enjoy faster productivity growth than more-developed areas, due to the scope for technological catch-up. The Italian evidence in this chapter clearly shows that, at a national level, this catch-up effect works very slowly.

In the case of Puglia, in 1980–95 (Appendix 7.2), relative productivity grew more slowly than in Italy (0.98). The intersectoral shift (IES = 0.31) was marginally smaller than in the centre-north, with smaller shifts out of manufacturing industries, and smaller shifts into services. In Puglia, the most important contributions were made by other market services (IES = 0.24), trade, hotels and public stores (IES = 0.14) and nonmarket services (IES = 0.12). Nonmarket services accounted for 23 per cent of regional jobs in 1995 (EM95 = 0.23). Of the intrasectoral productivity shifts, the most important contribution was made by agriculture (IAS = 0.36). In manufacturing, there were high relative productivity increases in ferrous and nonferrous minerals and metals (IAS = 0.12), metal products and machinery (IAS = 0.06), and food (IAS = 0.05). Note that, of the industrial sectors, only food (RP95 = 1.17), chemicals (RP95 = 1.09), which includes the Brindisi chemical complex (see Chapter 9), and ferrous and nonferrous minerals and metals (RP95 = 1.18), which includes the Taranto steel complex, had above average productivity. In some sectors, relative productivity declined. Examples include construction (RP80 = 0.94 to RP95 = 0.81), transport (RP80 = 1.13 to RP95 = 1.08), transport equipment (RP80 = 0.98 to RP95 = 0.82) and textiles (RP80 = 0.62 to RP95 = 0.56). In 1995, the relative productivity of textiles stood at just over one half of the national average.

In 1995–2000, the relative productivity of Puglia improved (RPG = 1.37), due mainly to real estate (RPG = 0.40), wholesale and retail trade (RPG = 0.32), and transport (RPG = 0.15). Three manufacturing sectors also made quite significant positive contributions, due largely to intrasectoral productivity improvements: machinery (RPG = 0.11, IAS = 0.09); textiles (RPG = 0.06, IAS = 0.08); and wood (RPG = 0.06, IAS = 0.05).

In contrast to Puglia, Basilicata (Appendix 7.2) saw its relative position improve in 1980–96 (RPG = 1.02). The major contributions to this improvement were transport (RPG = 0.21), transport equipment (RPG = 0.18), nonmarket services (RPG = 0.17), other market services (RPG = 0.16) and trade, hotels and public stores (RPG = 0.12). In Basilicata, employment in the transport equipment sector, which is closely related to the development of FIAT's Melfi complex, increased from 1 to 3 per cent of regional employment (EM80 = 0.01 and EM95 = 0.03), while productivity reached some 182 per cent of the national average (RP95 = 1.82).

In 1995–2000, there was a further relative improvement in the position of Basilicata (RPG = 1.01). Of the sectoral contributions, the most important were wholesale and retail trade (RPG = 0.45), real estate (RPG = 0.28), machinery (RPG = 0.17), coke, refined petroleum products and chemicals (RPG = 0.06) and agriculture (RPG = 0.13).

An advantage of this method is that it enables one to see that, in 1980–1995, the development of this transport equipment sector accounted for nearly 18 per cent of the increase of 1.02 in Basilicata's productivity relative to the Italian average. Also noticeable is the role of nonmetallic minerals in 1995–2000 (RPG = 0.07), which include the methane that underpinned an earlier wave of chemical investment in this area, and includes the recently discovered oil.

The analysis employed in this section can be developed further to ask what would have happened to, say, Basilicata's relative productivity, if there had not been a significant transfer of vehicle manufacturing to the region, because new investments were located either abroad or in another part of Italy. What is involved is the identification of a counterfactual. In 1980–95, employment in transport equipment increased from 1,300 to 4,900, while productivity increased from 84 to 182 per cent of the national average (Table 7.2). If, for example, the complex were located in another area, if the productivity of the Basilicata transport equipment sector had remained at 84 per cent of the national average, and if employment had changed at the national rate, declining to 831 full time equivalent jobs, the average productivity of Basilicata would have increased at 0.91 times the national rate rather than 1.02 times. The arrival of the car complex in Melfi added nearly 11 percentage points to Basilicata's relative productivity growth, transforming what would have been a further falling behind in terms of productivity into an actual catch-up.

What we have showed is that relative productivity growth in different industries has measurable impacts on overall regional productivity convergence, and GDP per head catch-up. An increase in relative productivity in less-developed areas contributes to GDP per head catch-up provided it is not offset by deteriorating employment rates (which requires either that it occurs in a context of overall growth or that it opens the way to an increase in market share). A relative increase in productivity in a sector contributes to overall productivity catch-up provided that it is not offset by a decline in the sector's employment share, or provided that the resources released are re-employed in sectors with higher relative productivity. The reason why is that regional productivity is equal to the sum of productivity in each industry weighted by the industry's share of total employment. Algebraically,

$$w^i = \sum_{j=1}^{n} e_j^i w_j^i \qquad (3)$$

Table 7.2 Relative regional productivity and shares of Italian employment, 1980–2000

Sector	Year	Relative productivity				Employment share			
		Piemonte	Lombardia	Puglia	Basilicata	Piemonte	Lombardia	Puglia	Basilicata
Chemicals, chemical products and pharmaceuticals	1980	0.88	1.07	0.93	0.77	0.08	0.38	0.03	0.01
Chemicals, chemical products and pharmaceuticals	1995	0.97	1.09	1.09	0.84	0.08	0.36	0.02	0.00
Coke, refined petroleum products, chemicals and chemical products	1995	1.02	0.93	1.28	0.97	0.07	0.40	0.02	0.00
Coke, refined petroleum products, chemicals and chemical products	2000	1.01	1.06	0.94	0.75	0.07	0.39	0.02	0.00
Transport equipment	1980	1.03	1.08	0.98	0.84	0.34	0.18	0.03	0.00
Transport equipment	1995	1.04	1.04	0.82	1.82	0.27	0.15	0.04	0.02
Machinery and mechanical, electrical and optical equipment; transport equipment	1995	1.01	1.04	0.79	1.19	0.16	0.28	0.03	0.01
Machinery and mechanical, electrical and optical equipment; transport equipment	2000	1.02	1.06	0.90	1.20	0.15	0.27	0.03	0.01

Source: elaborated from ISTAT, 1998; 2003b.

Table 7.2 summarizes the relative productivity differences and the shares of national employment in the regional economies and sectors containing the industries (vehicles and chemicals) that we plan to consider. In the case of chemicals, it reveals generally above average productivity in Lombardia and a remarkably large and stable share of national employment. The chemical poles in Basilicata and Puglia account for relatively small shares of national employment. Changes in the definition of the sector make interpretation of the productivity data difficult. What one can infer is that relative productivity increased in 1980–95 but declined in areas other than Lombardia in 1995–2000, and that it was significantly higher in Puglia than in Basilicata. In the case of vehicles, what stands out is the stability of relative productivity in Piemonte and its loss of employment share, compared with the strong rise of relative productivity and national employment shares in Basilicata. The next step is to explain these structures and trends. To do so, we shall start in the next two sections with brief accounts of the underlying changes in the geography of these sectors in Italy and the wider world.

The Changing Geography of Vehicle Manufacturing in Italy and the World

Alongside other advanced industrial countries, Italy and its regional economies have played a leading role in the development of the motor vehicle industry. Today the industry continues to make a major direct and indirect (motor vehicle sales, repairs, recycling, automotive equipment retailing, insurance and financing, fuel sales, driving schools, motor sport, transport services and road building) contribution to the creation of wealth, income and employment. According to successive Censuses (ISTAT, 2004b; ISTAT, 2004c) motor vehicle manufacturing employed 172,932 in 2001 compared with 275,306 in 1981. 52,304 of these jobs were in the province of Turin, 9,588 in Brescia, 8,451 in Milan, 7,955 in Naples and 7,788 in Potenza (Figure 7.5). In the first four of these provinces, employment was in decline (−71, 292, −1,786, −26,394 and −9,705 jobs in 1981–2001 respectively). In some areas, however, employment grew. Most striking were Potenza/Melfi (7,564) and Chieti/Termoli (3,875), Bergamo (2,592), Ferrara (1,702) and Avellino/Pratola Serra (1,439).

Output declined far less than employment. As Table 7.3 shows, Italian output stood at 1,612,000 vehicles in 1980, rose to 2,121,000 in 1990, but subsequently declined to 1,427,000 in 2002. At a global scale, output increased, although output growth varied with the economic cycle, declining in the early 1990s, increasing from 1994, declining with the 1997

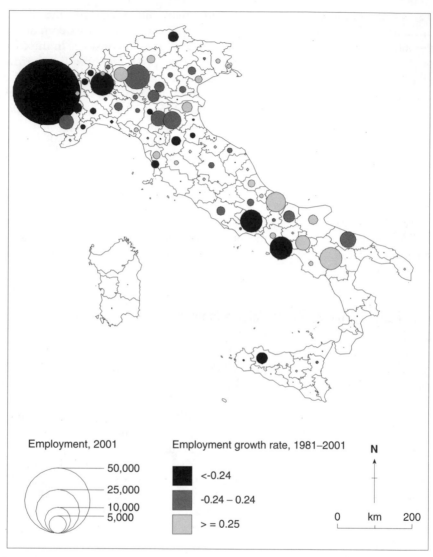

Figure 7.5   Motor vehicle manufacture (ISIC Revision 3.1: 34): employment (2001) and employment change (1981–2001) by province.
Source: elaborated from ISTAT, 2004b; 2004c.

financial crises and increasing again until 2000. Of the other EU countries identified in Table 7.3, only Sweden saw the volume of production decline. In all of the others, output increased or remained the same due to the relative success of national champions, the arrival of Japanese producers in

Table 7.3   Manufacture of motor vehicles by country and zone, 1980–2002

In '000s	1980	1990	1997	1998	1999	2000	2001	2002
EUROPE	14,547	17,920	18,130	19,452	19,771	20,191	20,044	19,899
WESTERN EUROPE	12,064	14,734	15,403	16,613	16,929	17,106	17,219	16,948
Germany	3,879	4,977	5,023	5,727	5,688	5,527	5,692	5,469
Belgium	929	1,252	1,101	1,065	1,017	1,033	1,187	1,056
Spain	1,182	2,053	2,561	2,826	2,852	3,033	2,850	2,855
France	3,378	3,769	2,581	2,954	3,180	3,348	3,628	3,693
Italy	1,612	2,121	1,817	1,693	1,701	1,738	1,580	1,427
Netherlands	113	151	232	271	307	267	239	231
Portugal	119	138	271	271	252	247	240	251
United Kingdom	1,313	1,566	1,936	1,981	1,974	1,814	1,685	1,821
Sweden	298	410	480	483	251	301	289	276
CENTRAL AND EASTERN EUROPE	2,483	2,977	2,383	2,495	2,544	2,654	2,554	2,604
Turkey	51	209	344	344	298	431	271	347
NAFTA	9,876	12,524	16,018	16,008	17,634	17,697	15,798	16,724
SOUTH AMERICA	1,387	960	2,326	1,892	1,682	2,087	2,115	2,006
Argentina	282	86	446	458	305	340	236	159
Brazil	1,165	915	2,067	1,573	1,351	1,682	1,817	1,793
ASIA-PACIFIC	13,141	16,403	17,576	15,308	16,871	18,071	17,954	19,926
AFRICA	405	335	383	327	301	329	393	286
TOTAL	39,397	48,201	54,434	52,987	56,259	58,374	56,305	58,840

Source: adapted from CCFA, 2003 and OICA, 2002.

the UK and inward investment in southern Europe. Also noteworthy is the fact that there was no net overall transfer of output to the CEECs, largely because of the catastrophic impact of transition on the manufacture of commercial vehicles.

Of the large producing countries, Italy was in a class of its own, losing substantial market share in the 1990s: in 1990–2002 sales of Italian vehicles declined from 52.3 per cent to 30.5 per cent of the Italian market, and from 13.6 per cent to 8.2 per cent of the European market. The decline in Italian output was essentially a result of the decline in the output of passenger cars from 1,445,000 in 1980 to 1,126,000 in 2002 (Table 7.4).[1]

Table 7.4   Manufacture of passenger cars by country and zone, 1980–2002

In '000s	1980	1990	1997	1998	1999	2000	2001[1]	2002
EUROPE	11,984	15,231	15,780	16,932	17,291	17,407	17,373	17,312
WESTERN EUROPE	10,401	13,062	13,451	14,522	14,844	14,779	14,939	14,815
Italy	1,445	1,875	1,563	1,402	1,410	1,422	1,272	1,126
CENTRAL AND EASTERN EUROPE	1,582	2,002	2,086	2,170	2,226	2,331	2,259	2,293
Czech Republic		188	321	369	348	428	457	441
Poland		284	295	380	547	482	336	287
Slovak Republic					127	181	182	225
CIS			1,031	909	994	1,010	1,067	1,021
Turkey	32	168	243	240	222	297	175	204
NAFTA	7,527	7,748	8,122	8,027	8,258	8,372	7,155	7,346
SOUTH AMERICA	1,136	703	1,890	1,493	1,355	1,650	1,722	1,674
ASIA-PACIFIC	8,797	11,910	12,421	11,271	12,637	13,573	13,326	14,571
AFRICA	277	210	240	202	218	213	251	212
TOTAL	29,721	35,802	38,453	37,925	39,760	41,216	39,826	41,116

Source: adapted from CCFA, 2003 and OICA, 2002.
Note
[1]As of 2001 some passenger cars were reclassified as commercial vehicles

In Italy the story of the motor vehicle and car industry is essentially the story of FIAT (Fabbrica Italiana di Automobili Torino) which today sells cars under three different brand names: FIAT; Lancia; and Alfa Romeo. In Italy, FIATs early development was in Turin and Piemonte. In 1922, it developed the Lingotto (Ingot) plant. Constructed on five floors with a futuristic test track on the roof, this plant is today the FIAT Group headquarters. The giant Mifafiori plant, modelled on Ford's River Rouge, Dagenham and Colonia plants, was opened by Mussolini in 1939, and was doubled in size in 1956–8. In 1967, the Rivalta works was opened (Table 7.5 which records total employment in FIAT Auto's evolving set of Italian plants in 1960–2001 and direct employment in 1997–2003). Until 1973, more than 83 per cent of the Group's Italian employees worked in the province of Turin (Dunford, 1988). 1969 marked a turning point.

Table 7.5 Evolution of FIAT Auto's plants in Italy: employment in FIAT Auto plants, 1960–2001 and direct FIAT Auto and FIAT-GM Powertrain employment, 1997–2003

Plant	Date of establishment/ acquisition-disposal/ Closure	Role	Employees at FIAT plant at end of year								Direct FIAT Auto and FIAT-GM Powertrain employees at end of year			
			1960	1971	1974	1977	1980	1992	1997	2001	1997	1999	2001	2003
In Turin and Piemonte														
Enti Centrali Torino			—	—	—	—	—	—	—	—	7,490	6,133	3,270	3,408
Costruzioni Sperimentali			—	—	—	—	—	—	—	—	834	778	607	311
Direzione Commerciale Torino			—	—	—	—	—	—	—	—	681	740	821	978
DVMI Torino		Italian marketing/sales			—		—	290	180	180	187	193	129	—
CDV Torino			—		—		—	300	108	115	107	115	96	—
Lingotto[5]	1923–82	Motor vehicles	5,090	5,100	7,629	7,500	9,300							
Mirafiori	1939–	Motor vehicles	16,535	31,000	60,000	49,000	37,400	33,341	26,313	21,188	24,277	19,189	14,784	12,395
Rivalta[6]	1967–2002	Motor vehicles	—		15,710	15,000	14,450	10,035	5,333	3,957	6,439	3,688	2,107	—
Torino Stampe	1975–	Press shop	—		—		1,170		—					
Villastellone	–1992 (disposal)		—		—			1,100	—		—			
Villar Perosa	1969–	Powertrain	—		185	350	400	—	—	—	—	—	—	—
Volvera Ricambi	1925–	Aftersales	1,028	2,500	1,689	1,600	1,850	—	—	—	—	—	—	—
Abarth	1975	Motor vehicles	—		—	—	330	250	106	80	—	—	—	—
Chivasso (Lancia)	1962/1969–1992	Motor vehicles	—		—	6,200[1]	7,400	4,180	—		—			
San Paolo (Lancia)	1906/1969–	Motor vehicles	—		—	3,500[1]	1,375							
Verrone (Lancia)	1974	Powertrain	—		—	1,700[1]	1,885	1,331	981	741	883	711	623	566
Balocco	1962	Proving grounds	—		—			—			42	55	67	87
In the rest of the Centre—North														
Arese	1963/1987	Motor vehicles	—		—	—	—		5,637[3]	—	5,457	3,729	2,282	1,180
Desio (Autobianchi)	1956–92	Motor vehicles	—		—	—	4,400	—						
Lambrate (Innocenti)[7]	1960/1992–1992	Motor vehicles	—		—	—	—	—						
Florence (Novoli)[8]	1942–84		513	—	1,092	1,800	1,775							
Vado Ligure	1970–	Powertrain	—		762	1,400	1,350							

(Continued)

Table 7.5 Evolution of FIAT Auto's plants in Italy: employment in FIAT Auto plants, 1960–2001 and direct FIAT Auto and FIAT-GM Powertrain employment, 1997–2003 (*cont'd*)

	Date of establishment/ acquisition- disposal/ Closure	Role	Employees at FIAT plant at end of year								Direct FIAT Auto and FIAT-GM Powertrain employes at end of year			
			1960	1971	1974	1977	1980	1992	1997	2001	1997	1999	2001	2003
Maranello (Ferrari)	1946/1969–	Sports cars	—	—	—	—	—	—	—	—	—	—	—	—
Modena (Maserati)	1940/1993–	Sports cars				—	1,175	—		—				
In the South														
Naples (Comind)	1956–	Components	—	—	—	900[1]	—	988[2]	1,096	767[4]	—	—	—	—
Termini Imerese	1970–	Vehicles		—	820	950	2,250	3,558[2]	3,066	2,798[4]	2,810	2,681	1,931	1,479
Bari	1970–2002 (disposal)	Industrial vehicles		—	—	2,800	—				—	—	—	—
Cassino	1972–	Motor vehicles		—	—	5,000	9,000	10,077[2]	8,014	7,028[4]	7,191	5,418	4,733	3,713
Sulmona	1972–	Suspensions		—	—	1,000	950	1,039[2]	1,133	971[4]	972	—	—	—
Termoli	1973–	Powertrain		—	—	2,900	2,830	3,292[2]	2,981	3,364[4]	2,985	2,547	2,315	2,388
Lecce (SASN Nardò)	1975–	Proving grounds				—	90	117[2]	125	115[4]	114	—	—	—
Val di Sangro, Chieti (Sevel Atessa)	1981–	Commercial vehicles						607[2]	3,248	4,179[4]	4,213	4,069	4,541	4,669
SATA—Melfi	1994–	Motor vehicles								6,336[4]	5,845	6,600	5,080	5,100
FMA—Pratola Serra	1994–	Powertrain								1,486[4]	1,304	1,920	1,450	1,726
Elasis (FIAT Auto)		Research								6,701[4]				
Pomigliano d'Arco (AlfaSud)	1971/1987	Motor vehicles		—	—	—	—	—	8,327	7,539[4]	7,808	5,616	4,942	5,046

Notes
1 1978
2 1991
3 1995
4 2000
5 Site redeveloped as conference and exhibition centre, auditorium, hotel, shopping complex and service centre and headquarters
6 FIAT Aviation
7 Site redeveloped as university premises, Courts of Justice, park, hotel, accommodation and commercial space
8 Innocenti (most famous for the Lambretta scooters) produced vehicles in an existing plant in Milan

In that year of unprecedented industrial and social unrest, FIAT lost 15 million hours of work through strike action. In these circumstances, FIAT decided to invest in the south. By 1982, 16 per cent of FIAT Auto employment was in the south, compared with 72 per cent in the centre-north and 12 per cent overseas. By 1997, 29 per cent was in the south, 33 per cent in the centre-north and 38 per cent overseas (Mariotti and Treves, 1999: 307).

The geography of FIAT's Italian operations was also shaped by a series of acquisitions. In 1969 FIAT acquired Lancia. In 1979, FIAT's car division was set up as an autonomous company with FIAT, Lancia, Autobianchi, Abarth and Ferrari as its brands. FIAT had already acquired 50 per cent of Ferrari in 1969. That figure subsequently rose to 87 per cent. In 1986, FIAT Auto took over Alfa Romeo and its associated companiesfrom the FINMECCANICA state holding company, after the Italian government had provided massive aid in the shape of capital injections, loss compensation and debt relief to prepare Alfa Romeo for privatization. This acquisition expanded FIAT's presence in the south in the shape of the Alfa Sud plant in Pomigliano d'Arco. Next FIAT added the prestigious Maserati sports car brand to its collection to complete the range of car brands.

As FIAT invested in the south, and pursued a strategy of globalization that we shall examine in the next chapter, existing establishments in FIAT Auto's place of origin were restructured, with significant repercussions for the regional economy of Piemonte. In 2001 in Turin, for example, the Mirafiori plant produced just 1,500 cars per day compared with 4,500 ten to fifteen years earlier. At that stage, it produced four models (Panda, Punto, Marea and Multipla) on two lines (Table 7.6). The Multipla was produced at a very low break-even point of 40,000 vehicles per annum by using a very labour intensive, space-room technology, though capital investment costs are only about one-half of those of traditional models. Orders for all models come mostly from dealers, although some come from individual customers. There is currently a 16-day ordering time, though there are also dealer networks and an ICT system with virtual stock, so that demand can be met from other dealers' stocks, obviating the need to produce to order. The plant remains integrated in that it is equipped to carry out all stages of vehicle manufacturing and assembly including gearboxes and engines, presswork, paint shop and body shop. The degree of corporate vertical integration within the plant is, however, much reduced. As a result of supply chain restructuring, manufacturing facilities on the site are run by other companies in the FIAT Group (Magneti Marelli and Comau) and by outside companies. (Table 7.7). In 1990, for example, FIAT Auto employed 46,650 staff in its Turin plants (Table 7.7). By 2002 total employment had fallen to 23,980. Of these

Table 7.6 FIAT Auto Italia: motor vehicle production in Turin and the rest of Italy, 1997–2003

Plant and model	Actual					May 2002 forecasts	
	1997	1998	1999	2000	2001	2002	2003
Totale Mirafiori	507,531	432,982	385,692	379,733	323,793	264,000	168,000
Panda	137,388	129,379	137,914	107,263	103,013	100,500	—
Old Punto	223,798	198,048	69,578	—	—	—	—
New Punto	—	—	89,494	195,413	145,182	87,000	90,000
Marea	146,345	99,855	54,722	42,149	39,861	15,000	—
Multipla	—	5,700	33,984	34,908	35,737	22,000	28,000
Lancia Lybra	—	—	—	—	—	25,000	25,000
Alfa 166	—	—	—	—	—	8,500	10,000
Lancia Thesis	—	—	—	—	—	6,000	15,000
Total Rivalta	60,837	48,557	73,644	77,040	50,586	—	—
Uno	13,632	—	—	—	—	—	—
Lancia Delta	13,831	9,263	4,493	—	—	—	—
Lancia Dedra	10,052	8,328	1,660	—	—	—	—
Lancia Lybra	—	230	27,923	51,037	37,934	—	—
Lancia K	17,322	16,037	12,597	6,460	—	—	—
Alfa 166	6,000	14,699	26,971	19,543	11,898	—	—
Lancia Thesis	—	—	—	—	754	—	—
Arese	16,754	14,442	22,975	18,287	13,000	11,270	—
Cassino	251,030	194,882	220,174	187,237	142,275	286,250	—
Pomigliano d'Arco	150,774	173,716	145,795	181,186	190,020	221,010	—
Melfi	408,777	384,247	376,819	364,882	347,272	344,050	—
Sevel	145,943	170,272	174,858	188,384	189,236	197,440	—
Termini Imerese	163,891	133,371	135,602	153,265	135,500	127,780	—
Production in Turin	568,368	481,539	459,336	456,773	374,379	264,000	168,000
Production in the rest of Italy	1,137,169	1,070,930	1,076,223	1,093,241	1,017,303	1,187,800	—
Production in Italy	1,705,537	1,552,469	1,535,559	1,550,014	1,391,682	1,451,800	—
Turin as share of Italy (%)	33	31	30	29	27	18	—

Source: elaborated from FIOM-Piemonte, 2002.

workers, only 14,019 were FIAT Auto employees. The rest were employed by other companies that had assumed responsibility for a substantial number of activities taking place in the FIAT Auto factories.

Increasingly the future of these manufacturing activities is uncertain. Already FIAT Auto has ended vehicle production at the Rivalta plant in Turin. Surviving models are to be transferred to Mirafiori, replacement models will go to Poland, and FIAT Aviation with its 2,000 employees is to transfer to the car manufacturing part of the Rivalta complex (FIOM, 2002). Also to close is the Alfa Romeo plant in Arese. In 2001, FIAT

Table 7.7 Automotive sector employment in FIAT Auto's Mirafiori and Rivalta complexes by company, 1989–2002: employment decline and outsourcing

Company	1989	1997	1998	1999	2000	2001	2002
FIAT Auto	49,512	31,334	27,179	23,043	16,474	14,673	14,019
Comau Service	—	—	—	2,009	2,041	1,680	1,681
Comau Stampi	—	—	390	360	670	416	324
Magneti Marelli (Suspension systems)	—	—	—	396	958	790	740
Fenice (Italenergia)	—	—	440	440	420	400	400
Sirio	—	—	—	340	320	310	300
FIAT-GM Powertrain	—	—	—	—	4,005	3,370	3,234
FIAT-GM Purchasing					448	427	411
Sava	—	247	287	319	320	315	310
Other FIAT companies	—	65	65	229	229	369	381
TNT Production Logistics	—	—	1986	1940	1990	2,015	1,810
Turinauto	—	—	—	—	395	380	370
Mirafiori and Rivalta	49,512	31,646	30,347	29,076	28,270	25,145	23,980

Source: elaborated from FIOM-Piemonte, 2000; 2002.

produced 374,379 cars in two plants (Mirafiori and Rivalta) in Turin. In 2003, in the absence of replacements for the Panda and Marea, output will fall to 168,000 cars (Table 7.6), and the shift of declining national output to the south will continue, with the Turin share declining from 33 per cent in 1997 to a forecast 18 per cent in 2002 and a yet lower share in 2003.

The decline in output was associated with a sharp drop in employment in FIAT Auto's vehicle manufacturing plants in Turin. Total employment in these two plants fell from 49,512 in 1989 to 31,646 in 1997 and to 23,980 in 2002 (Table 7.7). This declining figure was itself made up of three parts: a sharp decline in direct employment in the Mirafiori and Rivalta vehicle manufacturing plants, as FIAT Auto outsourced activities to other FIAT group or to independent companies; a consequent increase in the number of workers employed by these other companies in the two FIAT Auto factories from 312 workers in 1997 to 10,472 in 2001; and a decline in employment in many parts of the two establishments due to the decline in output, rationalisation and productivity growth.

As far as the city of Turin was concerned total car sector employment fell from 32,785 in 1997 to 26,270 in 2001 (Table 7.8). As the number of car sector employees at other sites in the province of Turin was relatively stable, this decline was almost entirely due to the decline of Mirafiori and Rivalta.

Alongside production-related employees, nearly 7,000 managerial, administrative and support staff are employed in headquarters-related

Table 7.8 Vehicle production and employment in Turin and Italy, 1993–2002

	1993	1994	1995	1996	1997	1998	1999	2000	2001	2002
Vehicle production in Turin ('000s) (FIOM)	571	572	465	401	568	482	459	457	374	—
Vehicle production in the rest of Italy ('000s) (FIOM)	593	832	1,040	988	1,137	1,071	1,076	1,093	1,017	—
Vehicle production in Italy ('000s) (FIOM)	1,165	1,404	1,506	1,389	1,706	1,552	1,536	1,550	1,392	—
Vehicle production in Italy ('000s) (FIAT Auto Bilancio)	1,165	1,404	1,506	1,389	1,628	1,476	1,480	1,478	1,272	—
Industrial production in Turin (Annual change, %)	—	—	—	—	—	—	—	−2.9	−6.1	—
Industrial production in Italy (Annual change, %)	—	—	—	—	—	—	—	−0.7	−1.3	—
Exports of components from Piemonte (€ million)	—	—	4,225	3,670	3,474	3,255	2,976	3,726	3,388	2,975
Exports of components from Piemonte (€ million)	—	—	2,590	2,850	3,451	3,631	3,507	4,273	3,905	3,952
FIAT Auto personnel employed at Mirafiori/Rivalta ('000s)	40.1	35.6	33.7	32.5	31.3	27.2	23.0	16.5	14.7	—

Staff of other companies employed at Mirafiori/Rivalta ('000s)	0.0	0.1	0.1	0.1	0.3	3.2	6.0	11.8	10.5	—
Staff employed at other car sector sites ('000s)	1.4	1.0	1.1	1.1	1.1	1.2	1.1	1.1	1.1	—
Total car sector employment in Turin ('000s)	41.4	36.7	34.9	33.7	32.8	31.5	30.2	29.4	26.3	—
Staff in supplier industries (112 per 1000 FIAT vehicles) ('000s)	—	—	—	—	—	—	—	—	42.0	—
Total car sector and supplier employment in Turin	—	—	—	—	—	—	—	—	68.3	—
Manufacturing employment in Turin ('000s)	323	318	324	313	307	301	306	302	293	298
Total employment in Turin ('000s)	882	868	870	886	879	861	894	916	916	912
Employment rate in Turin (% of population of 15 and over)	45.7	44.9	45.0	45.9	45.6	44.9	46.5	47.9	47.9	47.6
Male employment rate in Turin (% of population of 15 and over)	58.8	57.6	57.3	57.0	56.8	55.9	57.2	58.3	58.1	57.4

Source: elaborated from FIOM, 2002; FIAT, 2003; Unioncamere Piemonte, 2003; and IRES, 2003.

activities. Turin is, for example, a worldwide purchasing function centre, and is the site for a pilot factory, a design centre and a safety centre.

Additionally, Turin and Piemonte are the centre of a vast components sector. 1996 Census estimates (ISTAT, 1996) suggest that there were more than 105,000 jobs in the province of Turin motor vehicle value chain, of which some 70,000 were employed by companies other than FIAT Auto. In 2002, the Osservatorio provinciale sul mercato del lavoro estimated the size of the automotive sector 'indotto' at some 1,222 suppliers, employing some 67,000 people, and producing output valued at about € 16.2 billion. Most of these firms were in the metals and mechanical engineering sectors. Others were in the electrical engineering, rubber, plastics, glass, chemicals and services sectors. A sample survey showed that the degree of dependence of these firms on orders from vehicle manufacturers was extremely high, standing at 58 per cent for firms with up to 9 employees and 100 per cent for the largest firms (Table 7.9). A 2002 Confederazione Italiana della Piccola e Media Industria survey showed that the average degree of dependence of members on FIAT Auto stood at 45 per cent of turnover, and that more than 30 per cent of the turnover of 60 per cent of the indotto depended on FIAT Auto. This evidence was used to support demands for an industrial policy and a regional agency to support the vehicle components sector. Trades-union estimates suggested that if employment in FIAT Auto were to decline in proportion to the fall in output, employment at FIAT Auto sites could fall from 27,070 in 2001 to 14,244 in 2003, and employment in the FIAT-dependent supply chain could fall from an estimated 42,000 to 22,101 (FIOM, 2002).

The impact will, however, differ across companies and subsectors. Most of the multinationals present in the area are in the components field. Although these companies are first-tier FIAT suppliers, they also supply other automotive sector OEMs. As FIAT Auto has internationalized, these companies have increased their exports of components to FIAT's overseas assembly plants. This conduct helps explain the strong rise in component exports that accompanied the decline in motor vehicle exports: in 1993–2002 component exports rose from € 2,590 million to € 3,952, while vehicle exports declined from € 4,225 to € 2,975 million (Table 7.8). In the years to come, however, it is quite conceivable that these companies will relocate near the assemblers, and that the tendency for the location of assembly operations overseas to increase exports will weaken. Alongside these first-tier multinationals, a range of other companies are involved in the immaterial aspects of the automotive sector. Included are design houses, whose activity increased as manufactures made shorter runs of more diversified vehicles: Bertone; Italdesign-Giugiaro; and Pininfarina (which designs, develops and manufactures). For these design, engineering and prototyping houses, the market is increasingly outside of

Table 7.9 Motor components sector in Turin

			Number of employees				
	1–9	10–49	10–99	100–499	500–99	over 1000	Total
Production							
Firms ('000s)	263	427	109	132	16	10	957
Employees ('000s)	1,026	8,019	5,934	22,117	10,028	13,946	61,070
Turnover (€ '000s)	228	1,976	1,647	4,941	1,629	3,590	14,011
Services							
Firms ('000s)	89	122	26	26	1	1	265
Employees ('000s)	343	2,866	1,497	5,510	589	1,834	12,639
Turnover (€ '000s)	114	412	222	1,061	98	326	2,232
Total							
Firms ('000s)	352	549	135	158	17	11	1,222
Employees ('000s)	1,369	10,885	7,431	27,627	10,617	15,780	73,709
Turnover (€ '000s)	342	2,388	1,869	6,002	1,727	3,916	16,243
Share of turnover of surveyed firms dependent on automobile firms (%)	58	46	41	71	100	100	

Source: Osservatorio Provinciale sul Mercato del Lavoro della Provincia di Torino e IPL, 2002.

Italy. In the case of Italdesign, in 2000, 45.5 per cent of revenue came from the rest of Europe and just 35.1 per cent from Italy compared with 30.6 and 44.5 in 1999 (Italdesign, 2001). Insofar as design and headquarters functions remain in Turin, these jobs are likely to survive. Some second- and third-tier suppliers have sought to rationalize their activities, increase the scale of their operations, upgrade and diversify into new markets to reduce their dependence on FIAT Auto.

Although manufacturing employment declined from 323,000 to 298,000 jobs, although male employment rates have declined marginally (Table 7.8) and although the jobs lost paid relatively high wages to manual workers, in the 1990s as a whole, it is also the case that FIAT's restructuring has been accompanied by an overall increase in employment in the province of Turin. The scale of output decline in the next few years, however, significantly exceeds that of the recent past, is more concentrated temporally and is likely to have a much more negative short- and medium-term impact, especially if the skills and know-how found in the automobile district are not sufficiently strong to retain multinationals in the components sector.

The Changing Geography of Chemical Manufacturing in Italy and the World

The second industrial sector that we shall examine is the chemical sector. In 2001, the Italian chemical industry comprised slightly fewer than 6,000 companies, and employed more than 200,000 people (ISTAT, 2004b). The industry accounted for 1 per cent of total manufacturing companies, but for 4.2 per cent of employment, and 8 per cent of total manufacturing turnover, added value and investment (Federchimica, 2003b). Of the industry's main segments, primary chemicals and fibres were strongly represented, accounting for 34.7 per cent of total production. Speciality and fine chemicals accounted for nearly 30 per cent, followed by pharmaceuticals (22.7 per cent) and consumer chemicals (12.6 per cent) (Federchimica, 2003a).[2]

The Italian industry plays an important role at EU and international levels (Table 7.10). A 2002 turnover (sales plus exports) of € 65 billion accounted for 12.4 per cent of the EU total, and placed Italy third in the EU15, after Germany (25.1 per cent) and France (16 per cent). As an important EU15 manufacturer, Italy was also a major international player, accounting in 2002 for 3.5 per cent of world turnover, which stood at € 1,841 billion (Figure 7.6). In 1991–2002, however, Italian turnover increased at just 2.3 per cent per year compared with 4.3 per cent in the EU15.

Table 7.10   Chemical industry indicators, 1991–2002

	1991	1995	1998	1999	2000	2001	2002
Employment ('000s)							
Germany	717	536	485	478	470	467	462
France	264	246	237	238	235	240	236
United Kingdom	277	254	258	249	239	235	231
Italy	237	212	209	206	206	206	206
EU	2052	1743	1694	1680	1664	1664	1655
Rest of Western Europe	85	78	72	73	73	76	77
USA	1076	1038	1043	1035	1034	1022	1008
Japan	406	392	384	371	367	366	364
Turnover (= domestic + export sales) (€ billion)							
Germany	106.6	112.3	116.8	121.4	135.0	133.9	132.5
France	52.0	64.7	70.5	73.3	82.2	85.1	84.5
United Kingdom	37.4	40.7	45.8	49.1	53.6	55.0	53.3
Italy	50.9	48.9	55.8	57.0	62.5	64.2	65.5
EU	331.8	375.5	416.3	439.3	498.6	518.1	527.0
Rest of Western Europe	15.4	22.4	25.6	28.4	30.3	36.1	40.2
USA	256.7	295.2	373.7	403.2	497.3	510.4	485.1
Japan	145.3	190.1	158.9	188.3	240.1	213.6	192.7
Trade balance (€ billion)							
Germany	15.3	19.5	21.8	21.2	23.9	24.1	22.3
France	3.4	6.2	7.3	8.6	8.8	9.9	11.0
United Kingdom	4.0	3.9	6.1	5.8	5.9	6.7	5.7
Italy	−8.0	−7.6	−8.7	−8.6	−9.1	−8.2	−8.1
EU	16.2	29.9	43.9	50.0	60.1	71.1	77.0
Rest of Western Europe	4.8	7.0	8.0	8.9	8.9	10.6	11.9
USA	15.6	15.6	12.0	7.7	6.8	1.4	−5.3
Japan	0.0	4.2	5.7	7.0	8.7	5.9	7.9

Source: adapted from CEFIC, 2004.

Increased Italian turnover was to a significant degree a result of a 9.7 per cent average annual growth of exports (compared with 9.2 per cent in the EU15). As exports grew faster than domestic sales, exports accounted for an increasing share of turnover. (Chemical exports account for more than 10 per cent of total Italian manufacturing exports). Exports to countries other than Western Europe (EU15 plus Norway and Switzerland) grew yet more rapidly at 10.5 per cent per year, although the corresponding EU15 figure was 11.3 per cent. In 2002, western Europe remained Italy's most

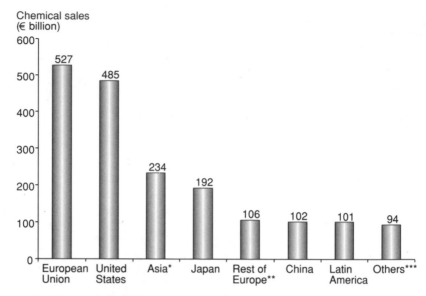

* Excludes Japan and China
** Switzerland, Norway, Central and Eastern Europe and Turkey
*** Includes Canada, Mexico, Africa and Oceania

Figure 7.6 The geography of world chemical sales in 2002 (€ billion).
Source: CEFIC, 2004.

important export market accounting for 65.3 per cent of exports. The corresponding figure for imports was a substantially higher 81.9 per cent.

The other striking feature of Italy's trade position was that, in contrast to other large EU Member States and the EU15 as a whole, Italy's trade was in deficit. In 2002, this deficit stood at 12.3 per cent of turnover. In absolute terms, the deficit increased in the 1990s, although it fell as a share of turnover, having stood at 15.7 per cent in 1991.

Employment was generally in decline, mainly as large companies reduced head-count. In 1991–2002, employment fell by 19.4 per cent in the EU15. In Germany 35.6 per cent of jobs were lost, compared with 10.4 per cent in France, 16.6 per cent in the UK and 13.1 per cent in Italy (Table 7.10). In Belgium, there was no change, while in some EU15 countries (Ireland, Denmark, Greece and Spain) employment increased.

In 1981–2001, Italian chemical employment declined by just over 30 per cent (ISTAT, 2004). The 2001 Census indicates that 88 per cent of chemical employment is in the centre-north (Figure 7.7). After north-Rhine Westphalia, Lombardia has the highest chemical sector employment in the EU15. Italy's chemical's capital comes first in terms of the number of companies (33 per cent of the total) and is one of the highest in terms of

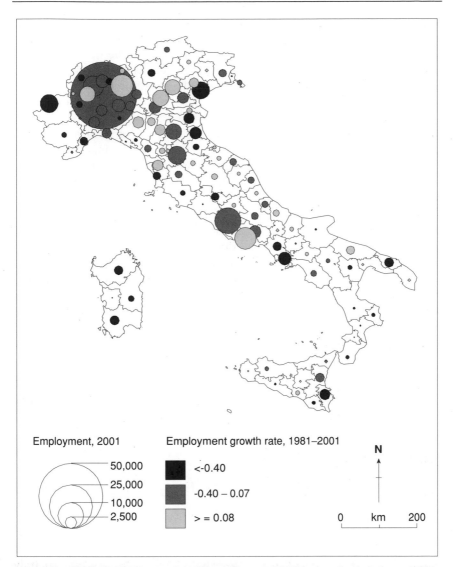

Figure 7.7 Chemicals (ISIC Revision 3.1: 34): employment (2001) and employment change (1981–2001) by province.
Source: elaborated from ISTAT, 2004b; 2004c.

specialization indices (chemical industry indicators as a share of the same indicator for all manufacturing industries). According to the 2001 Census, Lombardia accounted for 41.2 per cent of Italian employment. Milan itself, with 53,462 jobs, accounted for 26.1 per cent. The next most important provinces were Rome (10,050 jobs), Bergamo (7,957), Latina (7,526) and

Varese (6,930). Of the southern regions, the highest shares of national employment were for Campania (2.7 per cent), Sicilia (2.6 per cent), Sardegna (1.9 per cent) and Puglia (1.7 per cent). The development of these southern chemical poles was a result of new investments in the 1960s and 1970s. In Chapter 9 we shall consider their evolution.

An important characteristic of the Italian chemical industry is that, in contrast to commonly-held assumptions, it is primarily made up of a vast number of SMEs rather than a small number of large companies. The 2001 Census indicated that 64.1 per cent of companies had fewer than 10 employees. Only 1.2 per cent of establishments fell in the large company (500 employees or more) category, although large companies did account for a large share of employment (41.2 per cent).

In the 1990s, the volume of production (gross output) of EU chemical industries grew by 41 per cent (Figure 7.8). There were, however, wide national variations. Of the large economies, Italy performed the worst, with the volume of production increasing by 14 per cent. In Italy, gross chemical sector output increased more rapidly than aggregate output, and than industrial production, until 1988. After 1989, it fell behind (Figure 7.9). This aggregate figure was made up of a number of sharply differing components, with man-made fibres and primary chemicals growing slower than average. A finer product breakdown (Figure 7.10) indicates that in the

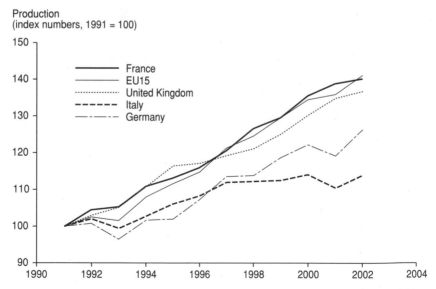

Figure 7.8   Gross output of the EU15 chemical industry and its largest economies, 1991–2002. *Source: elaborated from CEFIC, 2004.*

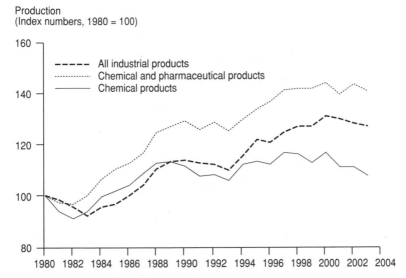

Figure 7.9   Gross output of the Italian chemical industry, 1980–2003.
Source: elaborated from Federchimica, 2004.

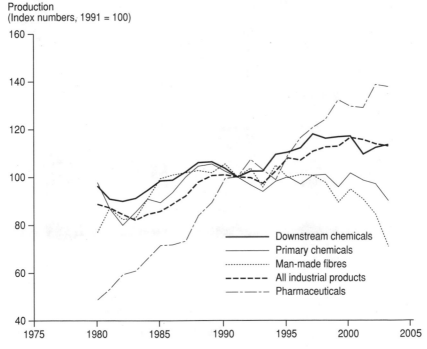

Figure 7.10   Gross output of the Italian chemical industry by product group, 1980–2003.
Source: elaborated from Federchimica, 2004.

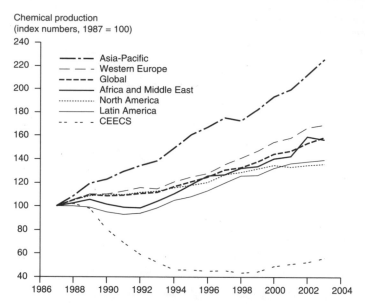

Figure 7.11 Evolution of chemical and pharmaceutical production by macroareas, 1987–2003. *Source: Federchimica, 2004.*

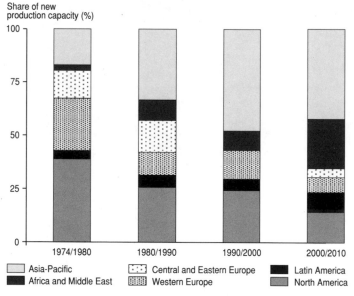

Figure 7.12 The changing geography of new petrochemical capacity. *Source: ACC, 2003.*

period since 1980 inorganic products, primary and intermediate organic chemicals and fertilizers declined in output while plastic materials, varnishes and paints, inks, detergents and cleaning products and cosmetics increased (Federchimica, 2004). Most remarkable was the growth of pharmaceutical products. The differing records of these components had important impacts on the relative development of different production sites.

An important cause of slackening growth was the weakening of demand and consumption in Europe relative to other parts of the world. The Chemical Industries Association (2001) estimated that by 2010 developing countries (excluding the CEECs) could account for 37.4 per cent of the world chemical demand, and 34 per cent of production, compared with 27 per cent and 21 per cent respectively in 1996. As Figure 7.11 indicates, in the 1990s Asia and the Pacific, in particular, recorded strong growth relative to western industrialized countries. A number of factors explain these trends. One is the interaction of demographic growth, increases in income and the diffusion of western lifestyles and modes of consumption. Another is the relatively rapid growth of textiles and clothing and other chemical-using industries in emerging economies. Figure 7.12, for example, indicates that new petrochemical capacity is increasingly located in Asia and the Middle East. The consequences for producing areas that grow less than average are declining market shares and declining employment.

Conclusions

In the first part of this chapter we identified the quantitative impact of relative productivity growth in different industries on overall regional productivity convergence and GDP per head catch-up, and we identified a number of significant movements in relative productivity and employment shares in all industries and, in particular, in the chemical and vehicle sectors in four regional economies. Essentially, we showed that the sectors we plan to examine have had impacts on the trajectories of Italy's regional economies that are measurable and that, while variable, are generally although not always small. As a result of this analysis we were able to show, for example, that without the creation of the Melfi motor vehicle complex the average productivity of Basilicata would have increased at 0.91 times the national rate rather than 1.02 times. The arrival of the car complex in Melfi added nearly 11 percentage points to Basilicata's relative productivity growth (as average productivity increased at 1.02 times the national rate rather than 0.91 times). As a result what would have amounted to a further falling behind in terms of productivity was translated into an actual catch-up.

In the second part of the chapter attention was redirected from the sectoral profile of regional development to the regional profile of industrial change. The main aim was to outline the underlying changes in the geography of the vehicles and chemicals sectors in Italy and in the wider world, as what happens in Italy is in part a consequence of the international redeployment of these industries. As we argued earlier, the evolution of relative productivity and employment and the changing geography of particular industries are a result of the context-dependent profit and upgrading strategies of enterprises. The aim of the next two chapters is to examine these restructuring strategies. In doing so, the aim is to exemplify an approach to the analysis of the microfoundations of productivity and employment changes that reshape the position of Italy's regional economies relative to one another and to the rest of the EU15.

Chapter Eight

Globalization, Industrial Restructuring and the Italian Motor Vehicle Industry

Introduction

In Chapter 7 we identified a number of trends in the regional profiles of output and employment in the Italian motor vehicle sector, including a movement into services, a decline of well-paid manual jobs in the northwest and a refocusing of Italian manufacturing operations in the south. As we shall show, these changes in Italy were closely associated with a reshaping of the international division of labour of FIAT Auto. The aim of this chapter is to explain some of these changes in the geography of the Italian motor vehicle sector. To this end we shall concentrate on a number of aspects of FIAT's value creating and upgrading strategies drawing on the categories introduced in Figure 3.1. The next section deals with the changing relative importance of FIAT Auto's functions in the motor vehicle value chain (method 3 in Figure 3.1) and of the chains in which the FIAT Group operates (method 4). In the third section a brief account of the changing market environment will open the path to an account of the ways in which FIAT Auto sought to globalize its operations in the search for new markets (method 2A) and cost advantages (method 1B). The following section is devoted to a consideration of organizational and technological upgrading (method 1A) and its impact on the changing geography of FIAT's operations in Italy. In the fifth section the restructuring of the supply chain and the implications of the recent joint venture with General Motors for the cost-driven restructuring of purchasing operations (method 1B) are discussed, while in the following sections we shall consider FIAT's latest restructuring plan, and the role of new model development (method 2A) in relaunching FIAT Auto.

The FIAT Group: Changing Functions in the Value Chain and Changing Chains

Although FIAT was founded, in 1899, as a car producer, it is not just a car manufacturer. After its founding, FIAT diversified horizontally into the manufacture of other types of transport equipment including trucks, trains, farm tractors, marine engines, aircraft and space launchers, and integrated upstream into metallurgy and components, and subsequently into production systems. Outside of manufacturing FIAT diversified downstream into a range of services. Some of these services were connected with the transport equipment sector including insurance and the financing of vehicle purchase. Also, however, it undertook conglomerate-style diversification into unrelated sectors such as publishing and communications. As a result the modern-day FIAT Group companies are organized into a number of operating sectors. In 2002, FIAT Auto accounted for just under 40 per cent of Group revenues (Table 8.1).

Over the course of time the actual profile of the group has changed as a result of a series of acquisitions, disposals and differential growth. Through these changes, FIAT has repositioned itself along the vehicle chain (Figure 8.1) which extends from research and design, through components and module manufacture, to sales and after sales activities, and has moved into/out of chains. In recent years, three developments were involved: a move into services; acquisitions and disposals designed to reduce the degree of horizontal diversification and to concentrate on core vehicle-related activities; and outsourcing of component and module manufacture, which also reposition the Group along the chain. The first two trends are considered in this section, while the third is considered below.

In the late 1990s, several Group sectors decided to make services, that were in some cases an integral part of their activities, a more central area of growth. What made services attractive were four factors: the growth in demand; their noncyclical character; the fact that minimal capital investment is required; and the overall profitability of service activities.

Two paths were pursued. The first involved the development of services to capture more of the part of the value added chain that connects vehicle manufacture with the final customers. The starting point was the sale of spares and the development of vehicle-related financial and insurance services. Today FIDIS, for example, provides financing for more than 40 per cent of the cars sold by FIAT Auto, while in 1997–2000 Toro Assicurazioni more than doubled its revenues. As new car purchases represent only a small part of the market associated with the use of motor vehicles, further material and immaterial service initiatives were pursued. Examples include: car rental

Table 8.1 FIAT Group net revenues by sector of activity, 1991–2003

(in € millions)	2003	2002	2001	2000	1999	1998	1997	1996	1995	1994	1993	1992	1991
Automobiles (Fiat Auto)	20,010	22,147	24,440	25,361	24,101	24,859	26,202	21,950	20,190	17,148	12,937	14,175	14,206
Agricultural and construction equipment (CNH Global)	9,418	10,513	10,777	10,770	5,246	5,126	5,284	4,420	4,209	3,925	2,949	2,368	1,093
Commercial vehicles (Iveco)	8,440	9,136	8,650	8,611	7,387	6,649	5,850	5,551	5,722	4,354	3,769	4,154	4,297
Sports cars (Ferrari)	—	1,208	1,058	—	—	—	—	—	—	—	—	—	—
Metallurgical products (Teksid)	844	1,539	1,752	1,873	1,682	1,165	1,190	1,013	1,037	865	646	600	616
Components (Magneti Marelli)	3,206	3,288	4,073	4,451	4,062	3,793	3,451	2,999	3,030	2,720	2,688	2,852	2,826
Production systems (Comau)	2,293	2,320	2,218	2,440	1,693	843	841	831	692	602	644	685	571
Aviation (Fiat Avio)[1]	625	1,534	1,636	1,491	1,361	1,361	1,263	901	851	805	862	636	606
Publishing and communications (Itedi)	383	360	347	354	413	437	406	360	326	242	225	245	209
Insurance (Toro Assicurazioni)[2]	1,654	4,916	5,461	4,363	3,922	2,959	2,016	1,773	1,226	1,085	1,175	1,030	903
Services (Business Solutions)	1,816	1,965	1,805	—	—	—	—	—	—	—	—	—	—
Rolling stock and railway systems (Fiat Ferroviaria)	—	—	—	216	375	389	384	378	243	194	165	146	147
Chemical-fibres-bioengineering (Snia BPD)	—	—	—	—	—	—	1,473	1,535	1,521	1,341	1,179	1,053	1,197
Civil engineering (Fiat Impresit)	—	—	—	—	—	—	—	475	374	328	—	—	—
Financial services (Fidis)[3]	—	—	—	—	—	—	—	—	1,542	1,482	1,495	1,534	1,474
Batteries (C.E.A.C.)	—	—	—	—	—	—	—	—	—	624	615	557	389
Retailing	—	—	—	—	—	—	—	—	—	—	—	2,324	2,129
Miscellanea and eliminations	−1,418	−3,277	−4,211	−2,375	−2,119	−1,812	−2,103	−1,942	−1,871	−1,710	−2,543	−3,197	−2,680
Total for the Group	47,271	55,649	58,006	57,555	48,123	45,769	46,257	40,244	39,092	34,005	28,176	30,526	29,174

Source: FIAT Group, Bilancio consolidato e d'esercizio and Relazione sulla gestione. FIAT Group, Turin various years.

Notes

[1] Aviation sector revenues until sale in 2003 (1 July 2003)

[2] Insurance sector revenues until sale (2 May 2003)

[3] Fidis was consolidated in the Automobiles Sector as from 1996

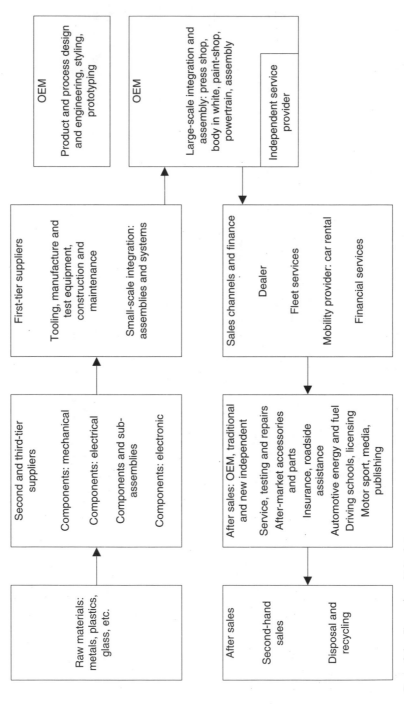

Figure 8.1 The motor vehicle chain.

and mobility services for FIAT Auto, CNH and Iveco; after-sales services and car maintenance and repair services for Magneti Marelli; and maintenance services for COMAU and FIAT Avio. In effect, the FIAT Group was seeking to make money not just by selling new vehicles but by providing services throughout a vehicle's life cycle. As a result, the occupational and geographical profile of employment and value added changed.

The second path involved more general service initiatives that were not directly connected with the car chain. To capitalize, for example, on its own internal services, FIAT Auto concentrated a number of administrative, infrastructural and industrial services in the hands of specialized companies. In 2000, the Group combined the units that provided shared producer services to Group Sectors (administration services, human resources management, facilities and utilities management, real estate services, information and communication technology services, and e-procurement) into a new operating Sector called Business Solutions. The aim was to service clients within the Group and external clients in Italy, Poland, Brazil and other countries where the FIAT Group operates. With revenues of about € 1.2 billion in 2000 and nearly 8,000 employees, Business Solutions was expected to expand by means of acquisitions, partnerships and commercial relationships. In its 2000 Annual Report the Group indicated that it expected combined service revenues to reach some 40 per cent of total sales within five years. In 2002, revenues increased to € 1,965 million. Over one-half came from noncaptive customers with the strongest growth in the Global Value joint venture with IBM Italia (3,000 staff) and the expansion of temporary employment agency services.

In July 2001, FIAT took another step forward in its strategy of enhancing and extending its services into the external market. On that date FIAT emerged as one of the main partners in Italenergia alongside, amongst others, Electricité de France (EDF). After two tender offers, Italenergia acquired exclusive control of Montedison. Montedison was the parent company of Edison, the most important private electricity operator in Italy. One reason for FIAT's interest in this venture lay in the profit-making opportunities offered by the deregulation of the Italian electrical energy market. (In 2002, the Edipower Consortium, in which Edison held a 40 per cent interest, made a successful bid for Eurogen, the largest of the three generating companies that state-owned ENEL was selling off). Another was the fact that FIAT itself had generating capabilities developed to supply its industrial operations. At the same time however FIAT's rivals (which included the steel producer, Lucchini, which itself had generating capabilities and was seeking to diversify into the electricity industry) argued that FIAT was acting as a Trojan horse to permit the entry of a state-owned French company into the Italian market. Certainly EDF was providing much of Italenergia's industrial capability and financial resources. Critics of FIAT argued that FIAT was settling scores stemming from a personal war between dominant figures in Mediobanca and FIAT. This conflict dated from 1980, when FIAT faced a number of financial difficulties, and when Mediobanca's president, Enrico Cuccia, put a former public sector manager, Cesare Romiti, in charge of the FIAT group in place of Umberto Agnelli, a grandson of FIAT's founder, who he considered incompetent (Interview data, 6th June 2001).

One year earlier, in 2000, FIAT had signed an agreement with the Detroit-based General Motors (GM). The agreement provided first for an exchange of shares between the two companies, with GM taking a 20 per stake in FIAT Auto in exchange for 5.1 per cent of GM's shares (valued at $2.4 billion). Capital efficient for GM, which was seeking to strengthen its position in Europe and South America, the exchange enabled FIAT to unlock value for its shareholders. As part of the agreement FIAT was also given the option of offering, three-and-a-half to nine years after the signing of the agreement, the remaining 80 per cent of its shares to GM. Together these moves were indicative of a transformation of the Italian model of family-controlled capitalism in the direction of a more Anglo-American model with professional managers and wider shareholding (see Chapter 3). In addition they opened the door to FIAT's exit from the automobile sector. The agreement also provided for a strategic industrial alliance involving the two group's automotive activities in Europe and South America. Two 50-50 joint ventures were created to combine their efforts in purchasing (GM-FIAT Worldwide Purchasing centred in Rüsselshein in Germany) and the production of engines and transmissions/gear equipment (FIAT-GM Powertrain centred in Turin). Also combined were their back office automotive financing functions. In addition, the two companies agreed to work together on a gradual convergence of their product platforms. In 2001, new joint centres for the design and development of common components and systems were established in Sweden and Turin.

The fundamental aims of this alliance were to reduce the costs of purchasing and of powertrain-related activities which accounted for 80 per cent of production costs. Through the development of common components and joint operations, fixed costs could be distributed over the larger combined car output of the two companies, while joint purchasing would enable them to secure larger discounts through the purchase of larger volumes. This agreement did, however, also permit FIAT Auto to concentrate on brand management, on the one hand, and distribution and customer services, on the other, reflecting its desire to alter the weight of its car making activities in its overall value chain.

At the time of this alliance the FIAT Group was on the verge of a major profitability and debt crisis, caused mainly by the mounting losses of its automobile operations. Table 8.2 sets out a series of indicators of the

Table 8.2 FIAT Group financial highlights, 1991–2003 (€ millions)

(€ millions)	2003	2002	2001	2000	1999	1998	1997	1996	1995	1994	1993	1992	1991
FIAT Group													
Consolidated revenues	47,271	55,649	58,006	57,555	48,123	45,769	46,257	40,244	39,092	34,005	28,176	30,526	29,174
Operating result	−510	−762	318	855	788	746	1,791	932	1,717	1,382	−433	145	342
Net result before minority interest	−1,948	−4,263	−791	578	506	916	1,550	1,420	1,329	—	—	—	—
Capital expenditures	2,011	2,771	3,438	3,236	2,712	2,418	2,398	2,746	2,918	2,351	3,439	3,061	2,160
Research and development	1,747	1,748	1,817	1,725	1,406	1,264	1,166	1,129	1,079	—	—	—	—
Net financial position	−3,028	−3,780	−6,035	−6,467	−4,031	1,420	1,340	−1,142	−1,341	−1,049	−2,710	−1,988	−139
Group interest in stockholders' equity	6,793	7,641	12,170	13,320	12,874	12,998	13,203	12,042	11,178	10,340	9,456	8,844	8,838
Net invested capital	10,522	12,459	19,642	21,676	18,798	13,700	14,122	—	—	—	—	—	—
Operating result/net revenues (ROS, %)	−1.1	−1.4	0.5	1.5	1.6	1.6	—	—	—	—	—	—	—
Operating result/Average net invested capital (ROI, %)	−4.4	−4.7	1.5	4.2	4.8	5.4	12.2	6.3	12.3	10.2	−2.7	1.0	2.8
Net result/Average stockholders' equity (ROE, %)	−26.3	−39.9	−3.5	5.1	2.7	4.7	9.9	10.5	10.3	5.4	−10.1	3.2	6.6
Number of employees	162,237	186,492	198,764	223,953	221,319	220,549	242,322	237,865	237,426	248,180	260,951	285,482	287,957
Automobiles (Fiat Auto)													
Net revenues	20,010	22,147	24,440	25,361	24,101	24,859	26,202	21,950	20,190	17,148	12,937	14,175	14,206
Operating income/loss	−979	−1,343	−549	44	−121	−108	758	243	420	221	−858	−281	83
Income before minority interest	−2,058	−2,739	−1,442	−599	−493	−258	—	—	—	—	—	—	—
Capital expenditures	1,100	1,115	1,331	1,412	1,464	1,373	1,341	1,677	1,700	1,589	2,616	2,027	1,115
Capital expenditures (% of Group)	54.7	40.2	38.7	43.6	54.0	56.8	55.9	61.1	58.3	67.6	76.1	66.2	51.6
Number of employees	44,563	49,544	55,174	74,292	82,553	93,514	118,109	116,144	114,386	119,618	120,338	125,378	128,925

Source: FIAT Group, Bilancio consolidato e d'esercizio and Relazione sulla gestione. FIAT Group, Turin, various years.

performance of the Group and of the performance of its automobile division from 1991–2003. In 1998, FIAT Auto's net result before minority interest was negative at € −258 million. In 2002 these losses reached € −2,739 million, declining in 2003 to −2,058 million. In 2001, the FIAT Group as a whole recorded a net result before minority interest of € −791 million. In 2002, these losses leapt to € −4,263 million. In 2003 another major loss of € −1,948 million was recorded. At the same time the Group was faced with massive debts: its net financial position turned negative in 1999. In 2001 it stood at € −6,035 million. As Table 8.2 also shows, the main indicators of financial performance were extremely weak: the return on sales (ROS), return on investment (ROI) and return on equity (ROE) all went negative, while the course of action the Group adopted saw the Group interest in stockholders' equity fall by nearly one-half in 2001–3.

The proximate causes of this crisis were a loss of market share in Italy and Europe, the difficulties of its internationalization strategy and excess capacity. Trends in market demand were one of the underlying causes. More important, however, was FIAT's relative lack of investment in plant and equipment, research and development and product development. This crisis precipitated a series of major restructuring plans. Involved were: a reorganization of the group's automobile operation into four separate units responsible respectively for the Alfa Romeo sports car brand, the FIAT and Lancia brands, all international activities, and services; capital increases; a decision to concentrate on the Group's core automotive and agricultural equipment activities and to divest itself of nonstrategic activities, generating cash, reducing debt and further helping fund associated industrial restructuring plans. As a result FIAT sold its shares in GM (incurring a net loss due to a fall in their market value), 34 per cent of its stake in Ferrari, a part of its equity holding in Italenergia, the Teksid aluminium operation, Magneti Marelli's after-market operation, the insurer Toro Assicurazioni, a part of the Fidis consumer finance operation, and the aerospace activities of FIAT Avio.

As will be seen, this evolving strategy first of conglomerate-style diversification and recently of reconcentration on core activities, as well as the goal of increasing the importance of service activities, had major implications for FIAT's more traditional activities, for the profile of employment in different parts of Italy and the wider world and for the geography of value added, earnings and incomes.

Globalization and Market-Seeking Investments

In Europe market demand for motor vehicles expanded at historically unprecedented rates until the mid-1970s. A relatively short slowdown in

1973–5 was followed by a longer slowdown in 1980–4 that prompted major structural change. After 1984 a phase of rapid growth lasted until 1989, a year of record registrations of new vehicles. After several years of uncertainty, 1993 witnessed a deep fall in car sales. Sales subsequently increased slowly until the onset of the recent recession which is leading to a further slump in sales.

A common remedy for flagging sales in what is increasingly a marketing-driven industry is the launch of new models to acquire a larger share of the market (see below), yet it requires a level of investment that is risky, is increasingly costly due to stricter environmental and safety regulation, and can exacerbate problems of excess capacity. In the case of FIAT Auto the decline in sales in EU and EFTA countries in the early 1990s was dramatic (Figure 8.2). After some growth in the middle of the decade, decline in EU and EFTA sales set in again. One reason why was FIAT's traditional specialization in small cars.[1] In the 1990s FIAT's position in this market segment came under threat from Japanese and Asian producers and from European producers of medium-range and upper-medium range cars who entered small car market segments.

Demand is not just cyclical in character but periodically grows less rapidly than capacity. In the 1990s overcapacity emerged in Europe, raising costs, encouraging overproduction, generating downward pressure on prices and profits and causing the less competitive producers to lose market share. A central part of FIAT Auto's response to this constellation of problems was to internationalize production and to seek new markets in

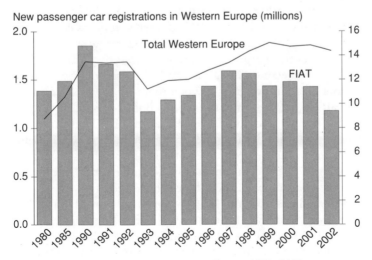

Figure 8.2 New passenger car registrations in Western Europe, 1990–2001. *Source: elaborated from CCFA, 2003.*

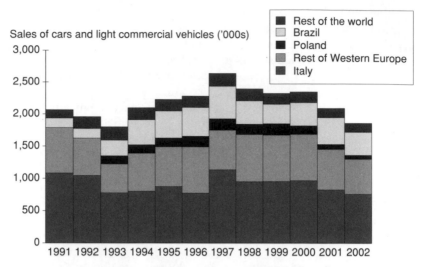

Figure 8.3 Sales of cars and light commercial vehicles, 1990–2000.
Source: elaborated from FIAT Group, 2001.

new market areas. However, other producers followed suit. As a result, the global economic downturn at the start of the new millennium saw excess capacity appear at a global scale.

The results of this strategy of internationalization in FIAT's case are reflected in Figure 8.3 which shows the way in which increased sales of cars and light commercial vehicles were achieved largely in Brazil, Poland and the rest of the world at least until the end of the decade.[2] As these areas are ones in which General Motors also operates, it helps explain the subsequent link-up.[3] A number of these new market areas came to suffer, however, from even sharper changes in economic fortunes than more-developed market economies as a result in part of the economic and financial liberalization designed to open them up for western investment.

FIAT's earliest overseas initiatives were confined to marketing, manufacturing under licence and some overseas assembly. These initiatives were driven by the small size of the domestic market and the consequent need to export in order to realise economies of scale in product development and manufacturing. (These economies play a fundamental role in explaining the oligopolistic structure of the motor vehicle sector).

The next step dates from the 1960s when FIAT moved in the direction of a multidomestic manufacturing strategy (Bélis-Bergouignan, Bordenave and Lung, 2000) starting to internationalize its production operations. In 1968 FIAT acquired a 15 per cent share in Citroën with whom it collaborated over production and sales. In 1973, this agreement was ended. Aware that production volumes and scale economies were important determinants of survival and success, in 1985 FIAT entered negotiations with Ford to create a car group that would have accounted for one-quarter of west European sales: at that time FIAT had 12.6 per cent of the west European market, Ford 12.7 per cent and Vokswagen 12.1 per cent. Agreement was not reached, however, due to differences over questions of control and governance. Not until the signing of a strategic alliance agreement with GM in 2000 did FIAT finally manage to link up with another major producer.

In the second half of the 1980s internationalization finally accelerated. After the chastening experience of the negotiations with Ford, FIAT chose projects involving financially and technically weaker partners and in which it played a lead role. FIAT's first decisions were to expand and consolidate its position in the CEECs (Poland, and Russia where a major project was finally cancelled due to the collapse of the Soviet regime) and in emerging economies (Argentina, Brazil, China, India and Turkey). A number of these countries were ones in which FIAT already had a presence. As early as 1960, FIAT had started to manufacture in Argentina. After the oil crisis FIAT progressively withdrew, so that in the 1980s cars were made by a local producer under licence. In 1995, however, FIAT decided to return to Argentina with the world car project. In Brazil FIAT Automovéis was established in 1973. Connections with Poland date from 1920. In 1931, production under licence commenced. In the 1970s, there was active industrial collaboration, leading to a 1979 agreement to export FIAT 126 cars made in Poland. In 1980, FIAT ended production of FIAT 126 cars in Italy, exporting to all its markets from Poland. In 1991, production of the Cinquecento started in a completely renovated FSM (Fabryka Samocho-dów Malolitrazowych) plant. In 1992 FIAT Auto Poland was established with the privatization of FSM and the acquisition by FIAT of 90 per cent of its share capital. As a result of these initiatives, dozens of Italian component suppliers followed FIAT into Poland. In Russia, a 1966 agreement paved the way for the Togliattigrad car plant. In this case FIAT was the prime contractor in a turnkey project and granted production licences. In 1987, FIAT agreed to participate in a joint venture to develop a manufacturing plant at Elabuga. As the economic situation in Russia deteriorated, how-ever, the project was scaled down, and the planned production of a new model was first transferred to Togliattigrad, and finally cancelled. Another failure occurred in the former Yugoslavia. In this case the presence of FIAT dated back 1954 when FIAT licensed Zavodi Crvena Zastava to manufac-ture some models. In the early 1990s, FIAT initiated negotiations to acquire a substantial share of Zastava. As the political situation deterior-ated, and the country fragmented, no progress could be made. In Turkey, collaboration dates back to 1960. In 1971, Tofas started to produce FIAT

Table 8.3   Number of passenger cars and commercial vehicles per 1,000 inhabitants at 1st January

	1985	1990	1995	2001	2002
European Union[1]	380	454	473	551	560
Poland	117	160	229	310	325
Turkey	27	37	65	90	89
Canada	559	617	562	571	572
United States	708	752	759	787	785
South Korea	25	71	177	255	273
Japan	375	456	527	573	576
Argentina	173	180	167	180	190
Brasil	86	87	89	115	117
China	3	5	8	12	12
India	3	5	6	8	8

Source: CCFA, 2003.
Note
[1]As of 1995 the EU comprises 15 countries

cars under licence in Bursa. In 1997, the same plant was chosen as a production site for FIAT's world car.

At the root of this internationalization strategy were several market-related factors. On the one hand, in existing market areas market expansion opportunities were limited for two sets of reasons. In Europe, FIAT faced a competitive challenge from East Asian manufacturers. At the same time there were high levels of car ownership (Table 8.3), so that demand was increasingly for replacement vehicles, was more irregular and required companies to manufacture a range of differentiated models. On the other, in developing country markets levels of car ownership were low, while their share of the world market was expected to increase from 18 per cent in 1994 to 33 per cent in 2000. FIAT therefore returned to its earlier focus on developing country markets, first to sell its established models, and then, in 1993, to sell a specifically-designed world car (Project 178) that could be adapted to a range of uses and a variety of different emerging markets.

The establishment of Project 178 was an important turning-point. Along with the earlier 1987 decision to produce the Cinquecento (and later the Seicento) in Poland for the entire European market, Project 178 amounted to a geographical shift from a multidomestic to a multiregional strategy, involving an intrafirm division of labour at a continental scale. The aim of the project was not just to increase FIAT's global role but also its global competitiveness: although FIAT was entering developing country markets in order to increase its manufacturing and sales in areas whose share of world markets was expected to increase, FIAT was also intent on achieving efficiency advantages in product design, purchasing and production. To meet these two sets of aims what FIAT proposed was the development of a modular global car for these non-EU markets made up of a family of five models, where a global car was understood as one that would cost $8,000 pre-tax. The development of a family of models was designed to enable FIAT to meet the different needs (different usage conditions, for example) and characteristics of different developing country markets (although at the expense of less standardization of components). As part of a quest for greater efficiency, a number of further decisions were made.

First, these models were to be developed on a common floorpan with a high degree of component commonality yet with five different bodies.

Second, FIAT sought to ensure absolute standardization of every version of each model even if these versions were planned for different markets, in part through the selection of global suppliers and their integration into the internationalization project (see Volpato, 1999:403–5). Each component manufacturer was to be responsible for the supply to every assembly plant of identical components, made using the same materials, technologies and manufacturing processes, either through delocalization, or through a joint-venture with a local firm where a transfer of know-how and quality standards or supplier training were involved. The aim was to put progressively in place a global procurement system that was globally optimized in the light of four factors: suppliers' production capacities; their just-in-time delivery capabilities to assembly lines and warehouses located in different countries; exchange rates and tariffs; and the coordination of the logistic chain. (Assembly, it was recognized, would have to take place inside market areas not least to reduce the risks of balance of payments crises in developing economies). Essentially a shrewd increase in the variety of models to meet the specific needs of different markets was to be combined with a careful sharing of the floorpan and components to create competitive advantage simultaneously in marketing (product mix) and cost (standardization and scale economies).

Third, FIAT planned to take advantage of the experience gained in the green-field Melfi plant in establishing and organizing new factories in developing economies.

Finally, FIAT aimed to put in place a process of organizational learning designed to permit the centralized design of products for markets in different countries. More specifically, the aim was to combine the skills of core country product development teams with local developing country knowledge of product use and market needs. As a result, FIAT Auto's first 'world car', the Palio, was developed on a world scale. (The range comprises the FIAT Palio, Siena, Palio Weekend and Pick-up). A 200-strong multinational, simultaneous engineering team worked with the Palio's 60 main components suppliers, who were involved as codesigners in the developmental stages of the project. Only body design remained outside the responsibilities of the codesign system. Model development time for the project was reduced to 36 months. Technicians from the countries where the models were to be manufactured were brought to the Turin pilot plant in Corso Orbassano. In addition, the development team was kept in place through the whole life cycle of the product, first, as the project was to be implemented sequentially in a set of plants, and, second, as part of a strategy of continuous improvement. What emerged was a matrix approach to global project organization, designed to combine the advantages of developed areas in high value added activities with the wage cost and market expansion advantages of less-developed countries. This approach implied the development of a new functional division of labour between cores and peripheries, involving functional upgrading in the cores in north-west Italy and the development of assembly and manufacturing operations in the peripheries.

Equipped with a versatile family of vehicles that offered it a chance of serving the needs of different developing country markets and of securing efficiency gains, FIAT had to decide where to invest. Three criteria were used. First, the countries chosen must offer considerable scope for market growth, implying that existing car ownership levels were low, that the population was relatively large and that rapid growth in the incomes of potential owners was anticipated. Second, they should be members of customs unions with high external tariffs that offered protection to produc-ers manufacturing inside the tariff walls. Tariffs vary by product and area of provenance and change over the course of time. At present they none-theless remain high. In 2003, the tariffs on passenger cars imported from the EU stood, for example, at 35 per cent in Brazil and Argentina, 10 to 135 per cent in Egypt, and 25 to 43 per cent in China (CEC, Market Access Database, http://mkaccdb.eu.int/). Third, the countries selected should offer significant fiscal and financial incentives for car makers and their suppliers,[4] and special import arrangements for vehicles or vehicle parts. Generally, these arrangements permitted imports by companies es-tablishing car complexes within the country provided that there were counterbalancing exports. As a result, the companies that chose to invest in these emerging economies were able to import, serve the host area markets and export to third countries.

Manufacturing of world cars started in Brazil in 1996 (Table 8.4) and was extended to Argentina in 1996–7. In 1997–9, world cars were assem-bled in Venezuela from Brazilian kits. In 1997, production started in Poland, and in 1998 in Turkey. In 1997, assembly started in Morocco followed in 1999 by India. Next to assemble world car models were plants in South Africa (1999), Egypt (2000), and China (2002). In 2001 Thailand was chosen as a site to assemble Alfa Romeo cars.

Table 8.4 Production of the Palio and Siena, 1996–2001

	1996	1997	1998	1999	2000	2001	1996	forecasts for 2001
Brazil	157,570	375,169	265,865	254,504	259,681	271,219		400,000
Argentina	—	48,141	60,544	26,139	25,736	31,315		100,000
Poland	—	9,006	28,521	21,957	10,945	6,168		50,000
Venezuela	—	9,766	8,858	1,867	—	—		23,000
Morocco	—	98	5,393	8,361	6,190	6,407		20,000
Turkey	—	—	20,095	23,964	43,892	13,835		110,000
India	—	—	—	3,766	2,701	8,990		100,000
South Africa	—	—	—	137	6,822	8,161		25,000
Egypt	—	—	—	—	3,201	3,375		15,000
Russia	—	—	—	—	—	—		70,000
China	—	—	—	—	—	—		100,000
Total	157,570	442,180	389,266	340,696	359,168	349,470		1,013,000

Source: Enrietti and Lanzetti, 2002.

Overall, however, sales fell well short of expectations. Output peaked in 1997 at 442,180 vehicles. In 2001, output stood at 349,470 world cars compared with expected sales of more than one million (Table 8.4). One reason for the shortfall was the economic failure of neoliberalism in developing economies: growth fell short of expectations, economic instability was endemic, income inequality grew and the conditions for a mass market for vehicles failed to emerge. In 1998, financial instability saw output decline dramatically in Brazil. In 1999 and 2000, the Argentina economy collapsed, forcing FIAT to scale down its operations to the minimum level consistent with retaining an ability to relaunch them were conditions to improve. Other difficulties were encountered in Russia. In 1998, FIAT had agreed on a joint venture with Russia's OAO GAZ car maker to produce its world car on a new greenfield plant at Nizhny Novgorod, once it had secured assistance from the European Bank for Reconstruction and Development (EBRD) and the Russian and Italian governments. In 1999, however, in the aftermath of the Russian financial crisis, the agreement was renegotiated. As a result, FIAT's investment was to be delayed and limited to an assembly operation. In 1999 instability reached Turkey, although the strong adverse effects it had on the passenger car market were partially offset by the expansion of commercial vehicle operations. In the same year, the Polish market declined to one half of its 1999 size due to an economic slowdown and competition from Skodas imported from the Czech Republic. It is these difficulties that help explain not just the shortfall in demand but also the rapid geographical reorientation of FIAT's efforts most recently towards China. In diversifying across countries, of course, FIAT is also able

Table 8.5 World production by manufacturer and economic area, 2002 and 1998 (% of total)

	North America (NAFTA)	South America	European Union	Other Europe and Turkey	Japan	South Korea	Other Asia, Pacific and Africa	Total ('000s)
2002								
European manufacturers	17.9	6.1	63.2	6.9	0.0	0.6	5.2	18,814
FIAT–Iveco–Irisbus	0.0	17.8	63.2	13.6	0.0	0.0	5.3	2,191
Daimler-Chrysler	64.3	1.1	33.2	0.1	0.0	0.0	1.3	4,456
PSA Peugeot Citroën	0.0	2.4	89.3	0.4	0.0	0.0	8.0	3,262
Renault-Dacia-Samsung	0.6	3.6	78.5	12.2	0.0	5.0	0.1	2,329
Volkswagen	6.6	10.7	59.1	13.9	0.0	0.0	9.7	5,017
American manufacturers	65.2	5.0	25.6	1.0	0.0	0.0	3.2	15,212
Japanese manufacturers	20.9	0.4	5.7	0.8	57.3	0.1	14.8	17,909
South Korean manufacturers	0.0	0.4	0.0	1.9	0.0	94.8	2.9	3,212
All manufacturers	28.4	3.4	28.8	5.0	17.4	5.4	11.6	58,840
1998								
European manufacturers	19.8	6.8	64.7	5.9	0.0	0.0	2.8	18,355
FIAT	0.0	19.5	62.5	15.5	0.0	0.0	2.4	2,696
Daimler-Chrysler	70.2	1.5	27.6	0.2	0.0	0.0	0.6	4,512
PSA Peugeot Citroën	0.0	1.5	94.8	0.7	0.0	0.0	3.1	2,247
Renault	1.4	4.7	84.0	9.5	0.0	0.0	0.4	2,283
Volkswagen	7.0	10.2	67.1	8.7	0.0	0.0	6.9	4,809
American manufacturers	66.2	4.3	26.7	0.4	0.0	0.0	2.4	14,361
Japanese manufacturers	18.9	0.3	5.4	0.6	66.6	0.0	8.3	15,089
Korean manufacturers	0.0	0.0	0.0	7.0	0.0	91.7	1.3	2,131
Other manufacturers	0.4	0.1	2.6	49.8	0.0	0.0	47.2	3,142
Total	30.2	3.6	31.4	5.4	19.0	3.7	6.9	52,987

Source: CCFA, 1999; 2003.

to spread risks across markets whose growth and contraction it hopes are not synchronous.

An important result, nonetheless, was the emergence of a hierarchical division of labour:

1   At the top was Italy where knowledge-intensive research and development, product development and technology, finance and strategic components were produced for the entire system, although steps were later taken to create an integrated international product development system involving Italy, Brazil and Turkey.
2   In Brazil, Argentina, Poland and Turkey integrated production complexes made and exported vehicles and components to the rest of the system. (Only some models are made in each market, although the entire range is potentially saleable in each market).
3   At the next level a series of developing economies (Morocco, Egypt, India, South Africa, and China) were the sites for assembly plants that imported components from Italy and the other production centres and served national markets.

For Italy multinational expansion resulted initially in an increase in net exports. The reason why was that increased exports of machinery, parts, components and other materials to overseas affiliates and markets exceeded in value the increase in vehicle imports. Over time, however, component exports from Italy declined as local content overseas increased, and component imports from overseas poles increased (Balcet and Enrietti, 2002).

The overall impact on the geography of the manufacture of passenger cars and commercial vehicles is recorded in Table 8.5. Two things warrant attention. First, most producers remain concentrated in their home markets: 63.2 per cent of the production of European manufacturers takes place in the EU, while 65.2 per cent of US producers' output is made in NAFTA. Second, setting aside the special case of the Daimler-Chrysler group, in 1998 the EU accounted for a smaller share of FIAT's output than that of any other company, indicating how FIAT had started to move earlier into new markets in the CEECs and South America. Other European producers did subsequently follow FIAT, however, most strikingly into the CEECs.

FIAT in Italy: Technological and Organizational Upgrading and Geographies of Production

In Chapter 7 and Table 7.5 we showed how FIAT moved from its initial home in Turin and the northwest to the south. At the root of relocation in the south were a different set of factors from the market-related consider-ations that initiated the globalization strategy discussed in the last section. Most important were considerations related to production technologies, the organization of work and the management of the workforce. Essentially the move to the south was related to decisions to replace an industrial system centred on large factories with mechanized production lines with smaller factories located in less urbanized areas: it was related in other words to the crisis of FIAT's variant of the Fordist model of mass produc-tion.[5]

The first wave of southern investment in vehicles and related sectors took place in the early 1970s with the expansion of the Termini Imerese plant and the construction of seven new establishments, manufacturing: fork lift trucks in Bari; vehicle assembly in Cassino; engines and gear boxes in Termoli; earth moving equipment in Lecce; steering boxes, steering columns and subsequently gearsticks and suspensions in Sulmona; aviation in Brindisi; and components in San Salvo. In 1977–80, new plants were added for the manufacture of coaches at Grottaminarda, diesel engines at Foggia and commercial vehicles in the Val di Sangro.

In the mid 1980s, FIAT decided on several further southern investments. The most significant was the San Nicola di Melfi plant in Basilicata. The area chosen had a strong agrarian tradition, high youth unemployment and no record of industrial conflict, yet was logistically near the Adriatic coast, Taranto and Naples. Work began at Melfi in September 1991. On 28 December 1993 the first Punto rolled off the line: just 28 months were required to build a 2.7 million square metre complex that would be capable of producing 450,000 cars a year once fully operational, and of providing work for 7,000 people. In the same year work also started on the Pratola Serra engine plant, near the epicentre of the 1980 earthquake in the province of Avellino (see Chapter 5). Together these plants contributed to a further rise from 29.6 per cent in 1990 to 49 per cent in 1998 in the south's share of FIAT Auto's declining Italian employment (Balcet and Enrietti, 2002).

The Melfi plant had important direct and indirect impacts on the region of Basilicata. First, most of the plant's employees were recruited in the region. The vast majority were young people with secondary school quali-fications starting their first job. Much of the workforce came from small towns and villages in a relatively large area surrounding the plant. Indeed, FIAT's desire to see the development of apartments for the workforce in the vicinity of the plant did not gain support (Interview data, June 2001), as the presence of a few people with industrial salaries in small villages, especially in upland and mountain areas, can play an important part in sustaining small communities. Second, the creation of new economic infrastructures created possibilities for further investments.

The development of the Melfi complex was an important step in FIAT's attempt to put in place a new productive model, in a new area, with new personnel, and with the aid of the Italian state. In 1986-92 FIAT made two planning agreements with the Italian government (see Chapter 5). The first involved the concession to FIAT of LIT 1,932 billion for investments associated with a restructuring and expansion of its existing plants in the south. Included were Cassino, Termoli,[6] Sulmona, Termini Imerese, FIAT Avio in Brindisi and Sevel in the Val di Sangro. Also supported was the creation of a network of seven research centres (ELASIS).[7] These investments were expected to amount to LIT 3,536 million and to create 1,200 new jobs. The second agreement provided for aid amounting to LIT 2,800 billion for investments of LIT 6,260 billion in new plants at Melfi and Pratola Serra. These two plants were expected to create 8,020 jobs (Franzini and Giunta, 1999: 283-6).

In the EU, general state aid rules require Member States to inform the Commission of any aids exceeding € 100,000 in a 3-year period (the so-called de minimis rule) unless covered by existing rules which apply to specific sectors including the agriculture, fisheries, steel, shipbuilding, textiles and motor vehicles sectors, while, in the case of regional aid, intensities must be within the ceilings set out in the map of Assisted Areas. In the automobile sector, state aids have since 1989 come under a specific EU framework designed to prevent a subsidy race and distortions of competition (Dancet and Rosenstock, 1995). The motor vehicle framework involved a stronger obligation to notify the Commission of subsides, and, in particular, of approved regional aid schemes, and established guidelines for the provision of state aid. In the case of regional aid, it involved the development of a cost-benefit method to compare the costs of investing and operating in a disadvantaged area with the costs in a selected non-assisted area in the EU. The Commission decided to accept an aid level that compensates investors for the net additional investment and operating costs they incur due to the less-developed area's structural handicaps. In this way assisted area are put in the same position as the nonassisted comparator region for attracting a mobile investment. In addition, the Commission decided to permit an additional 'regional top-up' up to a level of 3 per cent of the investment costs, if the project does not create any sectoral problems such as overcapacity.

The Melfi and Pratola Serra projects were one of the first three sets of investments subjected to the cost-benefit analysis methodology. Note, however, that this analysis is largely reliant on data provided by private companies, and depends, therefore, on absolute frankness about the relative costs of different locations. Only if state enterprises are involved in a sector can the state get direct knowledge of the real costs. Commission analysis of the Melfi and Pratola Serra projects did identify additional investment (infrastructure and machinery) and operating (freight, rejection rates, inventory and labour) costs. The Commission considered, however, that the top-ups were not entirely justified, and that aid intensities were excessive. As a result, the Italian government had to reduce aid levels. Commission approval of the aid package also required another step: FIAT was required to make irreversible capacity cuts, by closing three other plants, including Desio in Milan and Chivasso in Turin.

Aid is not a one-off event. The choice of the south made FIAT eligible for further subsidies as the years passed. In 1999, the European Commission decided to initiate six investigations into aid for an amount of approximately € 120 million that the Italian Government planned to grant to FIAT Auto under regional law 488/1992. These investigations were also conducted in the framework of the Community Framework for State Aid to the motor vehicle industry, and coincided with investigations into a Volkswagen assembly plant in the former German Democratic Republic and the Rover plant in Birmingham. In 1999, after a detailed investigation, the European Commission decided to approve regional aid of € 28.5 million granted to FIAT for its Termoli plant. In 2001, it approved a € 40.3 million investment aid for the production of the new Punto at Melfi.

A further advantage of the new investments was that new companies were set up to operate the new plants. In Melfi, for example, the new company was called SATA. The creation of a separate company meant that the wage rates were not subject to the FIAT Auto national agreements and could be set at lower levels (Interview data, June 2001).

Clearly, therefore, the choice of southern locations for these investments conferred a number of advantages. Any additional costs compared with alternative sites in the EU were more than compensated for by Italian state subsidies. A disadvantaged area location made further investments eligible for state aid. Also wage rates were reduced. As indicated earlier, however, FIAT's fundamental goals were to put in place, on greenfield sites, new manufacturing technologies and principles of work organization, involving factory integration, just-in-time supplies, to reduce buffer stocks and inventories, and team work.

The reason why a new productive model mattered is that, at the time of the the oil crisis and the social crisis in its factories in the 1970s, FIAT initially chose a path of automation and intensive use of computer technologies. In 1972, Mirafiori was using 16 robots to replace workers engaged in repetitive tasks. In 1974, robots figured in the new Cassino assembly plant. Next came intelligent machines that could respond automatically to the variability of complex situations, Computer Aided Manufacturing systems in which several machines were controlled and coordinated by a computer and Computer Aided Design. In 1978, the ROBOGATE system was introduced in Rivalta and Cassino. ROBOGATE was Comau's flexible, robotized body assembly and spot welding system. As a result Comau became the world leader in this field. In the 1980s, people started to speak of the automated, semi-staffless 'lights out' factory, with, in Italy, the FIAT plants at Termoli and Cassino setting the standard.

A major disadvantage of this system was high capital costs.[8] Another was the existence of rigid Taylorist hierarchies and strong demarcation. Yet another was the stockpiling of materials to cover unexpected demand. As with other car manufacturers FIAT wanted to find a new way of working, and to do so wanted a new green field development. FIAT's solution was the 'integrated factory', which sought to combine flexibility, just-in-time methods and improved accuracy and quality control as a result of greater reliance on intelligent, flexible and responsible human work and was implemented in Melfi.

In the leaner integrated factory the number of hierarchical levels was reduced, white-collar staff were relocated to the shop-floor level, and responsibilities were decentralized to a series of autonomous work teams (UTEs) to permit faster decision-making and to ensure that manufacturing problems were tackled and solved, as far as possible, as and where they arose. In its own area of competence, each UTE had responsibilities in relation to production technologies and equipment, maintenance, materials flow and supply, personnel management and quality. More specifically, each UTE was made responsible for self-checking, prevention, problem-solving and ongoing improvement. Each UTE leader was expected to motivate his or her team and delegate individual tasks to fall in with production needs. Quality was assured by making each UTE in the manufacturing cycle view the next UTE on the line as a final client who had to be satisfied. To help achieve these ends, a visual management system was introduced. The aim of this system is to regulate the flow of work from one UTE to another and to achieve full capacity working. Essentially, FIAT management define what is called the 'impostato'. The 'impostato' is the programmed number of cars that ought to be produced in a shift, given the capacity of the plant, the orders of FIAT's customers and the orders that need to be carried out to meet (as and when they are arise, in accordance with the pull principle of the kanban system) the needs of the next UTE along the line. This target is signalled on electronic displays, along with the volumes actually achieved in the course of the shift.

Once developed and implemented in Melfi and Pratola Serra, FIAT's version of the 'integrated factory' and 'lean production' were applied in other plants. An example was the Cordoba plant in Argentina. Subsequently, these principles were extended in the direction of a modular or digital integrated factory, while the UTE model was generalized to the whole of the FIAT Auto system.

A second important feature of the Melfi plant was the creation of an adjacent suppliers' park (see below).[9] Melfi's first-tier suppliers are required to locate next to the assembly plant, and are given direct access to the plant via a corridor called Via Roma, so that, at short notice, they can deliver small batches of the right components/modules to the right place at the right time, in accordance with just-in-time principles. Seat manufacture is, for example, synchronized with vehicle assembly, so that seats for the right model and of the right colour arrive from the adjacent park at an assembly-line work station just in time, and in a sequence that corresponds to the sequence in which cars are being handled on the line at that moment, avoiding the need for warehouses, and permitting rapid intervention if there are defects (Interview data, June 2001). Advantageous for the assembler, as it reduces lead times, avoids transport risks, reduces batch sizes, synchronizes supply with assembly and permits a rapid response to defects, this model is, however, costly for the suppliers, who must establish manufacturing establishments near each plant they serve.

Outsourcing, Redefining Corporate Boundaries and Restructuring the Supply Chain

The quest for new sources of competitiveness and higher rates of profit also involve attempts to identify and concentrate on core competences. This type of strategy rests on internal reorganizations, acquisitions and divestitures that also alter the boundaries of the firm. Of these changes, moves into services and changes in horizontal diversification were considered earlier in this chapter. In this section we shall concentrate on the changes associated with cost- and risk-reduction related reorganization of the supply chain.

Traditionally, car makers were responsible for a relatively large share of the activities that make up the motor vehicle manufacture and assembly chain. More recently, however, the degree of vertical integration has declined. Assemblers have sold off assets and/or transferred responsibility for a large range of activities to specialized suppliers, essentially to reduce risks and costs.[10] In the case of FIAT Auto these outsourced activities include activities that take place inside its own manufacturing facilities but which were transferred to other companies in the FIAT Group and outside companies (Table 7.7). Together, these transfers are one of the reasons for the dramatic recent fall in direct FIAT Auto employment from 128,925 in 1991 to 118,109 in 1997 and 49,544 in 2002.

Other major changes have taken place in the external supply chain itself. First, there have been drastic reductions in the number of suppliers to

European car makers. In 1990, European producers had between 723 (FIAT) and 2,500 (Mercedes) suppliers, compared to 162 to 270 for Japanese manufacturers. Four years later, FIAT had just 410 suppliers (Balcet and Enrietti, 2002). In 2001, it had just 330.

Second, a pyramid-shaped hierarchical structure has emerged, comprising first-, second- and third-tier suppliers (Figure 8.1). First-tier suppliers are mostly large and often multinational firms, and assume responsibility for the coordination of the activities of second-, third- and subsequent tiers of suppliers. A FIAT example is a project called guided growth. One aim of this project is to get second-tier suppliers to obtain ISO 9000 certification through first-tier suppliers (Interview data, June 2001).

Third, first-tier suppliers have assumed responsibility for increasingly complex modules and for entire subsystems. As cars get more complicated, and as divestiture proceeds apace, there are moves to identify core subsystems (such as complete dashboards or suspension systems), and to allocate responsibility for their development and manufacture to one external supplier (called modular manufacturing). As a result FIAT Auto's development activities will reside in the area of integration, while its manufacturing activities will involve the purchase not of numerous components for assembly but of modules that are sent directly to the final assembly line by a much smaller number of independent suppliers. These suppliers assume responsibility for the purchase of individual components, and often supply plants in a number of different locations. In this way, FIAT Auto streamlines its purchasing structure by creating a turnkey network (see Chapter 3), relying on sophisticated global suppliers, able to organize component manufacturing and sourcing on a global scale (Sturgeon and Florida, 1999), and able to reduce supply costs. FIAT used, for example, to make seats for its Mirafiori plant in Turin. This activity was sold to the Lear Corporation, as it can supply the same seats throughout the world, and can deal with second-tier suppliers.

As a result of these trends, employment in surviving independent suppliers should increase, with a substantial decline in the numbers employed directly by FIAT. Table 8.6, for example, indicates the way in which the number of suppliers is in decline, and the increasing extent to which suppliers are present in more than one of FIAT's main production poles (Italy, Poland, Turkey, Brazil and Argentina). The 100 suppliers present in more than one pole in 1998 accounted for about 70 per cent of global purchasing volume. Suppliers in Italy are one of the principal parts of this system.

This type of supply strategy is not without its disadvantages. Greater reliance on external suppliers and, in particular, on a single, large system integrator responsible for entire modules threatens a decline in the competence of assemblers, a loss of knowledge about the supply industry, including

Table 8.6   International FIAT suppliers and production poles

Year		1 pole	2 poles	3 poles	4 poles	5 poles	Total FIAT suppliers
			Suppliers present in				
1997	Number	903	63	17	14	6	1003
	%	90.0	6.3	1.7	1.4	0.6	100.0
1998	Number	766	55	19	16	10	866
	%	88.5	6.4	2.2	1.8	1.2	100.0

Source: Camuffo and Volpato, 2002.

its real costs, and a loss of power over the supply chain. In the years to come these trends are, however, expected to continue. Assuming an annual rate of growth of value added of 2.5 per cent per year, Figure 8.4 suggests, for example, that the share of value added of the assemblers will fall, while that of suppliers will increase, that the role of independent technical centres may increase, as codesign expands, and that a series of new downstream services will develop, as the assemblers move into activities closer to the final customer.

In the face of these new expectations, global first-tier suppliers concentrate on activities in which they have distinctive capabilities and a clear competitive advantage, outsourcing the manufacture of components to second-tier suppliers. To serve an OEM at a number of locations involves the development of strategies for reducing the associated risks, and securing economies of scale. To these ends, these suppliers seek greater volumes and increases in market power through mergers and acquisitions, or alliances

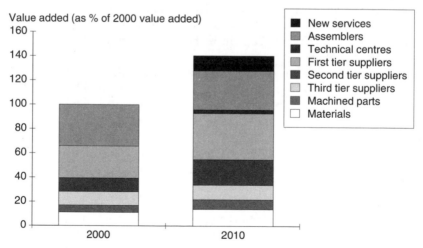

Figure 8.4   Trends in the distribution of value added.
Source: adapted from Volpato, 2002.

with local suppliers. An example is Breed Technologies which manufactures inflators, air bags, steering wheels, seatbelts and cockpit electronic equipment (Interview data, June 2001). Established in 1990, it has grown rapidly largely through acquisitions. The 50-year old Gallino plant near the FIAT Mirafiori complex which manufactures steering wheels was one such acquisition. The site has facilities for simulation, CAD, rapid prototyping and assembly, laser technology, testing and production. Of the 1.7 million steering wheels Breed Italia produces (compared with capacity for 2.7 million), 42 per cent go to FIAT Auto. The second most important group of customers is in the UK: 14.5 per cent goes to Landrover; 13.2 per cent to Jaguar; and 9.4 per cent to Rover. Seven years ago Breed Italia acquired a subsidiary in Hungary, and one year ago it acquired another in Romania. As wage costs in Romania are just $0.80 per hour, compared with $3.00 in Hungary, and $17.50 in Italy, Breed sends steering wheels that are to be leather-covered (410,000 per year) or wood-covered (60,000 per year) to the CEECs for trimming and finishing. The steering wheels for UK customers are subsequently shipped to UK warehouses, and a UK supplier that the Breed Group acquired in the Birmingham area will be downsized. (Note the logistic implications of these decisions). This delocalization strategy has coincided with other changes that have seen the workforce fall by 35 per cent in the recent past. In Gallino employment stands at 320. Including its subsidiaries, the company will employ just over 1,100. Of these employees, 350 work in Hungary, and 300 in Romania, although there are plans to raise employment in Romania to 500. Other recent economies include a 50 per cent reduction in scrap and a 70 per cent reduction of work in progress, while the number of suppliers declined from around 1,000 to 300, with plans to reduce this number to 100 in the near future,

To supply the 12 FIAT Auto operations (Italy, Argentina, Brazil, Venezuela, Poland, Morocco, Turkey, Russia, Egypt, South Africa, India, Thailand and China) and to manage the worldwide exchange of components that are purchased (as opposed to components exchanged within the FIAT Auto group) a Turin department of GM-FIAT Worldwide Purchasing runs global purchasing activities, with the assistance of representatives in the plants where FIAT Auto manufactures. There are several important aspects of this work (Interview data, June 2001). The first is the development of an information system that deals with orders, invoices and stock management. The development of a global system of this kind involves dealing with a large number of practical problems. An example is the problem of language, handled with the help of simultaneous translation systems to enable the translation of orders raised in, say, Turkish in Bursa into Polish, if the company that receives the order is in Poland. A consequence is the development of a database that can be used to select suppliers with the help of

information concerning previous transactions and performance with respect to cost, quality, reliability, service and so on. Another is the development of an electronic e-procurement system. The second important aspect
of this work is the creation of a global supply base. Two activities are
involved: getting existing suppliers to internationalize and locate in countries involved in the 178 Project; and scouting for new suppliers. The third
is the reduction in costs in part through guaranteeing cross-plant and cross-
market component uniformity, and through a systematic comparison of
prices and the performance of suppliers. By insisting on the same components being produced wherever the source, FIAT would have the ability, for
example, to switch parts made in Brazil to a car made in China. Clearly, a
system of this kind provides FIAT Auto with data to inform decision-
making and negotiation, although there remains the aforementioned risk
that the outsourcing of increasingly sophisticated modules and subsystems
will cause it to lose the knowledge that derives from learning by doing.

The global purchasing strategy recognizes a number of different situations classified according to two variables: the complexity of the part and
the extent to which suppliers require design capability or not; and the
incidence of logistic costs (which depend on transport costs, packaging
costs, costs of stocks in transit, exchange rates, tariffs and local content
rules) relative to sales value (Figure 8.5). For example, supplies of com-

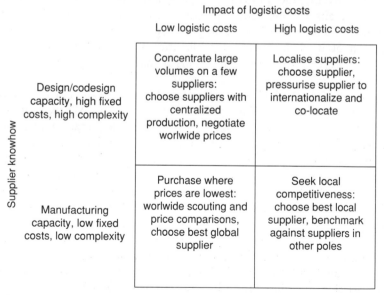

Figure 8.5   Global purchasing policy matrix.
Source: interview data, June 2001 and Camuffo and Volpato, 2002.

ponents requiring a high level of know-how and high fixed costs, yet involving relatively low logistic costs, are concentrated in the hands of a few large suppliers, who do not need to locate near to the OEM plant. Conversely, in the case of components requiring high knowhow and high logistic costs (such as dashboards where transport and packaging costs are high or seats which must be delivered just-in-time) there are few suppliers and these suppliers locate near the OEM plant, insofar as volumes are sufficiently high to justify the investments involved. In cases where components are simple, which of course can include the components used to make dashboards or seats, cost is the key consideration, with worldwide sourcing where logistic costs are low, and local sourcing where they are high.

The first step, called localization, involves scouting to identify potential suppliers, negotiation and supplier start up. Normally, 50 to 90 per cent of components are purchased in the countries in which assembly takes place. Next, the global sourcing activity involves the development of a global information system that enables comparison of the prices of each part in each country and permits either import from the cheapest source, allowing for transport costs and duties, or a realignment of prices in each country with the best world price (Figure 8.6). Clearly, price differences can reflect differences in capacity utilization. Differences can also result from productivity differences, or cost differences between developed and developing

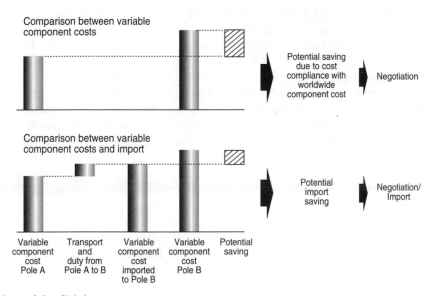

Figure 8.6   Global sourcing support system.
Source: interview data, June 2001.

areas, and can result in substantial pressure on high cost suppliers. Where differences do exist, purchasers can select the cheapest source, or renegotiate conditions with existing suppliers. At present, exit (switching to an alternative supplier) is much less likely than voice (negotiating in order to improve the performance of suppliers, and to realign prices in countries that are more expensive with the cheapest country) for several reasons. One is that, where there are fixed costs, exit is inefficient in the medium-term. Another is the need to preserve cooperative relationships with suppliers, reflected for example in the fact that FIAT Auto shares the advantages of cost reductions with suppliers. (The system also permits strikes or other factors that disrupt production in one pole to be overcome by using another plant to meet output or supply needs, and, more generally, permits multiple sourcing, which reduces dependence on suppliers, increases availability and increases pressure on suppliers). In 2000, it was estimated that this global purchasing operation resulted in savings of 13 per cent (€ 32 million).

As in the case of outsourcing, the changes in the geography of purchasing operations that result from these mechanisms play an important role in changing the geographies of employment and output that influence regional economic performance.

Crisis, Markets and Models

Crises are not new for FIAT. In 1980, there was an industrial relations crisis. In the early 1990s there was another crisis associated with the increase in the influence of Mediobanca at the expense of the Agnelli family. In 2000, FIAT faced new difficulties due to the impact of its loss of market share and flat market demand on profitability and indebtedness (see above).

In December 2001, the FIAT Group accelerated existing cost cutting and inventory reducing measures, and approved a more far-reaching restructuring plan providing for an increase in share capital, divestitures and an industrial reorganization programme. The latter involved the closure or restructuring of 18 plants (two in Italy and sixteen abroad, including the closure of Rivalta and the transfer of its facilities to Mirafiori), a reduction in the number of production lines in existing plants (from seven to four, for example, in Mirafiori), the loss of 6,000 jobs outside Italy and temporary layoffs funded by the CIGS (a special temporary layoff wage compensation fund) in Italy. In the car and light commercial vehicles sector the aim was to reduce capacity and related costs and to increase capacity utilization rates from just over 60 per cent in 2001 to nearly 78 per cent in 2003 and in the longer term to 90 per cent (Figure 8.7).

In May 2002, a new strategic recovery plan (Stage 1 of a FIAT Auto restructuring plan) was announced. The plan aimed to address issues

European car and light commercial vehicle volumes ('000s)

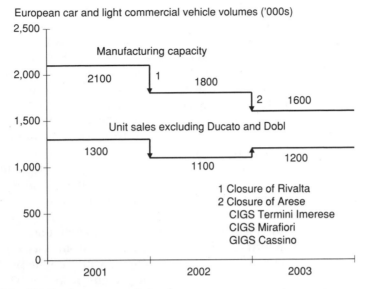

Figure 8.7   Matching European capacity to sales.
Source: FIAT 2002.

related to the distribution network and channel mix (increasing fleet and retail sales) and to reduce overhead costs. The plan involved financial measures, aimed at rescheduling debts, and an industrial reorganization project. The industrial side of the plan involved further active workforce reductions through the use of schemes requiring government authorization such as the CIGS and MAP (early retirement mobility allowances applicable to employees who would otherwise be eligible to receive a pension within a maximum of three to four years). Managed by the INPS, these programmes are funded by payroll contributions from companies and workers. (The Group drawing on them has to make a contribution).

In June, the Chamber of Deputies' committee for productive activities drafted a report on the condition of FIAT and of the Italian automobile industry. In July, after reaching an agreement with the government and all of the trade unions (FIM-CISL, UILM-UIL, FISMIC and UGL) except FIOM-GIGL, FIAT announced the dismissal of 2,887 employees, with severance payments and mobility allowances to cushion the effects on the workers involved.

In October 2002, a new plan (Stage 2 of the FIAT Auto restructuring plan) was announced. To reduce overcapacity, FIAT asked the government to declare a state of crisis in FIAT and in certain Magneti Marelli and

Comau factories to permit the group to draw on the CIGS. Also announced were plans to accelerate product range development. The announcement led to widespread industrial unrest. The trades unions declared several general strikes. In the course of the year work stoppages averaged more than 100 hours for the Group as a whole, and as many as 300 hours at FIAT's Termini Imerese plant in Sicily. In December 2002, a government plan (Framework Agreement) was was accepted by FIAT but rejected by the unions. Amongst other things, the government authorized GIGS funding for the temporary layoff of up to 5,600 employees at FIAT Auto, Comau Service and Magneti Marelli (1,000 in Mirafiori, and another 350 in Turin, 1,000 in Arese, 1,200 in Cassino, all 1,800 in Termini Imerese and 250 others) as of December 9, 2002 and of a further 2,000 from July 2003. In addition, Magneti Marelli and certain service sector companies were authorized to dismiss 500 additional employees with MAP mobility allowances by the end of 2003. After local discussion with unions early in 2003, most agreed on further measures permitting the dismissal by the end of 2003 of up to 2,400 FIAT Group employees with 'long-term' MAP mobility allowances (approved by the Italian parliament in April 2003 to provide early retirement benefits to workers otherwise eligible for a pension within up to seven years) and another 700 'short-term' mobility allowances.

In June 2003, FIAT announced a new relaunch plan. Designed to restore profitability, the relaunch plan confirmed and pursued further the drive for cost reduction via capacity cuts and layoffs (12 plants were to close in 2003–4, while Group employment was to decline by 12,300 in 2003–6, though with new recruitment of 5,400 staff) on the one hand and industrial cooperation with General Motors on the other. Costs were also to be reduced through improvements in quality. In addition, the plan was designed to raise funds internally and through a rights issue to pay for restructuring charges, and to address weaknesses in research and development, product development, marketing and distribution.

At the root of this crisis, and the consequent plans to reduce capacity and employment, were a number of factors. The first (see above) was the failure of FIAT's internationalization strategy. The second was a decline in market share in Europe and Italy. The third was the costs of the resulting excess capacity. The fourth was the relatively high costs of nonquality, reflecting the relatively high number of defects, and the consequent costs of warranty schemes (themselves related to outsourcing, the frequent changing of components and suppliers and the small margins of many suppliers which prevented significant upgrading to meet FIAT's demands). (High stocks were a further cause of excess costs).

The decline in market share reflected FIAT's overconcentration on the small car segment, and its failure to anticipate the evolution of the market. In western Europe the rate of growth of car ownership was in decline, as

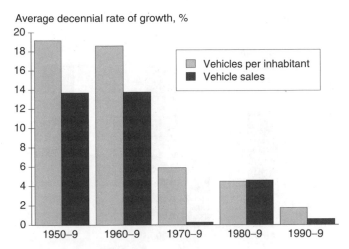

Average decennial rate of growth, %

Figure 8.8 Growth of car ownership and sales in Western Europe, 1950–2000.
Source: adapted form Volpato, 2002.

was the rate of growth of sales (Figure 8.8): sales increased at 13.7 per cent in the 1950s and 1960s, 0.3 per cent in the 1970s, 4.6 per cent in the 1980s and 0.6 per cent in the 1990s. Increasingly, demand was for replacement vehicles, while excess capacity created a buyer's market in which product range and the rate of new model development were important determinants of sales volumes. FIAT's investment and expenditure on research and development lagged behind those of its rivals. After 1993, investment declined in part due to a decline in the share of group investment in the automobile sector. To protect its market share FIAT sold a relatively large proportion of vehicles in the zero kilometre and rent-a-car channels. In these channels margins are small, while the resale of these cars created excess supply of semi-new cars, putting downward pressure on prices in the retail channel. At the same time, to sustain retail sales, FIAT incurred high advertising costs, and offered discounts to dealers, sacrificing its own profit margins.

Addressing these product development and marketing weaknesses was one of the aims of the relaunch plan. Essentially, FIAT decided first to concentrate on the retail and fleet channels, to reduce sales to the rent-a-car and zero kilometre channels, essentially to stop unprofitable sales, and to reduce inventories. Second, it decided to strengthen its fragmented distribution system and to invest in marketing, concentrating in particular on 50 metropolitan area markets that accounted for 40 per cent of sales. Third, it decided to accelerate the rate of renewal and reduce the average age of its products (vehicles and powertrain) in the A (city compact), B and C, and

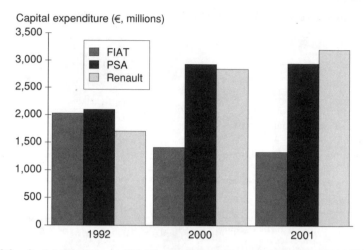

Figure 8.9   Capital expenditure by FIAT Auto, PSA Peugeot Citroën and Renault, 1992–2001.
Source: elaborated from FIAT, PSA and Renault Annual reports and accounts and Volpato, 2002.

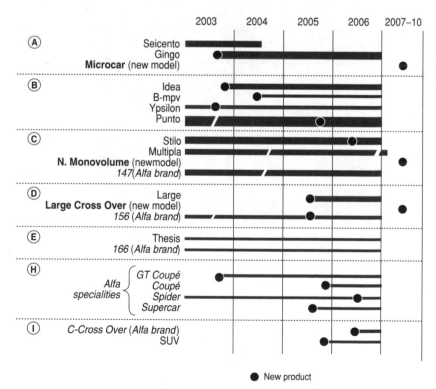

Figure 8.10   The game plan: new FIAT/Lancia and Alfa product development by market segment.
Source: FIAT, 2003.

Table 8.7 The game plan: platform convergence and component standardization

Platform convergence	2004	2008	2012
Total number of platforms	13	11	6
Number existing in 2002	12	5	0
New platforms	1	6	6
Joint platforms with partners	1	4	3
Number of models per platform	2	3	5
Annual number of vehicles per platform (including partners) ('000s)	140	250	450
Component standarization			
Families of petrol engines			
FIAT stand-alone	5		1
FIAT and partners	3		5
Average volume per family ('000s)	300		600
Families of diesel engines			
FIAT stand-alone	0		0
FIAT and partners	5		6
Average volume per family ('000s)	380		580
Families of transmissions			
FIAT stand-alone	4		4
FIAT and partners	6		6
Average volume per family ('000s)	410		535

Source: FIAT Auto, 2004.

light commercial vehicle market segments and to extend its range into the high margin segments (D, E, MPV and SUV) while exploiting a common architecture and shared components with GM to order to contain development and manufacturing costs (Figure 8.10).

FIAT's relaunch plan draws, in other words, on a number of paths to upgrading. Amongst these paths the development of its product range is seen as a central determinant of market share, and of the destinies of the plants to which these new models are assigned in a world of replacement demand in developed market economies. Accompanying these actions are a number of others. One is a restructuring of the distribution system and a concentration on strategic geographical areas. A second is platform convergence and component standardization to achieve critical mass and scale economies (Table 8.7). A third is a more concentrated geographical footprint with higher rates of capacity utilisation and a movement in the direction of manufacturing one platform per pole.

Conclusions

In the last two chapters we have identified a number of trends in the geography of motor vehicle output and employment. In core areas in Piemonte there is an emphasis upon the development of highly qualified professional and managerial jobs in services in and outside the vehicles sector. These vehicle-sector jobs for executives and clerical workers may also disappear, however, if finally FIAT sells out its automobile interests to GM, and the centre of decision-making moves away from Italy, although the recent re-commitment of the FIAT Group to its vehicle-related activities makes this outcome less likely in the immediate future. At the same time manual jobs are in decline, weakening the demand for manual workers and their relative incomes. Within Italy, car production and the manual jobs it creates are increasingly located in the south, which, in spite of the development of a research arm, is characterized by a relative under-representation of more highly qualified and paid jobs, captured in some ways by the virtual absence of offices in the Melfi complex. At the same time, there are clear signs of the creation of a new hierarchical international division of labour with market-seeking and cost-reducing investments leading to a relocation of employment and wealth creation in less-developed countries, though these investments are associated with high degrees of risk due to the extraordinary instability of market-led models of development in the developing world.

The rise and decline of individual establishments, the changing geography of wealth creation, income and employment and their regional impacts are results of an underlying set of generative mechanisms that operate in the context of an inherited set of resources and in specific market and institutional conditions. At the root of these mechanisms are corporate profit strategies which themselves involve cutting costs, raising revenues and reducing risks. In the case of FIAT Auto, these strategies involve a combination of product development, globalization to seek new markets, global sourcing to reduce costs, greater concentration on core competences, a shared use of turnkey production networks, innovation and productive organization to increase competitiveness, functional upgrading in the value chain towards design and distribution and an albeit fluctuating movement into new chains, as is illustrated by the movement of the FIAT Group into services. These mechanisms have significant direct and indirect effects on plant performance and on regional performance: decisions as to whether to source in Italy or elsewhere in the light of logistic and transport costs alter the geography of income and employment, while the failure of projects to enter new markets or to develop new products will have negative impacts on group income that will adversely affect value creation, employment and regional performance across the firm's set of establishments.

Afterword

In February 2005, GM agreed to pay the FIAT Group € 1.55 billion to void the put option in the contract agreed in March 2000. This agreement involved a swap of a 6 per cent stake in GM for a 20 per cent stake in FIAT Auto, and the contentious put option that required FIAT to sell the rest of FIAT Auto to GM between January 2004 and July 2009. In 2005, the fact that FIAT Auto continued to make very large losses and was faced with debts of more than $10 billion made the put option increasingly attractive for FIAT and increasingly unattractive for GM which was struggling to deal with its own loss-making activities in Europe and was in the process of eliminating 12,000 European jobs to save $600 million per year. As early as 2003, the relationship between FIAT Auto and GM had deteriorated as a result of FIAT Auto's decisions to sell 51 per cent of its consumer-finance operation FIDIS to four creditor banks and to recapitalise FIAT Auto with an injection of € 3 billion that reduced GM's holding to 10 per cent. Faced with a FIAT group threat to exercise the put option, GM agreed to return its 10 per cent stake and to the dismantling of two major cost-saving joint ventures in the fields of European purchasing and powertrain activities, although some co-operation will continue especially in the diesel engine area. This agreement will permit FIAT Auto to reduce its debt, although its continuing losses and remaining debt place its future in jeopardy.

Chapter Nine

Reconfiguring Industrial Activities and Places: The Italian Chemical Industry

Introduction

The main aim of this chapter is to outline and explain the rise and the fall of the primary petrochemicals sector in Italy and its impact on some of the localities in which it developed. In the phase of expansion this sector played a major role in driving regional performance. As economic and institutional circumstances and corporate strategies changed, these and other areas that figured prominently in the map of Italian chemical industry activities were required to confront difficult and politically mediated reconversions and redundancies, which also help explain why certain regional economies lost economic momentum.

To understand these stories, the evolution of these localities must be placed in the wider context of the evolution of the subsectors, the companies and the governance mechanisms that determined their destinies. Accordingly, the first part of the chapter will deal with the performance of the industry and some of its main characteristics.

A particularly important characteristic of the chemical industry is that it produces a wide range of intermediate goods that are used by the chemical industry itself (intraindustry demand) and a variety of end-user/customer industries (interindustry demand), as well as final products for consumers. Figure 9.1 shows that in the EU15 30 per cent of chemical sales are made to the chemical industry itself. Another 50 per cent of chemical industy sales is made up of sales of chemicals as intermediate goods to other industries, such as manufacturing, agriculture, and the service sector, with final demand accounting for just 20 per cent of sales. As a result the growth of the industry depends to a significant extent on the growth of the industries it supplies, including in particular the rubber and plastics, metal manufacturing and service sectors.

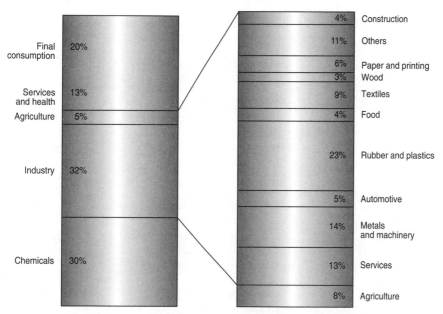

Figure 9.1   Composition of chemical sales and intermediate demand circa 2002.
Source: Federchimica, 2003a.

A second important feature of the industry is its internal differentiation. Table 9.1 identifies four main segments: primary chemicals and man-made fibres; fine and speciality chemicals; life sciences; and consumer chemicals. Given the characteristics, technologies and competences of the life sciences sector, it should be considered a separate industry, while the consumer chemicals sector is distinguished in part by a smaller use of chemicals and a higher marketing content. In these two areas, intermediate demand represents a small share of sales. Conversely, in primary and speciality chemicals and especially in petrochemicals, it is much more important, as Figures 9.2 and 9.3 suggest. Figure 9.3 for example shows how a barrel of oil, that is not used for energy, is first transformed into primary manufactures and then into final products.

The Italian Chemical Industry and its Changing Position in the Wider European and World Context

The Italian chemical sector faces sustained competitive pressure, as does the west European industry in general. There are several reasons why. In 1989–93, European production fell by 11.3 per cent. As this downturn

Table 9.1   The segments and main subsegments of the EU chemical industry in 2002

Segment	Subsegment	Share of EU15 production in 2002 (%)
Primary chemicals and fibres		38.2
	Petrochemicals	
	Plastics and synthetic rubber	
	Man-made fibres	
	Inorganic chemicals	
	Industrial gases	
	Fertilizers	
Speciality and fine chemicals		27.8
	Paints, dyes and inks	
	Crop protection	
	Materials for pharmaceuticals	
	Other speciality chemicals	
	Other fine chemicals	
Life sciences		23.4
	Pharmaceuticals	
	Biotechnologies	
Consumer chemicals		10.6
	Perfumes and cosmetics	
	Soaps and detergents	

Source: adapted from CEFIC, 2004.

occurred just after a worldwide wave of investment in the late 1980s' phase of industrial growth, the whole chemical industry suffered from conditions of overcapacity (Federchimica, 2004). A recovery started in 1993 only to be interrupted by the Asian crisis (Federchimica, 2001), while the more recent competitive performance of the EU sector was adversely affected by the strength of the € relative to the US $.

Another factor affecting EU industry costs is increasingly strict environmental standards and regulation. From 1990 to early 2003, the European Parliament passed nearly 530 new pieces of legislation: 35 per cent concerned plant security; 33 per cent the use of hazardous substances; 18 per cent air pollution; 9 per cent waste management; and 5 per cent water pollution (Federchimica, 2003a). Of major importance was the White Paper on Chemical Products. Approved by the Commission in 2001, it opened the way to the establishment of the REACH (Registration, Evaluation and Authorization of Chemicals) system. Amongst other things, the REACH regulation will require EU enterprises that manufacture or import more than one tonne of a chemical substance per year to register it in a central database, to assess the risks associated with its use and to take steps

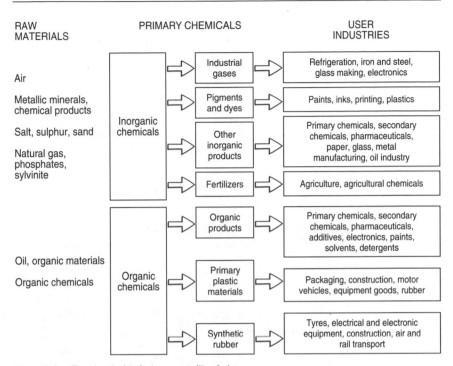

| RAW MATERIALS | PRIMARY CHEMICALS | | USER INDUSTRIES |

Figure 9.2   The chemical industry commodity chain.
Source: adapted from Bunel and Duclos, 2001.

to manage any risks they identify. Safety information on substances will be passed down the commodity chain. The aims of the proposed new regulation are to improve the protection of human health and the environment and to increase public confidence in the EU chemical industry, while preserving its competitiveness and enhancing its innovative capabilities. Increasingly strict environmental standards have led to massive investments in clean technologies. Trade associations such as Federchimica are more critical. Federchimica (2003b) claims that the new regulation will increase SME production costs and could drive them out of the market, and that close regulation of EU manufacturing activities exposes the EU chemical industry to environmental dumping by non-European producers.

The recent difficulties of the chemical industry are a part of a more general decline in rates of growth which dates from the 1980s. Until the late 1970s demand for chemicals in western industrialized economies grew far more rapidly than GDP. At the same time, rapid product and process innovation especially in petrochemicals, plastics, fibres and pharmaceuticals transformed the profile of chemical production. In the 1980s, a number of factors modified market conditions. The first was

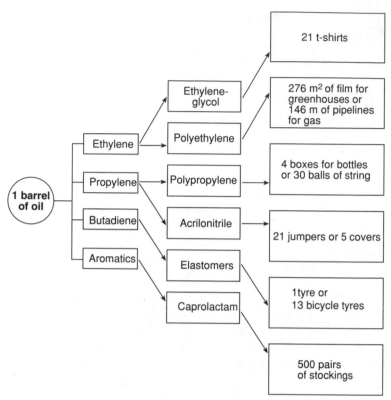

Figure 9.3 The intermediate nature of petrochemical products.
Source: Federchimica, 2003a.

the repercussions of the two oil shocks (which served to increase the importance attached to feedstock and to efficiency in its use) and over-optimistic demand forecasts (which indicated a failure to recognize the structural character of the growth slowdown). The second was the entry of new producers especially in oil-rich countries. Together these factors put pressure on prices and profits. In subsequent years the chemical premium (the difference between the rate of growth of demand for chemicals and GDP) gradually declined and eventually disappeared. At the same time there was a decline in the overall economic growth rate and an amplification of cyclical movements.

In these new circumstances, the chemical industry entered a state of maturity, characterized by slower product innovation. This loss of momentum places a question mark over the capacity of large parts of the industry to retain their frontier roles and to raise further the share of high unit value, speciality/fine chemicals relative to large-volume, low unit-value commod-

Table 9.2   Chemical industry research intensity in several industrialized countries

	Chemical industry R&D expenditure as a % of chemical industry value added			*Chemical industy R&D expenditure as a % of manufacturing industry R&D*		
	1990	*1999*	*Difference 1990–9*	*1990*	*2000*	*Difference 1990–2000*
Japan	14.2	15.2	1.0	9.7	8.5	−1.2
Germany	12.6	14.3	1.7	15.1	10.9	−4.2
France	10.3	7.2	−3.1	9.3	6.1	−3.2
USA	8.5	6.6	−1.9	8.7	5.9	−2.8
UK	8.5	6.6	−1.9	6.4	4.2	−2.2
Italy	4.6	3.2	−1.4	6.4	5.6	−0.8

Source: Federchimica (2004).

ity chemicals. These concerns get some support from evidence suggesting a worrying decline in research intensity outside of the pharmaceutical industry (Table 9.2). In the period 1990–2000, in all countries the share of chemical industry R&D in the total manufacturing R&D fell, while in 1990–9 in a number of countries chemical industry R&D fell as a share of value added.

The response of western chemical companies was twofold. The first was the quest for scale economies and increased market share. The aim was to reduce costs and increase profit rates. Especially in heavy chemicals manufacture, however, investment in new capacity and increased production in economic upturns resulted at a global scale in overcapacity and cyclical profitability crises with phases of expansion and accumulation followed by periods of restructuring and down-sizing (Chapman, 1991; Financial Times, 1997).

The technical characteristics of the petrochemical industry[1] encouraged large oil companies such as Standard Oil and Shell to assume a dominant role. At first the movement of oil companies into petrochemicals was part of a strategy of diversification. In the 1980s these companies increased their shares of petrochemical capacity at the expense of chemical companies (Chapman, 1991). To defend their market position, chemical companies opted for backward integration into oil refining and joint ventures with oil companies. ICI, Union Carbide, BASF and others were made into vertically integrated companies, characterized by complete production cycles, and the maintenance of a range of in-house support services.

The second was diversification. Chemical companies entered a variety of chemical businesses, from industrial to speciality and fine chemicals, where the presence of the latter had an anticyclical function, allowing the companies to make profits in troughs in the petrochemical cycle.

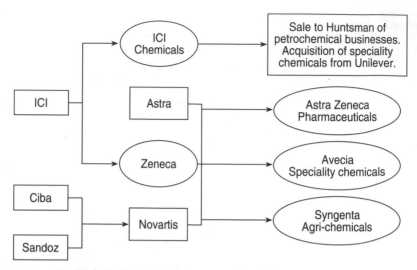

Figure 9.4   The restructuring of ICI.

After the mid-1980s the pressure to upgrade increased. Chemical restructuring intensified, although it did not follow uniform patterns and, in many cases, there were high degrees of continuity in production programmes and processes (see Bathelt, 2000). Especially radical were the strategies of companies such as the British ICI[2] and the German Hoechst,[3] that abandoned the production of industrial chemicals to seek worldwide competitive advantage in less volatile, higher value added markets (Figure 9.4). Conversely, Du Pont and BASF maintained but strengthened their core activities, sold noncore activities (focus as opposed to diversification) and sought to increase their market shares in the sectors in which they were market leaders. The watchwords that dominated boardrooms in the late 1980s and 1990s were 'specialisation and disinvestment'. This new corporate convention underpinned a season of large-scale, selective, sell-offs. In addition, it gave rise to asset swaps and mergers designed to reduce market competition (Chapman and Edmond, 2000; Greco, 2002). A case in hand was that of Basell (Figure 9.5).

In the vast majority of cases, rationalisation involved a move towards speciality and fine chemicals, where economies of scale and company size are less important than in the development of industrial chemicals; petrochemical production remained primarily an activity of oil companies. As a consequence, in the last decade some of the major chemical corporations have disappeared, while new groups and activities have emerged from the reconfiguration of old ones. Table 9.3 shows the changes that occurred at

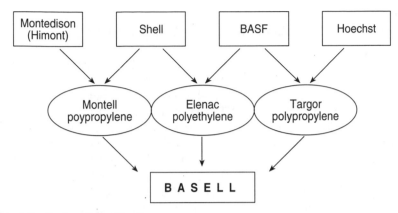

Figure 9.5 The formation of Basell.

Table 9.3   The main chemical companies in the world

	Turnover 2001 ($ billion)		Turnover 1992 ($ billion)
Merck	47.7	Hoechst	29.4
BASF	29.7	BASF	28.5
Glaxosmithkline	29.6	Bayer	26.4
Dow	27.8	DuPont	21.7
Bayer	27.7	ICI	21.3
Pfizer	26.9	Dow	19.0
DuPont	24.7	Ciba-Geigy	15.8
Aventis	21.0	Rhone-Poulenc	15.4
ExxonMobil	19.3	Elf Aquitaine	13.2
Novartis	19.0	Sandoz	10.3
Atofina	17.9	Shell	10.1
Roche	17.2	Merck & Co	9.7
AstraZeneca	16.5	Akzo	9.4
Mitsubishi Chemicals	14.7	Mitsubishi Kasei	9.3
Shell	13.8	Roche	9.2
Akzo Nobel	12.9	Exxon	9.1
L'Oreal	12.6	EniChem	9.1

Source: Federchimica, 2003a.

an international scale. As we shall show, this restructuring of industrial activities involved not only changes in production but also in the geography and economy of places.

History and Characteristics of the Italian Chemical Industry

As in the case of their rivals, in the 1980s and 1990s, large Italian companies were profoundly restructured in ways that had important implications for the regional economies in which they were located. At that time three major chemical groups (SNIA, Montedison and ENI) accounted for one-half of national chemical production (Federchimica, 2001).

The dramatic industrial events involved in the transformation of the Italian chemical industry in those two decades cannot be detached from an account of the political events that profoundly affected the industry. The peculiarity of Italian chemical manufacturing compared with its European rivals lay in the politically-driven character of its development. As Zamagni (2003:16, our translation) argued: 'it [the story of the Italian chemical industry] amounts to an account of the ambitions of the Italian entrepreneurial class and political elites of the years from the 1950s to the 1980s to be a part of world development trends, such as those involving petrochemicals, without a clear conception of the instruments with which they might successfully achieve that objective'. As indicated in Chapter 5, post-war governments accepted the view that economic modernization required the development of strong intermediate goods industries (Castronovo, 1995). In the late 1950s these goals were coupled with the objective of industrial development in southern Italy. The chemical industry was accordingly seen as an instrument of national economic growth, and a development opportunity for the south. In the words of Enrico Mattei (speech delivered on 4/02/1962, cited in Perrone, 2001, our translation): 'the modernization of the economic and social structures of regions abandoned to secular stagnation will receive a decisive push from these activities.'

At the start of the 1900s, there were two major chemical companies, SNIA and Montecatini. Established as a shipping and shipbuilding concern, SNIA later entered the chemical and man-made fibres businesses. By 1968 it dominated the man-made fibres market, accounting for 30 per cent of domestic sales, while in the 1970s it experienced its greatest expansion. Montecatini was originally established to extract copper from the hills of Toscana, moving later into the production of fertilizers and, after WWI, of ammonia. Montecatini encountered industrial and financial problems during the Great Depression of the 1930s. After WWII its situation worsened. According to Amatori and Brioschi (1997) Montecatini was reluctant to adopt the emerging petrochemical technology paradigm of the time, despite the fact that it had built Italy's first petrochemical plant in Ferrara in 1950. The main obstacle to effective modernization was its conglomerate structure.

In the 1950s, Montecatini's market leadership was challenged by Edison and by ENI and ANIC. Edison was an electricity company that entered the chemical sector with the financial resources it acquired from electricity nationalisation in 1962. ANIC was created in 1936, and started fertilizer and subsequently petrochemicals production at Ravenna in the second half of the 1950s. Other petrochemical plants followed at Gela (Sicilia) in 1962, Pisticci (Basilicata) in 1963, and Sarroch (Sardegna) in the late 1960s. ANIC's entry into the fertilizers market, which Edison had also entered, resulted in an immediate reduction of 15 per cent in the prices of the main fertilizers (Zamagni, 2003). In 1966, Montecatini and Edison merged to form Montecatini Edison (shortened to Montedison in 1969).

The new company was created without a clear market strategy from two companies with financial and production problems (Barca, 1997). At the root of these problems were excessive diversification (Montedison was a conglomerate with 173 plants of which just 44.5 per cent were chemical plants) and harsh internal competition that prevented it from exploiting the opportunities offered by the rapid growth of world petrochemical production. In the 1970s, Montedison's CEO, Eugenio Cefis, sought to deal with these problems that had plagued the company. A partial rationalisation of the company had, however, to await the 1984 implementation of the plan of another manager, Mario Schimberni.

Alongside these groups there were two other private companies that relied heavily on public money: SIR and Liquichimica. SIR was founded in 1931 to manufacture bakelite. In the 1950s, it entered the polyester and polystyrene markets, and in 1962 with the help of government subsidies it entered the petrochemicals sector at Porto Torres (Sardegna). SIR was the first Italian producer to use oil as a raw material. In the late 1960s, it diversified into man-made fibres. Liquichimica manufactured caustic soda. As in the case of SIR it launched plans to invest in the south (at Augusta in Sicilia and Ottana in Sardegna) although at a point in the 1970s when market growth was slackening.

The competitive process of chemical expansion involving these private and public companies was subsequently referred to as the chemical war (Table 9.4 and Dunford, 1988:185–98). Autarchic corporate cultures, intense rivalry and an absence of industry-level planning resulted in strong vertical integration and anarchic competition that delivered a duplication and lack of coordination of investments. For instance, Montecatini's decision to construct the huge Brindisi plant (covering 5 square kilometres with 45 kilometres of internal roads, 20 kilometres of internal railway lines and a dock pier capable of receiving medium-size ships) was a response to ANIC's construction of its fertilizer and petrochemical plant in Ravenna. At the same time, state action did not entirely help. On the one hand, the system of incentives for southern development had perverse effects. Although partially tailored to

Table 9.4   The main Italian chemical groups in 1971–2

| | Capacity (1972, %) | | | | | |
	Montedison	ANIC	SIR-Rumianca	SNIA	Others	Total ('000 tonnes)
Organic chemicals	54	20	16	—	10	7,179
Inorganic chemicals	53	15	8	1	23	9,432
Plastic materials	54	12	18	—	16	2,245
Man-made fibres	38	9	3	38	12	653
Turnover (1971, LIT billion)	2,023	226	222	368	—	—
Employment (1971)	173,500	15,791	9,892	43,196	—	—

Source: Zamagni, 2003.

serve the investment strategies of chemical groups, the incentives encouraged a proliferation of plants, the choice of less than optimal scale plants and a splitting up of investments in the case of companies such as SIR that were seeking to maximize the level of financial support. (The owner first of Liqui-chimica and later of SIR and some of his associates were subsequently accused of fraud against the Italian state). On the other, state attempts to put the sector in order and to get not just public but also private companies to implement government industrial decisions in the 'name of the superior interest of the nation' (Zamagni, 2003:19) were painful failures. More generally, state aids and incentives were seen by critics as contributing to a culture of corporate unaccountability and political interference. Trigilia (1992), for example, argued that the influence of political parties on management undermined efficiency and created a situation in which state holding companies came to symbolize economic inefficiency and political productivity (as the state sought to grapple with the crises of southern industrial establishments, the chemical industry, political difficulties and social conflict).

The chemical war itself had several important consequences. First, the struggle for market share and increased profits was confined to the creation of new capacity in large plants. Second, there was an associated concentration on petrochemical production despite the fact that secondary and speciality chemicals could provide higher returns in a situation of steady prices. Third, petrochemical production technologies were relatively mature and replicable in emerging economies capable of contesting the markets of Italian producers. Finally, in order to take advantage of state incentives for the industrial development of the south many of these investments were, as indicated earlier, located in the Mezzogiorno. As Ranci and Vaccà (1979) argued, at the end of the 1970s the Italian chemical industry was suffering from the consequences of a model of quantitative development centred on the growth of inputs and outputs and the use of constant-returns-to-scale technologies. Insufficient research and insufficient management capacity resulted in insufficient process innovation, product development and product differentiation (La Malfa and Coppola, 1974; Zamagni, 2003). A strong domestic market orientation and the location of plants in places that were remote from European markets helped limit Italy's share of international markets. Capacity created by separate, warring companies was dispersed. According to Chemical Week (1981:31) 'for a decade, the Italian government . . . has subsidized many of its [chemical industry] projects. World scale plants were built, but they lacked the necessary worldwide sales and distribution systems. The result has been rampant overcapacity and a constant flow of red ink.' In fact the rate of plant utilisation for ethylene production in Italy decreased from 83 per cent in 1976 to 33 per cent in 1982. In the same period the comparable figures were 81 per cent and 69 per cent for Germany, and 74 per cent and 62 per cent for France (Giannetti, 1996).

Trajectories of Restructuring

1979 saw a sharp downturn in international petrochemical profitability. Italian companies had already failed to modify their business strategies (Giannetti, 1996). Many managers considered the 1970s crisis conjunctural rather than structural and were in any case locked into earlier investment

Table 9.5 The performance of the main international chemical groups, 1980–1

	Income before tax/ turnover (%)	Income after tax/ turnover (%)	Interest and financial expenses/gross operating profit (%)	Interest and financial expenses/ turnover (%)	R&D/ turnover (%)
Montedison	−6.6	−6.3	154.9	11.3	2.2
ANIC	—	−17.2	101.3[1]	5.1[1]	1.0
Rhone-Poulenc	−3.4	−3.8	79.8	5.5	6.1
Atochimie	−8.4	−1	45.6	2.2	1.7
BASF	4.3	1.2	—	—	3.5
Bayer	4.5	1.8	40.6	4.2	8.2
Hoechst	4.2	1.3	29.2	3.2	5.4
ICI	5.0	1.2	18.3	2.0	3.6
DuPont	8.9	5.1	9.8	1.6	3.2
Monsanto	6.6	4.3	n.a.	0.8	10.7
Dow Chemical	8.9	6.3	20.6	3.4	3.3

Source: Federchimica, 1991.
Note
[1]1980

Table 9.6   Net operating income of the main Italian chemical companies during the 1980s ($ million)

	1982	1983	1984	1985	1986	1987	1988	1989
Montedison	−717	−315	−26	243	464	527	490	362
ENIChem	—	−651	−159	−387	—	131	496	—

Source: adapted from Federchimica, 1991.

decisions made without a prior reorganisation of the sector. As a result, Italian capacity continued to rise as international demand grew more slowly.

In 1978, Liquichimica collapsed. Next to face insolvency was SIR. Montedison was making major losses, while ANIC, although less affected, was also in a difficult position. In the absence of restructuring, the situation of Italian companies was substantially worse than that of their competitors, especially in terms of their degree of indebtedness and their research intensity (Table 9.5). To start to deal with this situation Law 784/1980 opened the way to the creation of a duopoly comprising a public pole around ENI, made up of ANIC and the surviving SIR-Rumianca and Liquichimica plants, and a private pole around Montedison. In 1981, ENI created a new company called ENIChimica to manage its petrochemical interests, a name changed to ENIChem in 1985. At the same time, ENI indirectly controlled 50 per cent of Montedison shares.[4] The next step was the only partially successful 1982 National Chemical Plan developed when Gianni De Michelis was Minister for the PP.SS. To rationalize chemical manufacturing, ENI and Montedison were encouraged to swap assets and to specialize in their core activities. As a result of the subsequent 1983 ENI-Montedison agreement, ENI consolidated its presence in primary chemicals (ethylene, polyethylene and polyvinylchloride), plastic materials and elastomers. Montedison retained polypropylene and polystyrene and concentrated its efforts on light chemicals and pharmaceuticals. This rationalisation of industrial production did not involve, however, a rationalisation of plants, which remained small, scattered and insufficiently connected. Once market conditions improved in the mid-1980s, Italian companies started to make profits again (Table 9.6).

Montedison

At the root of Montedison's problems lay its conglomerate nature. In the first half of the 1980s, after privatization, and under Schimberni's chairpersonship, a complete reorganisation was pushed through. 80 per cent of the top managers were replaced by international managers. Montedison integrated its polypropylene interests with Hercules of the US in the

Himont joint venture. Exploiting Montedison's new Spheripol technology, Himont quickly secured over 20 per cent of the world market. Montedison failed, however, to reorganize its chemical activities sufficiently. Instead an attempt was made to deal with its financial difficulties through an 'aggressive policy of growth' (Amatori and Brioschi, 1997: 141) and through financial mechanisms. Schimberni also conceived the idea of getting away from Mediobanca, which controlled Montedison through Gemima, and establishing Montedison as a public company (Zamagni, 2003), but was prevented by Raul Gardini after the Ferruzzi family takeover.

The Ferruzzi group took over in 1986. In 1987, Schimberni left. As new head of the group, Gardini reorganized Montedison around three poles: chemicals, pharmaceuticals and energy. The company's financial position remained, however, very weak. To avoid bankruptcy, a new 1988 joint venture agreement was made with ENI. At first called ENIMont, the new joint venture was renamed ENIChem in 1991. Montedison and ENIChimica/ENIChem had 40 per cent of the shares each. The remaining 20 per cent were sold on the market. For the first time, a single company controlled the whole of the petrochemical sector plus fertilizers, plastics and elastomers. The new company, which did not include Montedison's pharmaceutical and polypropylene interests, was one of the ten largest chemical companies in the world, although it inherited large debts and Italy's productive weaknesses (dispersed, underdimensioned plants, insufficient research, limited internationalisation and weak links with downstream activities). Conflicts between the managements of the two parent companies quickly emerged. In 1990, ENI assumed complete control of ENIMont, with Montedison receiving a payment of LIT 2,805 billion which was considered to exceed by a substantial margin the value of its interests. As for Montedison, in the 1990s it withdrew from the chemical sector to pursue a role as an energy company. The particularly successful Himont polypropylene operation was transferred first to a joint venture with Shell, called Montell, and subsequently to the Shell and BASF-owned Basell.

SNIA

SNIA's upgrading strategy involved specializing in high added value activities. In the 1980s, SNIA sought to diversify. Most important was the acquisition of Sorin Biomedica, as it permitted the SNIA Group to enter the medical technology sector. A number of 1990s' market operations sought to strengthen SNIA's position in this market. In 1992, it acquired the cardiovascular division of the US Pfizer Group making it the European and joint world leader. In 1997, it sold its medical diagnostics business, concentrating its resources on the development of its cardiovascular and haemodialysis operations. The acquisition of US Cobe Cardiovascular and of the French ELA Medical further enhanced the group's competitive advantage. In 2002, medical technologies represented 75.8 per cent of Group net revenues of € 886.1 million.

The share of revenues from chemicals was in decline (21.2 per cent in 2002). In the early 1990s, SNIA entered into the Novaceta (cellulose acetate filaments) and Nylstar (polyamide filaments) joint ventures with Courtaulds and Rhône-Poulenc. Other joint-ventures were set up for technopolymers (NYLTEC) and fibres for carpets (NOVALIS). In 2001, SNIA divested its agrochemicals operations to Isagro to strengthen its medical technology activities.

2002 employment reflects SNIA's specialisation: 84 per cent of its worldwide workforce of 5,043 people were employed in the medical technology sector. 687 were employed in the chemical sector. About 8 per cent of the total workforce are involved in research and development.

In 2003, SNIA demerged Sorin because of the different strategies pursued by the two businesses. SNIA retained its speciality chemicals and man-made fibres activities. Sorin took over all medical technology operations and managed to attract foreign attention making it 'an atypical Italian success' in cutting-edge technologies (*The Economist*, 2004).

Re-engineering a corporation: ENIChem's complex restructuring

With the 1998 withdrawal of Montedison, ENIChem was left with the task of putting the whole of the Italian petrochemical industry in order. The task was not an easy one: ENIChem inherited an extremely difficult financial, industrial and social situation, a result of decades of mismanagement. The new company could rely on some strong businesses but it also had 40 production sites, some of which were technologically backward and only partially exploited. To deal with this situation ENIChem embarked first of all on a period of intense rationalisation. A decision was made to concentrate on primary petrochemicals, polymers and elastomers and to sell off its least profitable and nonstrategic activities (man-made fibres and agricultural chemicals).

The degree of down-sizing was dramatic. Employment fell from nearly 55,000 at the time of creation of ENIMont to some 21,000 in 1995 and continued its downward path to stand at just over 7,000 in 2003 (Table 9.7). After a number of years, however, it reaped its reward for the company, in the shape of the restoration of profitability, and a reduction in indebtedness, although an important role was also played by improved

Table 9.7 The performance of ENIMont, ENIChem and ENI petrochemical interests, 1988–2003

	Net sales revenue	Total net profit/loss	Net operating profit/loss	Net operating profit/net sales revenue (%)	Medium-long term financial indebtedness	Net financial indebtedness	Capital expenditure/ ENI capital expenditure (%)	Employees
LIT billion								
1988	14,662	817	1,843	12.6	9,606	—	—	54,715
1989	15,347	716	1,665	10.8	6,950	6,186	23.4	52,656
1990	15,060	−88	743	4.9	9,362	8,341	24.9	49,000
1991	13,424	−722	77	0.6	9,506	7,004	16.6	37,017
1992	10,964	−1,542	−308	−2.8	9,723	7,391	12.2	32,963
1993	10,654	−2,668	−819	−7.7	9,725	8,356	6.2	28,913
1994	11,556	−840	399	3.5	5,363	4,849	3.7	23,501
1995	13,942	1,095	2,116	15.2	2,957	1,735	3.4	21,358
€ million								
1995	7,200	549	1,093	15.2	1,527	896	3.6	21,358
1996	5,282	117	127	2.4	1,367	529	4.2	17,131
1997	4,986	124	187	3.8	1,375	492	4.3	16,839
1998	4,048	−229	−1	−0.0	1,251	589	6.5	15,316
1999	4,096	−611	−362	−8.8	—	838	5.5	14,567
2000	6,018	1	18	0.3	—	832	4.7	13,343
2001	4,761	—	−332	−7.0	—	—	5.5	11,022
2001[1]	5,108	—	−415	−8.1	—	—	5.9	12,479
2002	4,781	—	−347	−7.3	—	—	3.1	11,691
2002[2]	4,516	—	−126	−2.8	—	—	1.8	7,258
2003	4,487	—	−176	−3.9	—	—	1.6	7,050

Source: Zamagni, 2003 ; ENIChem, 1999; and ENI, various years.
Notes
[1] 2001 including a full consolidation of Polimeri Europa
[2] 2002 after reclassification of some activities

economic circumstances: in particular, the 1994 and 1995 results reflected a rise in exports above all towards China, a rise in product prices and the devaluation of the Italian currency (ENIChem,1995).

Alongside this strategy of rationalisation, ENIChem pursued a number of other upgrading paths. Capital investment increased strongly in 1989–92, as the company pursued a strategy of technological upgrading and work reorganisation.

On the technological front, ENIChem adopted second-generation technologies to improve plant performance. For instance, at the Brindisi plant, production was concentrated in one plant rather than in three. Commencing in 1993, annual output doubled to reach 400,000 tonnes in less than a decade, while employment dropped from 150 to 70, energy costs declined by 50 per cent and maintenance costs fell by 25 per cent (Interview data, May 2000; Greco, 2002).

As far as work organisation was concerned, two sets of changes were implemented in order to improve efficiency and reduce cost. First, ENIChem decided to deploy its core workforce on core activities and to contract out a number of in-house support services. Maintenance in particular was contracted out, although, in contrast to other companies such as ICI, ENIChem decided to retain most of the specialized activities. Second, ENIChem adopted new management models, automated some jobs and placed greater reliance on multiskilling.

A final component of ENIChem's upgrading strategy was the pursuit of a number of strategic joint ventures with foreign companies to exploit company-based competences and to reduce competition in certain segments of the market. The most important were EVC, the result of an agreement with ICI for the production of vinyls, and Polimeri Europa, established in 1995 as a result of a joint venture with Union Carbide for the production of polyethylene and olefins.

In 2001, ENIChem concluded an asset swap with Dow Chemical, exchanging its polyurethane activities for the 50 per cent of Polimeri Europa that Dow had acquired as a result of its takeover of Union Carbide. After acquiring 100 per cent control of Polimeri Europa, in January 2002, ENIChem transferred all of its viable chemical activities (olefins and aromatics, intermediate products, styrenes and elastomers) to it. In 2002, Polimeri Europa with its 11 production sites in Italy and 8 in Europe (Table 9.8), 7,200 staff and a turnover of € 4,516 million was all that remained of Italy's large-scale chemical industry (Table 9.9). How much longer it will survive is in question, as its creation was a step marking the desire of ENI to withdraw from chemical manufacture (which was loss-making again from 1998) and to return to its core vocation as an energy company (Greco, 2004). The ENI Group was privatized in 1995–8 when the Italian government sold most of its stake in four tranches of some 15 per cent each, and

Table 9.8   Polimeri Europa's production sites

	Site capacity ('000 tonnes per year)			
Production site	Basic chemicals	Polyethylene	Elastomers	Styrenics
Italy				
Brindisi (Puglia)	815	550	—	—
Ferrara (Emilia-Romagna)	—	110	85	—
Gela (Sicilia)	410	230	—	—
Mantova (Lombardia)	780	—	—	895
Porto Marghera (Veneto)	920	—	—	—
Priolo (Sicilia)	2,495	200	—	—
Porto Torres (Sardegna)	1,337	140	33	—
Ragusa (Sicilia)	—	170	—	—
Ravenna (Emilia-Romagna)	150	—	350	45
Sarroch (Sardegna)	488	—	—	—
Settimo Milanese (Lombardia)	—	—	—	40
Europe				
Feluy (Belgium)	—	—	—	195
Champagnier, Dunkirk (France)	590	340	40	—
Grangemouth, Hythe (UK)	—	—	225	60
Oberhausen (Germany)	—	140	—	—
Neiva (Portugal)	—	—	—	5
Szazhalombetta (Hungary)	—	—	—	115
USA	—	—	—	50
TOTAL	7,985	1,880	733	1,405

Table 9.9   The sales of the main Italian chemical companies, 2001

Company	Net sales revenue (€ millions)
EniChem	4,761
Polimeri Europa	1,455
Gruppo Snia[1]	1,280
Radici Group	1,130
Gruppo Mossi and Ghisolfi	948
Gruppo Bracco[2]	871
Mapei	730
Montefibre	530
Aquafil	383
Gruppo Colorobbia	338

Source: Federchimica, 2003a.
Notes
[1]Includes medical technology sales
[2]Includes biomedical sales net of pharmaceuticals

sees its future as lying in: (1) oil and natural gas exploration and production; (2) natural gas supply, transmission, distribution and sale; (3) electricity generation; (4) the refining and marketing of petroleum products; and (5) oil-field services and engineering.

One of Polimeri Europa's major achievements was the modernization of the Brindisi plant. A recent investment of € 200 million was added to the € 400 million spent in 1992–4 to put in place Union Carbide's celebrated Unipol PE technology for manufacturing polyethylene. The gas-phase plant in Brindisi combines high-pressure solution and slurry polyethylene technology to maximize the effectiveness and versatility of production for different commercial uses, while respecting environmental standards. The new process has improved the quality of output, ensures batch-to-batch consistency and permits major efficiency savings (Provincia di Brindisi, 1999). This attempt to enhance its competitiveness is however weakened by the peripherality of the Brindisi site. As Federchimica (2001) has indicated, production sites are progressively moving eastwards and northwards, towards the UK, Germany and Holland, and towards places in the vicinity of oil pipelines and platforms.

The Role of SMEs

As large chemical companies have declined in number and employment share, SMEs have increased in importance (Table 9.10), although in chemicals they generally tend to be larger than in other sectors. In 2001, nearly 94 per cent of chemical and man-made fibre companies had fewer than 100 employees. These SMEs accounted for nearly 31 per cent of chemical workers compared with 22 per cent in 1981. Conversely, companies with more than 200 employees accounted for 3.2 per cent of the 2001 total but for 57.6 per cent of employment. In pharmaceuticals the corresponding large-firm employment share was much higher (76.3 per cent) (ISTAT, 2004).

Although their employment share remains small, SMEs make a substantial contribution to Italian chemical industry turnover and exports. Chemical sector SMEs also make an important contribution to the specialisation, innovativeness and capacity to adapt to clients' demands of a number of industrial districts and Made-in-Italy sectors. Some of the districts in which chemical production is especially important are listed in Table 9.11.

More generally, SMEs play an important role in the development of innovations and prototypes and in fine, speciality and niche markets, although they also suffer from a number of disadvantages especially in Italy where SMEs tend to be smaller than their EU15 counterparts. Included are: access to finance; the high costs of research and development,

Table 9.10 Size distribution of Italian chemical firms, 2001

	Firms			Employment		
	1981	1991	2001	1981	1991	2001
Chemicals and man-made fibres						
1–9	3,790	3,382	3,783	14,107	12,204	12,244
10–19	1,011	819	844	13,566	11,255	11,552
20–49	663	604	633	20,302	18,969	19,560
50–99	311	245	286	21,332	17,144	20,183
100–99	208	195	169	28,883	27,594	24,050
200–499	132	131	117	41,809	40,835	34,055
500 and more	101	79	72	169,722	119,214	85,075
Total	6,216	5,455	5,904	309,721	247,215	206,719
Pharmaceuticals						
1–9	153	171	269	740	738	798
10–19	75	74	66	1,040	1,049	887
20–49	97	92	98	3,170	2,920	3,255
50–99	69	57	65	4,779	3,974	4,710
100–99	43	65	49	6,052	9,375	7,047
200–499	50	46	38	16,006	14,382	11,264
500 and more	32	38	35	36,917	43,747	42,395
Total	519	543	620	68,704	76,185	70,356

Source: elaborated from ISTAT, 2004.

Table 9.11 High chemical-using industrial districts

Industrial districts	Sector
Biella and Prato	Textiles
Como	Silk
Castel Goffredo	Womens' stockings
Cadore	Spectacle frames
Pesaro	Furniture
Montebelluna	Sports shoes
Bologna	Packaging
Fermo	Footwear
Bari	Furniture – sofas
Rossano Veneto	Saddles for bicycles

Source: adapted from Federchimica, 2001.

immaterial investment and marketing required to avoid the trap of produ-cing standardized commodity-type chemicals; the difficulties of inter-nationalizing; the threat from informal microenterprises that avoid the costs of regulation;[5] and the tendency of large specialized groups to enter profitable fine and speciality chemical markets, perhaps through acquisition. Federchimica (2001) has, for example, noted a tendency for foreign firms to acquire chemical sector SMEs that were growing in size.

Another Aspect of the New International Division of Labour: Foreign Companies In Italy

The restructuring of domestic and international chemical companies has altered well-established relations of production. As a result of the acquisition of Italian SMEs and the divested activities of large Italian companies, the presence of foreign companies in Italy has increased. On the basis of data from the Reprint database, Federchimica (2001) identified 214 foreign companies, employing almost 50,000 workers. These chemical sector companies accounted for 10 per cent of total manufacturing employment in foreign companies, and for nearly one-third of chemical sector employment. Foreign chemical companies suffered less from the fall of employment in the sector than indigenous companies (−7 per cent in 1995–2000 compared with −20 per cent). Size-wise these foreign companies included large groups (30 per cent of total turnover is generated by the 10 largest companies), but many were SMEs. 63 per cent of these foreign companies were European, although American and German companies accounted for 50 per cent of employment (Federchimica, 2001).

The foreign sector has grown largely through acquisitions. Only 4.4 per cent of employees work in new, green-field plants. In many cases, however, acquisitions were essential for companies seeking to develop their existing activities, and to enhance their export orientation. A Federchimica survey of foreign companies found that production activities were more important than commercial activities. In 1995–2000 production in Italy by foreign or foreign controlled companies increased by 26.4 per cent (the national average was one-half of this figure) to reach around 35 per cent of the total. 36 per cent of their gross output was for export. Also highlighted in the report was evidence indicating that what attracted inward investment to Italy was Italy's role as a trend-setter market, and, especially, the presence of competitive SMEs as clients and/or as quality suppliers.

The promotion of SMEs and the attraction of inward investment are important aspects of an industrial strategy, designed by the Ministry of Industry in 2000, and implemented by the 'Osservatorio per il settore chimico' (established in 1997). Although the Observatory has acted as a mediator, as, for example in the case of the Accordo di Programma for Porto Marghera, its main responsibility is the environmentally-sustainable regeneration of existing chemical poles. To achieve these ends the Observatory opted for a territorial approach (Figure 9.6 which plots the provinces

Figure 9.6 Geography of Osservatorio per il Settore Chimico local observatories.
Source: adapted from Osservatorio per il Settore Chimico, 2004.

in which local observatories were established). The aim was to draw upon local conditions and productive traditions, to mobilize local actors and policies and to coordinate them with national actions. A specific programme entitled 'Chimica 2000' involved, for example, 13 provinces with substantial chemical capacity, and was designed to map territorial investment opportunities and to encourage inward investment. A second project aimed to make environmental audits easier. A third sought to diffuse innovations to SMEs through seminars and presentations. Indeed, for the Ministry of Industry, research and innovation are the policy priorities for these chemical poles. Accordingly, two laws have been proposed. The first envisages financial incentives for expenditure on R&D. The aim is to encourage companies to develop technologies and patent them rather than purchase licenses. The second involves incentives for the improvement of human capital and for encouraging the exchange of researchers between private and public institutions. Also related to innovation capacity are measures to encourage the internationalization of Italian chemical companies, although these measures aim also to deal with a lack of information about external opportunities and a lack of visibility.

Experiences and Regional Impacts of Restructuring: The Disengagement of the Chemical Industry in Puglia

In the context of programmes for southern industrialization, the Puglia region received LIT 1,258 billion from the state for investment in 1950–73. Almost 60 per cent of this total was devoted to the development of the province of Taranto and its new state-owned steel complex. The province of Brindisi received LIT 171 billion (almost 14 per cent of the total). 93 per cent of resources were allocated to the chemical sector and primarily to the Montecatini plant (Mele, 1975). ENIChem later established a plant at Manfredonia (Foggia). Manfredonia was less important than Brindisi, although, as an area dependent on a single plant, it suffered when the plant was closed. At present, Manfredonia is the subject of 'Contratto d'Area', with funds for regeneration and for new initiatives, especially in cooperation with entrepreneurs from the Veneto region.

At the start of the 1960s, Brindisi was chosen as a 'growth pole'. Although justified in terms of expectations concerning the expansion of Mediterranean markets, the selection of Brindisi as a chemical manufacturing location was a result of the competition between Montecatini and ANIC. The Brindisi plant could refine one million tonnes of oil per annum, and the first cracker produced 160,000 tonnes of ethylene a year. Other products were chlorine, bromine, polymers, PVC and aromatics. Early 1960's employment stood at more than 1,000 workers. In 1965, this figure reached more than 5,000. From the very beginning, however, the plant's suboptimal production arrangements were having disastrous effects on Montecatini's budget (Amatori and Brioschi, 1997). The production of ethylene was, for example, spread over three plants rather than concentrated in one (Interview data, May 2000). To raise additional capital, and to increase capacity to meet booming demand, in 1964 Montecatini entered a joint venture with Shell (Table 9.12). New Dutch managers were critical of the productive arrangements and the 'excessive' size of the workforce employed at the Brindisi plant. Their proposed production and work rationalization project was turned down by Montecatini, and in 1966 Shell withdrew.

In the 1970s and 1980s, the petrochemical complex faced falling demand, rising costs and increased competitive pressure. In 1975, the 'topping' plant, in which crude oil was processed to obtain virgin naphtha and diesel oil, was halted due to technical problems. Then, in 1977, the P2T ethylene cracker blew up. A smaller cracker capable of producing 80,000 tonnes of ethylene per annum survived. Without the topping plant and the main cracker, however, the few remaining active production lines

Table 9.12 Changing ownership of the Brindisi petrochemical plant

Year	Owner
1929	Montecatini (fertilisers)
1959	Montecatini (petrochemical production)
1964	Monteshell (Montecatini-Shell joint venture)
1965	Montedison (Montecatini-Edison merger)
1966	Shell leaves the company
1987	ENIChem-ANIC (EniChem-ANIC 75%, Montedison 15%, Himont, 10%)
1988	ENIMont (ENI 40%, Montedison 40%, market 20% joint venture)
1990	ENI purchases all the company's shares
2002	ENIChem passes all its activities to Polimeri Europa

(the VCM plant) required polyethylene, butadiene and ethylene imports. Not surprisingly, the viability of the plant was seriously compromized: reduced production brought a large rise in unit costs, and fixed costs reached 28–32 per cent of annual turnover compared with less than 10 per cent in normal times. Costs per unit of product exceeded revenues by up to 20 per cent (Greco, 2002). Montedison decided, therefore, to close the plant in spite of the contrary indications of the prevailing national Chemical Plan. In 1978, the first 800 redundancies were implemented. Massive union-led demonstrations in Rome forced the government to intervene. Montedison was required to revise its decision, and to negotiate with the unions on a basis that ruled out plant closure. The result was a plan to rebuild the ethylene plant, invest in new technology, downsize the workforce and reorganize production with the help of public money (*Chemical Week*, 1981), Another 700 out of a total of roughly 4,000 workers were made redundant in 1983–4. The wages of the laid off workers were guaranteed by the extensive use of CIG, and special laws were approved to allow the early termination of employment. Workers with families and children were given preference, when deciding who should keep their jobs. The negotiated solution to this crisis did not, however, conclude the eventful history of the plant. At the end of the 1980s ENIChem assumed control the plant, whose manufacturing activities are summarized in Table 9.13.

In the 1990s, three important companies operated in Brindisi: Polimeri Europa which assumed control of ENIChem's viable assets; Basell which produced polypropylene; and EVC. EVC closed in December 1999, but later became PowerCo for the production of VCM and MDI. At present, after recent ownership changes, the chemical industry in Brindisi consists

Table 9.13   ENIChem's production in Brindisi in the 1990s

Activities	Activity description and main products	Examples of application	EniChem's position
	Olefins and aromatics (ethylene, propylene, butadiene, and benzene)	Intermediate products for the plastics industry, solvents, fine chemicals, tyres and antifreeze products	European leader for ethylene, second producer of benzene, and third producer of xylene
Petrochemicals	Intermediates (phenol)		Second European producer of phenol
	Intermediates for polyurethanes (chlorine, caustic soda, MDI and TDI)	Vehicle components, furniture, footwear, construction materials and decorating	Third European producer of TDI
Polymers	Styrene	Industrial and food packaging, vehicle components, and semi-conductor and micro-electronic packaging	Second European producer of styrene
Elastomers		Tyres, footwear, adhesives and building products	European leader in SBR and BR tyres

of large companies (Figure 9.7) such as Polimeri Europa, Dow Chemical, Basell, Chem-Gas and Aventis Bulk (which resulted from a merger involving Hoechst and Rhône-Poulenc) along with EniPower which provides electrical power, Exxon Mobil and other locally-owned companies, some of which received funds from the Territorial Pact (Table 9.14 which lists the main chemical companies operating in Brindisi and some locally-owned companies that received funds under the Territorial Pact).

The restructuring of the Brindisi chemical sector has had significant implications for its socioeconomic fabric. Over the period 1981–91 nearly 40 per cent of jobs in refining, chemicals and plastics were lost (Greco, 2004). In 1991–6 a further 32 per cent were lost compared with 15 per cent in manufacturing industry as a whole.

As early as the first employment crises, many observers noted that the model of industrialization pursued in Brindisi was flawed as a model of development. The creation of a vertically integrated and capital-intensive

Table 9.14   Chemical companies in Brindisi

Company	Activity	Employees
Polimeri Europa	Ethylene, polyethylene, butadiene	574
Basell	Polypropylene	163
ChemGas	Industrial gases	30
Dow Chemical	MDI	180
PowerCo	MDI and VCM	35
Aventis	Pharmaceutical products	210
Exxon-Mobil	Cling film in polypropylene	160
Europlastic Sud	Plastic materials	n.a.
Europlastica	Plastic materials	n.a.
San Marco Sub	Diving suits	2
Icem	Waterproof products	13
Ri.Be.	Audio and video cassettes	9
Areta	Plastic furniture	27
Camassa	Plastic pipes	20
A.G.Scavi	Plastic materials	12
P.P.E.	Polyethylene containers	19
Polyman	Plastic packing	11
Giano Plastica	Plastic materials	30

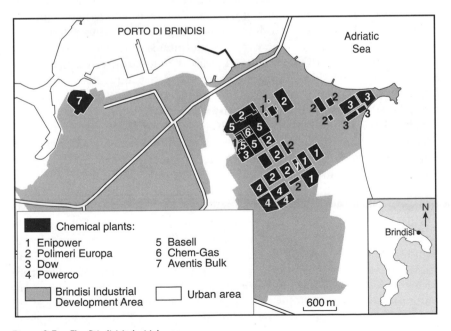

Figure 9.7   The Brindisi industrial area.
Source: adapted from Consorzio SISRI Brindisi, 2004.

plant had limited multiplier effects, and did little to stimulate indigenous industrial development. The plant failed to act as a growth pole (CER-PEM, 1990; ECOTER, 1993) and remained a cathedral in the desert. The outcome should not have been surprising. One of the reasons for choosing upstream intermediate goods industries for the south was that these industries did not require a developed industrial environment (Dunford, 1988). Not surprisingly, these industries did not create what they did not require.

Over the years several successive waves of redundancies added pools of middle-aged and semiskilled male workers to the local labour market. Many of them were unable to find similar work or work with similar rates of pay, and ended up in the local informal economy. As a result, however, of the dependence of local employment and local incomes on the direct and indirect employment the plant provided, first Montedison and later ENI-Chem were not allowed simply to walk away. The state intervened in order to sustain employment, incomes and consumption (Provincia di Brindisi, 1999). As indicated earlier, state intervention as a whole came at a cost: it delayed the restructuring of major industries, and it transferred costs to the state budget.

As far as the future is concerned, however, these sectors are assigned a very limited role. The 2000–6 POR for Puglia has three objectives: more equal development; improving the quality of life; and increasing competitiveness, innovation and the creation of employment. More specifically, the competitiveness and economic growth objective is concerned with the growth of local systems of production/industrial districts, integrated development and the creation of new innovation-intensive activities. To achieve these objectives, the POR identifies a strategy that is addressed exclusively at SMEs (setting aside Territorial Pacts and other agreements involving large companies) and aims at gradually reducing the size of capital incentives. The POR reflects, in other words, a decision to adopt a model of local development. To implement it, the POR draws on PITs, planning agreements (programmazione negoziata) and horizontal measures designed to improve economic conditions (such as support infrastructures, logistic arrangements and international cooperation) for all sectors.

The aim of PITs is to mobilize and upgrade local resources, and to strengthen local systems of production. The 2000–6 POR identifies three PITs for Brindisi. The first aims to develop a logistic and distribution system to expand intraregional trade. This PIT applies to the province of Brindisi alone. The second PIT is designed to support agricultural innovation, diversification and development and the integration of agriculture with the food processing industry. This PIT involves the provinces of Taranto, Lecce and Brindisi. The third PIT is directed at the textile and clothing sector and aims to strengthen it, improve its innovative capacity and create an integrated local economic system. This PIT applies to the provinces of Bari, Taranto and Brindisi.

The 2000–6 POR therefore makes no reference to existing poles of development centred on chemicals and steel other than to mention the implications of their activities for the environment. This oversight is quite deliberate, reflecting the prevailing view that the promotion of local systems of production is the way forward for Puglia, and that large-scale industries have nothing to contribute (see CUM, 1995; and the Special Report on the Mezzogiorno in Il Sole-24 ore, June 2001). This decision is not a recent one. A regional government regional planning document for the period 1994–6 opted for a district model (in the Marshallian sense and as a simple concentration of SMEs) as the linchpin of the region's medium- and long-term development strategy. The aim was to ensure that regional government action did not benefit individual companies, but rather aided a plurality of companies located in industrial districts. Adoption of the district model involved turning away from large-scale industries and large firms and was seen as a clear alternative to models involving large companies. As far as the Puglia regional government is concerned, chemicals, steel and similar industrial sectors have only a marginal role in government strategies for future growth and development.

From Growth Pole to Industrial Cemetery! The Disengagement Of The Chemical Industry from Basilicata

In 1978, in the Basento Valley ASI, 18 manufacturing firms provided employment for 4,935 people (Figure 9.8). Employment in these companies was expected to reach 5,868 and the arrival of a further nine enterprises was expected to raise the total number of jobs to 6,870 jobs. Of these jobs 3,500 were in the ANIC synthetic fibre plant established in 1964. A further 20 and an anticipated 800 were in the Liquichimica caustic soda and PVC plant.

Just over 20 years later, at end of 1999, the Consortium that managed the ASI reported that there were 41 firms and 1,881 jobs. The deindustrialization of an industrial complex, that was one of the most important in the Mezzogiorno, and whose development had played a significant role in the improvement of the relative economic fortunes of Basilicata in the 1960s and 1970s, was in part a consequence of the crisis and restructuring of its core chemicals sector. However, it was not the only cause, as a remarkable report, published as a result of an enquiry conducted by a Commissione Regionale d'Inchiesta (CRI) established by the regional government of Basilicata, showed (Regione Basilicata, 2000).

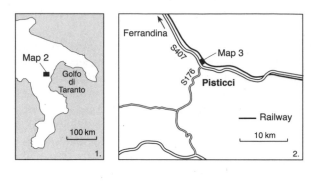

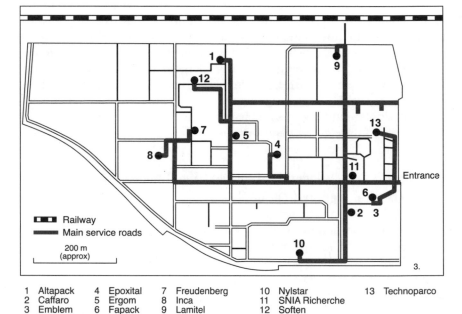

1	Altapack	4	Epoxital	7	Freudenberg	10	Nylstar	13	Technoparco
2	Caffaro	5	Ergom	8	Inca	11	SNIA Richerche		
3	Emblem	6	Fapack	9	Lamitel	12	Soften		

Figure 9.8   The Basento Valley ASI.

The first agreement between ENI and the trade unions to restructure and reconvert the chemical pole was in 1981. In 1985, the 'lodo Darida', named after the Minister of PP.SS., envisaged pacts to attract new externally-controlled industries. In a December 1987 decree of the President of Council of Ministers, applying Article 7 of Law 64/1986, these principles were implemented in a Planning Agreement. The Agreement envisaged the sale of ENIChem Fibre, ENIChem ANIC and SNAM infrastructures to a new company, Tecnoparco Valbasento. The role of the new company was to provide infrastructural (energy and water), technological and producer services. The target of these services was a range of (1) consolidated activities, (2) activities restructured into higher value added segments, and (3) new and more diversified activities (chemicals, technical polymers, fibres, textiles, manufacturing and food processing) to be established by ENI and others. To aid the reconversion programme, the Minister for EI agreed to authorize Finanziaria Meridionale to participate in the share capital of the new service company, and to authorize the Agenzia to award financial aid. Once completed, these industrial initiatives would guarantee the Val Basento some 2,900 jobs. These commitments were confirmed in 1990.

The initial Planning Agreement, which rested on investment in excess of LIT 500 billion envisaged the realization of these objectives within five years. The end date was extended several times until April 1996, while the rephasing of the programme was formalized in a protocol approved in November 1994 by CIPE.

At the end of 1992, 240 staff had been transferred to the management company and ENI had started 15 new initiatives, of which 8 were with other private companies and 1 (Metapontum Agrobios concentrated mainly on research in molecular biology and genetic engineering) was with the Basilicata region. An asset swap with SNIA BDP had also resulted in the transfer of some of its staff to this group, which was also involved in some of the other new initiatives. Table 9.15 lists the companies present in the early 1990s.

At that stage there was a shortfall of 1,500 jobs. The shortfall was attributed to (1) the government, for the uncertainty surrounding refinancing of Law 64, which had delayed the approval and payment of subsidies, (2) the Regione Basilicata and the CSI, for delay in the provision of infrastructures and services and (3) general sectoral conditions. As the Regione Basilicata report commented: 'further tasks and duties weighed on everyone except ENI which seemed to have nothing further to do other than bank . . . some further tens of billions of Lire and abandon the Basento Valley to its destiny as an industrial cemetery' (Regione Basilicata, 2000: 3082). The case prepared by the Comitato di Coordinamento for permitting ENI to walk away was, however, not accepted. The 1994 amendment confirmed ENI's responsibilities in terms of employment. Indeed, the amendment spoke of the presentation of initiatives capable of creating 5,021 new jobs, and indicated the importance of maintaining the employment levels identified in requests for subsidies. Nonetheless, unlike the Brindisi case, ENI managed to avoid a sharp social and economic confrontation.

At the end of the day, the programme fell far short of its targets, and ENI was successful in its attempt to extricate itself from the area. As the report of the Regione Basilicata (2000) indicated, the creation of Parco

Table 9.15 The Valbasento industrial zone in the early 1990s[1]

Company	Partners	Start date	Sector	Activity	Employment
Altapack	100% Altapack	1986	Rubber and plastics	Air bubble film for industrial packaging	36
Blucover	100% ENIChem	1986	Textiles	Upholstered textile products	4
ENIChem Fibre	100% ENIChem	1963	Chemicals	Acrylic and Polyester Fibres (closed)	29
Epoxital	50% Planasia[3]–50% Dow Chemical	1986	Chemicals	Epoxy resins	42
Filteni	100% private	1986	Textiles	In liquidation	
Gruppo SNIA	100% SNIA BPD	1990	Chemicals	—	588
Fapack	100% SNIA/Caffaro	—	—	Nylon film for wrapping food	56[2]
Emblem Europe	100% SNIA/Caffaro	—	—	Biaxially oriented film	67[2]
Nylstar	50% SNIA, 50% Rhodia	—	—	Polymer and nylon fibre	281[2]
SNIA Richerche	100% SNIA	—	—	Technical polymer R&D	
Inca International	100% ENIChem	1986	Chemicals	PET polymer and food containers	128
Industrie Resine Speciali	100% ENIChem	1987	Chemicals	Special resins	44
Lamitel[4]	100% Planasia[3]	1987	Rubber and plastics	Multi-layer laminates for printed circuit boards	75
Metapontum Agrobios	50% ENIRicherche–50% Regione Basilicata	1986	Chemicals	Agro biological R&D	69
Nuova Chimica Ferrandina	100% Planasia[3]	1964	Chemicals	Former Liquichimica industrial area	21
Orla	100% ENIChem	1992	Textiles	Needles and threads	149
Safiplast	100% ENIChem	1986	Rubber and plastics	Reinforced resin products	17
Soften	100% private	1991	Textiles	Upholstering products	12
Technoparco Valbasento	40% ENIChem, 40% CSI, 20% FIME	1990	Chemicals	Service management	198
Terbond	100% Locatelli	1987	Chemicals	Textiles and nonwovens	95
Total					1,507

Source: Tecnoparco Valbasento, 1994.

Notes

[1] Other 1980's initiatives included Alta (ENIChem: anticorrosive coatings); Sinel (ENIChem: electoluminescent systems); and Comav (ENIChem: prepregs)

[2] 2000

[3] Planasia was a 100%-owned subsidiary of Enichem

[4] Lamitel was acquired by MEW Electronic Materials (itself owned by Matsushita of Japan) in 2000 and production capacity was tripled

Tecnologico, which was to cost LIT 226,989 million was a slow and tortuous process that was indicative of the inadequacies and inefficiency of the CSI at least up to the deadline for the completion of the project. Not until 1991 did the CSI launch a competition for the design of the technology park. The proposals that resulted from the consequent contract with an association comprising Arthur Anderson, FIAT Engineering and Basica, that cost nearly LIT 2 billion, was thrown away due to a rescaling of SNIA BDP's activities. Only after another two years did the CSI approve a redefined project (Regione Basilicata, 2000: 3092).[6]

The record of the Comitato di Coordinamento that sought to argue that the ENI had met its obligations by the early 1990s was also criticized. In particular, the report documented that, in 1987–96, 11 members of the Comitato di Coordinamento, excluding two representatives of the Regione Basilicata who waived payments to themselves, misused large sums of money for personal purposes.

Most deficient was its record on the employment front, where there was a large shortfall. In responding to the CRI, ENIChem claimed that its commitments were to spend a specific sum of money and not to create a determined number of jobs. Also it claimed that, with the rephasing of the agreement, its task was to promote projects by third parties. It claimed that there were 1,552 jobs in ENIChem, SNIA and new initiatives (Table 9.16) and that it submitted to the Italian government a list of 71 initiatives. The view of the CRI was that there were in fact fewer jobs (1,106 in part as 106 staff were employed simply to guard and maintain plant that was no longer operational) and that ENIChem's commitment was not simply to spend LIT 410,837 million but instead to help enterprise and employment creation.

As for the list of 71 initiatives and 5,019 jobs, just 25 initiatives creating 1,606 jobs were included by the Ministero del Bilancio e della Programmazione Economica in a ranked list approved for further consideration by the banks. Of the 25, only one, Ergom which produces parts for FIAT

Table 9.16   Employment in the Basento Valley, 31 December 1992

	ENIChem's claim			Regione Basilicata			
	Actual employment	CIG	Total	Real employment	CIG or mobility	Total	Job loss
Enichem	756	848	1,604	606	892	1,498	−998
SNIA	465	—	465	465	—	465	
Other	331	214	545	329	214	543	−216
Total	1,552	1,062	2,614	1,400	1,106	2,506	−1,214

Source: elaborated from Regione Basilicata.

Auto, and whose owners recently diversified by acquiring the Torino foot-ball club, was operational in 2000 (Interview data, Pisticci, June 2000). Ergom had 40 employees, compared with an earlier target of 166 in three initiatives. In their report, the CRI called for an enquiry into the relations between ENI and the investors in the original list and in the short list of firms that did not complete negotiations with the banks.

The CRI also queried the payment of LIT 9 billion by the CSI to the ENI Group for some of its assets. The value of these assets was established in October 1987 by American Appraisal Italia. The valuation at replacement cost less depreciation rested on the assumption that the context was an industrial one. The failure to reindustrialize the area rendered some of these fixed assets obsolescent, and implied a massive reduction in their value. At an early stage, the Italian government had raised questions concerning the characteristics of the assets and the valuation methodology, while the valuation of the thermal power station, which was considered obsolete yet accounted for 40 per cent of the total valuation, was con-sidered unacceptable. Nonetheless, the sale went ahead (at LIT 90 billion against a replacement cost of LIT 176 billion) perhaps, as the report suggests, as ENI was committed to reinvest the proceeds of the sale in new projects in the area. As in relation to other matters that it considered of concern, the CRI asked for magistrates and the Guardia di Finanza to investigate this case, looking forward to a more effective set of measures after an agreement between the regione Basilicata and the national govern-ment to spend LIT 212 billion on future planning agreements designed to regenerate this part of Basilicata.

Although the context was one of decline, some new investments did occur. Examples include Freudenberg Politex nonwoven (formerly Ter-bond), Japanese Matsushita's MEW Electronic Materials (formerly Lami-tel) and Nylstar. which made the largest European textile investment for 30 years in its Pisticci plant. Nylstar designs, manufactures and markets nylon threads/yarns used to make socks, stockings, lingerie and sportswear, and works closely on fibre development with fashion designers (Figure 9.9 indicates the geography of Nylstar, including its presence in the CEECs, where it has transferred some production lines). The second largest produ-cer in the world, Nylstar was established in 1994 as a joint venture when SNIA combined its polyamide filaments activities with those of its main European competitor Rhodia. (SNIA itself concentrated on nylon, after a 1990 agreement with ENIChem under which the latter took control of polyester and acrylic fibre manufacture).

In 1994–8, Nylstar added a new one-step technology plant to the original 1964 two-step plant (Interview data, Pisticci, June 2000). Also involved was an automation of process control and handling. The old plant was capable of producing 3,000 tonnes of POY and 4,200 tonnes of FDY per

Figure 9.9   Nylstar in Europe.

year, while the new plant had an annual capacity of 9,000 tonnes of FOY. The new technology led to a large increase in productivity (Table 9.17), although the use of the more complex Nylon 66 required higher manning levels than Nylon 6. As the products have different uses, output increased (Figure 9.10), and the old lines went down slowly, while at the end of 1999 booming sales saw the plant operating at full capacity. After 2001, however, there was a sharp and sustained fall in the demand for man-made fibres as the economic cycle swung downwards, and delocalization accelerated. Nylstar was in financial difficulties, SNIA was seeking to withdraw and a major restructuring programme was announced. In the case of Pisticci, a decision to close the older plant led to the announcement of layoffs.

Table 9.17   Nylon plant productivity, 1994–8

Tonnes per person year	1994	1995	1996	1997	1998
FDY6/POY	36	41	36	33	33
FOY6/FOY66	17	65	67	79	72

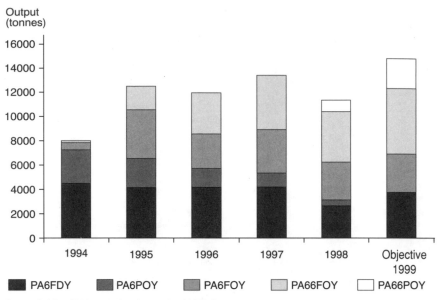

Figure 9.10   Nylstar production trends, 1994–8.

This announcement coincided with other job losses. Most significant were Dow Chemical's decision to close the former INCA International PET plant (although PET had enormous growth potential), and the earlier SNIA decision to close the polymer facility that supplied its nylon film activities (Fapack and Emblem Europe) at Pisticci.

The justification given for the 2003 closure of Dow's PET plant and the loss of 70 jobs was an enduring deterioration in market conditions, and a need to concentrate production in larger and more efficient plants. PET is a polymer used extensively for plastic bottles that are progressively replacing glass and aluminium cans as drinks containers (Figure 9.11). PET is derived from PTA which itself is employed to manufacture polyester for use in fibres and films. PET is therefore one example of the way in which textile polymers can be put to new uses, offering the technical possibilities of the industrial reconversion envisaged for the Basento Valley. Dow entered the PTA and PET sectors in 1996, when it acquired INCA International and its plants at Pisticci and Ottana in Sardegna. In 2000, Ottana had a PTA capacity of 175,000 tonnes per year and a PET capacity of 85,000, while Pisticci had a PET capacity of 60,000 tonnes per year and was supplied with excess PTA by Ottana. Italy is the largest consumer of PET in Europe: in 1999 Italians consumed 6,711 grams per head, the French 4,311, the UK and Ireland 3,498, Spain and Portugal 2,323,

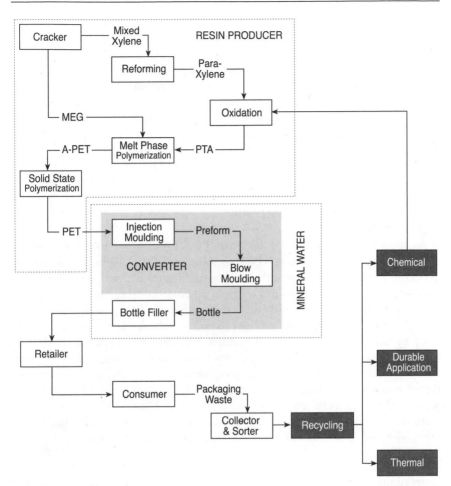

Figure 9.11 The PET chain.

Germany and Denmark 939 and the CIS 413. In 1995, Dow started the reconstruction of the Buna Sow Leuna Olefinverbund chemical complex in Schkopau in the former GDR. Included was a 155,000 tonne PET facility that used state-of-the-art technolgies and was supplied with some of the PTA it required from Ottana. In 2003, Dow announced its intention to construct a second grant-aided PET production train in Schkopau with an annual capacity of 175,000 tonne. The aim was to serve the growing market for PET in Germany and the CEECs, in relation to which the East German complex is well-positioned. Soon after this announcement Dow announced the closure of the Pisticci site: its capacity fell far short of that of newer plants, while Dow could fully serve its customers from its two remaining sites. Pisticci's production costs were higher due to differences in plant efficiency and required manning levels, although in 2000 wages were one-half of those in Germany, and the logistic costs of supplying southern markets were far smaller. The fundamental problem, however, was downward pressure on prices, especially as a result of Asian imports: when it was established INCA International employed 151 people and earned LIT 4,000 per kilogram; by 2000, although it had reduced its staff to 70 and its costs to LIT 1,200 per kilogram, the price it received for its PET scarcely covered its costs (interview data, Pisticci, June 2000).

The rise and the subsequent fragmentation and fall of the ANIC complex had a significant impact on the local economy, as the role of the chemical industry in the regional economy increased and subsequently decreased. The decline phase was thrown into relief by the rise of other parts of the regional economy.

As in the case of Puglia, the Basilicata POR for 2000–6 does not mention the chemicals/plastics sector. The development of the regional economy is expected to involve the strengthening of traditional sectors including green tourism, archaeological/cultural tourism and agriculture, which is expected to develop stronger links with food processing. As far as industry is concerned, most importance is attached to motor vehicles and related subcontracting activities and to the furniture industry where the lead firm is Natuzzi. Services are also an area of growth. The document acknowledges that regional economic growth is geographically uneven, as it is concentrated around the borders of the region: FIAT, Barilla and Parmalat are driving growth in the north, while the province of Matera is looking towards Bari. The declining areas are the formerly privileged regional capital and, especially, the area that once hosted the region's chemical pole.

Conclusions

In this chapter we have outlined the rise and the fall of the primary chemicals sector in Italy and the complex restructuring of the industry as a whole. The specific circumstances that shaped the development and structure of the industry delayed the restructuring of large corporations and weakened the industry's performance at international level. The processes of upgrading, consisting essentially of a progressive specialization in core competences and activities, occurred only when the spheres of influence of chemical corporations were more clearly defined and when their economic rationales changed.

The evolution of the chemical sector testifies that the model of large-scale capital intensive industrialization was a costly one for the state as well as for the regions chosen as the locations of chemical investments. The model of development was neither sustainable, nor did it create conditions for further growth. In the phase of expansion it brought permanent and well-paid jobs, which raised local incomes and integrated local economies into wider economic relations. As the industry went into decline, however, these areas were required to confront the difficult path of reconversion and waves of redundancies that left a great number of workers unemployed. In this chapter we have also shown that the processes of disengagement differed geographically. In the case of Brindisi, ENI's exit from chemical production occurred through asset swaps and mergers which brought other companies to the area. In the case of Basilicata, restructuring meant downsizing, plant closures and the decline of the industry, in spite of state intervention which was seemingly designed to provide alternative chemicals-related employment, yet which in effect helped ENIChem walk away.

Chapter Ten

Conclusions and Further Remarks

Introduction

The aims of this volume were to describe and explain the character and trajectories of Italy's regional economies, and to develop a set of conceptual tools applicable more generally to the study of economic geography, moving beyond existing representations of Italian development and existing theoretical approaches to regional dynamics and industrial change.

Geography as a Spatial Expression of a Social Order

At the centre of the conceptual framework adopted in this volume is the idea that geography examines the spatial expression of a social order (including its relation to nature). In economic geography, analyses of the the structure and dynamics of regional and urban economies emphasize the dynamics of the market sector and the geographies of the production of goods and services. An understanding of the trajectories of advanced economies and societies requires, however, a more complex model which recognizes the structure, role and interactions of the three fundamental orders of contemporary mixed economies: the economic, political and domestic orders (see Théret, 1994). Each of these orders involves a set of practices, that are structured by a set of material/technical and social relations. These social relations are simultaneously economic, political and cultural in character, and characterized by different and contradictory motives.

The economic order in its widest sense is the set of practices through which human beings use the material resources they derive from inanimate and animate nature to meet human needs. In capitalist societies, a more restricted conception of the economic order confines it to those economic practices oriented towards the accumulation of wealth, where wealth is understood as material and immaterial goods and services that possess exchange value, and signs that represent/command goods and services, of which the most important are monetary titles. The domestic order is the realm in which households/individuals live their lives, and in which social reproduction takes place, while the political order is a realm in which are established rights of sovereignty and citizenship, on the one hand, and a set of fiscal obligations and social rights/expenditures, on the other. In order to regulate their internal contradictions and to improve their mutual correspondence, these three orders are regulated and articulated via a set of monetary, legal and ideological/discursive systems of mediation. Of these systems of mediation and regulation, most attention was paid, in this volume, to those that shape the economic order, and articulate it with the political and (to a lesser extent) domestic orders. Of particular importance was the further conception of the economic order as a complex combination of activities of production, distribution, consumption and exchange. More specifically, in mixed capitalist economies the economic order comprises (1) a set of processes of labour, capital valorization and production and exchange of goods and services in the market sector, and a set of noncapitalist processes of production in the welfare state; (2) an organization of the distribution of income to the agents who own productive assets, are self-employed or are employed as wage earners; (3) a system that embraces the redistribution of income, the expenditure of wages and other incomes on commodities, the consumption of goods and services and the reproduction, in the domestic order, of individuals as owners and wage-earners and as citizens. Of these elements, most attention was paid, in this volume, to production, distribution and their articulation. More specifically, we examined contrasts in the work people do, what they get paid for it and how much wealth they create in different parts of contemporary Italy, and the ways in which work, employment and income are changing.

Geography and Development Models

A second key idea underpinning this volume is that areal differences in what is produced and in the rewards people receive are a reflection of different yet interdependent development models/development pathways/worlds of production, and, more generally, of processes of combined and uneven development. A concept of combined and uneven development is important for two reasons. The first is that it deals with the development of an entire social order across different territories. The second is its emphasis on the interdependence of different parts of the division of labour and of different localities.

These ideas are associated with one of two possible interpretation of synchronic differences in development. The first interpretation, with which we do not agree, suggests that there is a single development path and that differences between places stem from the occupation of different positions, or the attainment of different stages, along this development path. (An example is Rostow's stages of economic growth). Associated with these ideas is the view that there is an end state, that more advanced localities are closer to this end state and that the most developed areas offer less developed areas a vision of their own future. The second interpretation sees differences as stemming from the performance of different, and un-equal, but interdependent roles by different groups of people and localities. In this case, even if areas that are less developed move in directions that make them more similar to areas that are already advanced, inequalities may not disappear. The reason why is that advanced areas also change, that the context of development differs and that elites can reproduce existing/create new types of unequal interdependence. More radical versions of this thesis suggest that in capitalist societies development and underdevelop-ment are two sides of the same coin.

The adoption of this second stance puts us in general agreement with those approaches which identify multiple, context- and path-dependent development trajectories, and contrasting and constantly changing worlds of production. In this volume, the dimensions on which these pathways and worlds were seen to differ fell into two groups. The first comprised different sectoral realities and different functional roles in the division of labour. The second included the inherited regional context, and the associated modes of regulation and varieties of capitalism. However, established concepts of multiple development paths, path dependence and worlds of production pay insufficient attention to the reality of interdependence. Each role in the division of labour and in different value chains and each distinctive world of production presuppose the existence of other roles and other worlds on which they depend. A specific mode of organization appropriate for one part of the division of labour is not necessarily appropriate or applicable for others. The development path of one locality is conditioned by, depends upon and can only be explained in the light of its relations with others.

Contemporary Perspectives on Industrial Change and Regional Economic Performance

After identifying our general object of analysis, our attention turned to the interpretations and theories found in the literature. As we pointed out, a number of theories dominate recent analyses of regional dynamics and the geography of the economic order: neoclassical theories of growth and convergence; the new geographical economics; supply side theories of clusters, districts and regional economies; and theories that emphasize the importance of regional social, cultural and institutional conditions.

In Chapter 3, we argued that neoclassical theories of trade and growth and the new geographical economics have several merits. The first is that these theories seek to answer why questions. The new geographical economics, for example, draws on and develops an earlier locational tradition in geography to derive economic landscapes from analyses of profit-seeking. The second is that endogenous growth models and the new geographical economics recognize the coexistence of centripetal and centrifugal forces affecting the geography of economic activities, as did earlier theories of cumulative causation. An important implication is that trends in territorial inequality depend on the relative weight of these two sets of forces. The third is that recent models of location and development recognize that the geography of economic activities is a result of a dialectical process in which causes are consequences and consequences are causes (see Dunford and Perrons, 1983) and in which an area's resource endowments are an endogenous consequence of its development path. Explanations are, therefore, almost always historical in character (Krugman, 1999). The reason why is that, while there are general mechanisms at work, what happens at each stage is in part a result of what happened earlier.

At the same time, neoclassical theories of trade and growth and the new geographical economics have several limitations. The first concerns the specification of causal mechanisms. Neoclassical theories of growth and convergence (as opposed to endogenous growth theories) dwell solely on catch-up mechanisms. The models of the new geographical economics involve at least two specification errors. First, they employ a narrow concept of corporate strategies, industrial organization and of the relevant external environment. Second, no attention is paid to the functional/occupational divisions of labour and the factors that shape them.

The second limitation of neoclassical theories of trade and growth and the new geographical economics is that the quest for parsimony runs the risk of determinism and of a reductionism that ignores the significance of social, cultural and institutional conditions.

The third limitation of most of these models is that they are models of fictitious worlds. Any model is a simplified representation of reality. Good models aid our understanding of reality. If, however, simplification involves ignoring fundamental features of the real world, difficulties arise. Most problematic in this respect is the common neoclassical assumption that adjustment to economic shocks involves processes that are smooth and automatic. In the real world, resources released as a result of structural change frequently remain unemployed, and markets do not automatically lead to full employment. Neoliberal economists often argue that what is wrong in these situations is not their models but those features of the real world that prevent the adjustment of prices: the real world, they argue, should be made to correspond more closely with their fictional world. Other social scientists rightly contest this claim. According to Keynesians, for example, the initial response to a decline in demand is not a price adjustment but a quantity adjustment. Adjustment falls on quantities as a consequence of two fundamental features of real money economies that neoclassicists assume not to exist: the costs and time involved in the acquisition of information; and the fact that money is the medium of exchange, so that excess supply in the labour market diminishes the effective excess demand for goods and services (see Dunford, 1997).

The new economic geography provides richer accounts of the economic and institutional determinants of the structure and development of individual clusters, districts and regional economies. Conceptually, it concentrates on the roles of economic and institutional resource endowments, corporate strategies, interfirm relations and innovation and learning. Methodologically, it relies on deep description and region-centred approaches. The difficulties with these ideas are severalfold. First, insufficient attention is paid to earlier explanations of the changing map of relative development in terms of economic interdependence and changing spatial/territorial divisions of labour (Massey, 1979; Lipietz, 1977; Hymer, 1975). Second, too much attention is paid to wealth creation (who does what, where and when and in what relationships they stand to other people and other economic activities) and to governance at the expense of questions of distribution (what rewards people receive) and uneven development.

Theorizing Industrial Change and Regional Inequality: Profit Strategies and Value Chain Upgrading

In the light of this assessment of the strengths and weaknesses of recent research, an alternative approach to the analysis of regional inequalities and the underlying territorial division of labour was presented. This alternative approach involves a synthesis of the political economy traditions of the 1970s and 1980s with elements of approaches that emerged in recent years, and is justified by the fact that the complexity of the phenomena under investigation requires an approach that recognizes the complementarity of different narratives.

The starting point for this alternative approach was an empirical account of the creation and reproduction of regional inequalities, seen as involving relative overdevelopment at one pole and relative underdevelopment at the other. Emphasis was then placed on the ways in which these inequalities reflect the interaction of equalizing/centrifugal and unequalizing/centripetal tendencies in an interdependent world. Centrifugal mechanisms, which include the relocation of factor-cost sensitive activities in low cost peripheral areas and technological catch-up, were seen as playing an equalizing role. Centripetal mechanisms, which include scale economies, cumulative increases in knowledge and skills, infrastructural improvements, flows of investment and people, and increased market size, were, conversely, seen as creating new differences and increasing inequalities. Whether inequalities get larger or smaller depends on the relative weight of these centrifugal mechanisms and the countervailing centripetal mechanisms.

The next step was to argue that an analysis of the microfoundations of regional dynamics and regional interdependence implies an analysis of industrial/sectoral performance, of the evolution of the underlying territorial divisions of labour and of the mechanisms that shape them. Of particular importance was the territorial division of labour defined as who does what, where and when, what rewards they receive and in what relationships they stand to other people and economic activities in other places.

Against the new geographical economics' preoccupation with questions of location, we argued that the evolution of the territorial division of labour depends not just on where activities are located but also on the answers to what and when questions. Analyses of location should therefore be integrated into a more general explanation of the evolution of capitalist enterprises.

To analyze the territorial division of labour, a distinction was made between the residential economy, that is driven by the expenditure of regional incomes and serves a region's inhabitants (and perhaps its visitors), and those economic activities than serve wider markets (perhaps excluding sales to visitors as they involve local consumption). In this volume, attention was concentrated on some of the industrial activities that comprise the latter. The approach adopted considered the profit-seeking firm as the central economic actor in explanations of industrial performance and change. Considerable attention was also paid, however, to the impact of a firm's external environment on its actions, and of its actions on its environment. One aspect of this environment is a firm's network relations with other firms. Networks exist at a regional level. In contrast to the new regionalism, however, the approach adopted in this volume paid more attention to national and international networks and organization, and to interregional interdependence. In addition, whilst emphasizing the fundamental profit-seeking nature of enterprises, it recognized that an institutional environment also affects corporate economic performance, and that this institutional environment comprises not just other firms/market organizations but also a set of social, political and cultural conditions.

These concepts offer some of the foundational elements of an account of regional performance and the territorial division of labour. Although the starting point is the inherited territorial division of labour, its trajectory depends on the profit and upgrading strategies of individual enterprises. A number of paths to profitability and upgrading were identified: (1) cost reduction via technological upgrading/cheaper inputs/scale economies; (2) the development of new commercially relevant products or the penetration of new markets; (3) changes in the relative weight of different functional roles to increase the share of functions commanding higher returns; and (4) disinvestment and a transfer of resources from one value chain to another.

A profit/upgrading strategy depends upon the quality, appropriateness and coherence of the internal and external resources the enterprise can mobilize, and on the governance compromise it puts in place in order to implement its chosen profit strategy. Clearly, therefore, it depends on the resources it derives from its network relations with suppliers and customers, and with the wider social, scientific and educational environment. Moreover, these conditions in which development strategies unfold and by which they are shaped are, as cumulative causation approaches make clear, the result of previous phases of development.

Areal Differentiation and Uneven Development In Italy: From the North-South Divide to the Three Italies and After

As indicated earlier, the second aim of this volume was to draw on this conceptual framework to analyze Italian regional economic development, and some of its sectoral foundations. This analysis provided a new and more complex picture of Italian areal differentiation and regional economic development. This new picture distanced itself from the traditional north-south dichotomy, and moved beyond the concept of the Third Italy in two ways: first, it emphasized that the district system is a context-dependent part of a wider interregional division of labour and not simply a replicable model; second, it presented at a number of points a new account of the rise of the Third Italy, and emphasized some of its overlooked weaknesses.

As we showed, the north-south divide[1] in particular and regional inequalities in general were reproduced. Measured by GDP per head, contemporary Italy is still characterized by large development gaps essentially between its four main territorial units. At the same time, it is important to note that gaps in income per head are far smaller than gaps in GDP per head, largely due to the impact of state-organized redistribution to Italy's weaker regional economies.

Although Italian regional inequalities were reproduced, the scale of these inequalities nonetheless diminished rapidly in the second half of the

30-year Fordist Golden Age, which saw rapid economic growth in the centre-north, and yet more rapid state-sponsored modernization in the south. In the subsequent 22 years (1974–96), Italy experienced increases in territorial inequality, reversing earlier processes of catch-up and convergence. The main proximate reason for this reversal was a widening of employment rate differentials: growth in the aggregate demand for labour and, therefore, of the economy were not sufficiently fast to absorb all of those seeking employment as productivity increased (more rapidly in less-developed areas than in more-developed areas). In those years, an economics and politics of excellence rewarded and strengthened successful regional economies and weakened and discouraged unsuccessful economies. In the recent geographical literature, this central feature of Italy's economic geography was frequently overlooked, and ceased to be an object of explanation.

In the first post-Fordist cycle of slower growth (1974–89), the Italian economy performed relatively well compared with other European economies. In no small measure, this achievement was due to the relative dynamism of regional economies in the centre-northeast, and of nonmetropolitan areas in the northwest. As their growth was connected with a model of industrial development centred on SMEs, industrial districts and traditional industries, many commentators were led to the conclusion that this Made-in-Italy or flexible specialization model of industrial development offered a replicable alternative to the mass production system, and a way out of the crisis of Fordism.

Although its growth achievements were significant, the recent downturn in the fortunes of the Made-in-Italy system suggests that these claims were excessive. As we showed, the model relates to just one set of roles in a wider division of labour, and to one of several mutually interdependent worlds of production. More specifically, a flexible specialization model is appropriate in particular sectors (the manufacture of clothes, knitwear, textiles, ceramics, and shoes where fashion and/or seasonal factors lead to rapid product changes), in particular conditions of income distribution (as it is oriented towards design-conscious, middle and upper income markets), in economic conjunctures in which instabilities predominate (due to its small fixed costs and flexibility), and in specific cultural and institutional contexts (that include a strong tradition of microentrepreneurship). Alongside it, there were, therefore, other worlds of production that comprised equally important parts of a wider division of labour. Included were: the world of large-scale industries of the kind considered in the sectoral chapters of this volume; the world of diversified producer and consumer services; and the welfare sector. Concentration on the Third Italy model drew attention away from other territorial realities associated with other parts of an interdependent division of labour. Most significant were: Italy's metropolitan areas; and the south which, after years of falling behind, saw some improvement in its relative performance, as well as striking internal differentiations, that separate growing districts from earlier poles of development.

Economic Decline and the Limits of the District Model

The case for transcending existing concepts of the Third Italy also stems from the recent deterioration in its economic performance, and from the Italian economy's loss of momentum. In the 1990s, Italy found itself with a strong dependence on sectors and market areas whose vulnerability to international competition had increased. At the end of that decade, it was clear that Italy had suffered from successive decennial declines in its rate of growth. In the last few years, average annual growth rates have not exceeded 0.4 per cent. Worst affected were the 'Made in Italy' sectors and districts: in 2003, Carpi suffered a 14 per cent reduction in the number of enterprises, a 15 per cent drop in employment and stagnant turnover. In Como, turnover has declined by 20 per cent since 2000. For Biella, 2003 was the third year of crisis. Greater international competition, accelerating technological change and the imitative capacity of new competitors have helped place the district model in crisis (Bersani and Letta, 2004). At the same time, one-third of 25–40 year-olds have fixed-term contracts and no employment-related pension rights, while individual redundancies are widespread. The Italian population is ageing, and fertility is declining.

Apparently, the very internal factors that had made the remarkable growth of the 'Made in Italy' industries and areas possible were under strain (external diseconomies, a degree of individualism that constrained collective action, a transformation of family structures and revised popular expectations). An identification of weaknesses also permitted a new explanation of the rise of the districts, and of the three decades of growth of microenterprises in traditional sectors outside of the main industrial areas of the country. As these developments profoundly changed the industrial geography of the country, and underpinned some of the regional development trajectories identified earlier in this volume, this new explanation is also an explanation of important aspects of Italy's model of territorial development. At the centre of this new explanation are two factors which were given little attention in the literature. The first was the special role the Constitution attributes to craft enterprises and the legal status of SMEs, which played an important part in permitting decentralization as a reaction to social conflict in large plants in established industrial areas. The second was the downward movement of Italy's exchange rate. Together these factors encouraged defensive adaptations at the expense of weaknesses in the shape of small average firm size, a reliance on price competitiveness and a disadvantaging of more capital- and research-intensive sectors.

As evidence of the deterioration in Italy's economic performance mounted, a string of new publications started to speak of the country's economic decline (Fazio, 2004; Toniolo and Visco, 2004; Turani, 2004). All of these publications identified similar concerns. In his 2004 annual speech (Considerazioni Finali), the Governor of the Bank of Italy, Antonio Fazio, spoke of Italy's disease. At the centre of the problem was a decline in competitiveness, reflected in a decline in Italy's share of world exports and, Turani (2004) would add, slow growth. Fazio identified several causes. The first was Italy's extremely low productivity growth. After growing at 0.9 per cent per year in 1991–5, productivity grew at just 0.5 per cent per year in the second half of the 1990s, and it declined by 0.7 per cent per year in 2000–3. At the root of this dismal productivity growth record lay, Fazio argued, insufficient investment in new technologies, insufficient exposure to international competition and the predominance of SMEs. The predominance of SMEs was related to the second cause: the small average size of Italian companies. In 2001, the average number of employees in industrial and service companies was less than four. If the 2.4 million companies with just one employee were excluded, average company size stood at just 8 people compared with 13 in France, and 15 in Germany and the UK. Nano-companies have greater difficulties than larger ones in carrying out R&D, innovating, increasing productivity and finding new markets. As Turani (2004) pointed out, the other side of the coin is the fact that just five Italian companies are in the Fortune top 500, and just three in the top 50 in Europe (Assicurazioni Generali, FIAT and ENI). In the aeronautics, defence, chemicals, information and communications technologies and food and drinks industries, Italy had no companies in the top 500. The third factor identified by Fazio was the fact that the adoption of the € prevents Italy from devaluing in order to offset its economic inefficiency. Also identified as a weakness was Italy's specialization in mature, traditional industries that account for a small share of world trade. Almost all of these arguments were developed in Chapter 5 of this volume in our account and critique of Italy's specialization, enterprise and district model, and in our subsequent accounts in Chapters 8 and 9 of the national and regional trajectories of the vehicles and chemicals sectors.

Turani's (2004) diagnosis of Italy's malaise, provocatively entitled 'Why we have the worst capitalism in the world', is similar but more politicized: insufficient R&D; the decline of the chemical, microelectronics and perhaps of the automobile industries; a concentration on cost-sensitive, low-technology Made-in-Italy industries, whose activities are easily replicated; a capitalist class that relied on the protection of politicians and protected markets; the predominance of family-owned companies with strong kinship ties; and the small average size of companies, which constrains investment in R&D, distribution and marketing.

Another influential critique was that of the well-known economic journalist, Guido Rossi. In an interview (Rampini, 2004) Rossi attributed the decline of Italy's industries to three pathologies: the flaws of family capitalism; the flaws of state capitalism; and the role of an indebted state. All three, he argued, prevented the development of the country's industrial apparatus, through their effect on the relationships between companies, financial institutions and the state.

An evident implication of these ideas is the need for a reappraisal of Italy's industrial and regional trajectories. The conclusion of Bersani and Letta (2004) is, for example, that, if Italy is to be more than 'the environmental and archaeological Disneyland', its districts must be revitalized. For Fazio (2004), an Italian renaissance depends on the growth of Italy's medium-sized companies (with 50–500 employees) and their capacity to drive the creation of consortia capable of overcoming the disadvantages of the small-size of many Italian companies.

Also required in our view are three other elements. The first is a reconsideration of the potential role of more technology- and research-intensive parts of the division of labour in which Italy is weakly specialized (energy, transport, environment, health, and information and communication technologies), the development, perhaps with European partners, of new long-term technological upgrading and investment programmes, and a related reorientation of the profile of the national economy. Essentially, the diminished dynamism/crisis of the Made-in-Italy sectors suggests that in economic development terms more attention should and should have been paid to the strengthening of high-tech and modern industrial sectors that are concentrated in other parts of the national territory, and that suffered from the conditions that favoured the rise of microenterpises and the remaking of Italy's economic geography. To achieve these ends, Italy clearly needs an industrial policy. Ideally, this policy should be articulated with its regional policies and with substantially higher expenditure on research and development. The second is a strategy that can effectively counter the climate of illegality, informality and a lack of respect for rules that prevails to different degrees throughout the country and has a large negative impact on individual enterprises and on the production system as a whole. The third relates not to the industrial system but to the job market and the need for full employment which depend in part on the rise of the residential economy that we have not considered explicitly in this volume. On this score there were important initiatives including the socially useful work (lavori socialmente utili) and the publicly useful work (lavori di pubblica utilità) programmes that, at their peak in the late 1990s, provided 160,000 jobs for the long-term unemployed and unemployed young people especially in the south.[2]

Industrial and Regional Performance

Not only does this recent debate support a number of the arguments developed in this volume. It also suggests that a revised appreciation of the trajectories of all of its industries and regional economies is required to make sense of Italy's past, present and possible future development. An analysis of relative regional economic trajectories, and of the respective roles of centripetal and centrifugal mechanisms, requires, however, a study not just of aggregate trends but also of the parts of which regional economies are made up. The reason why is that aggregate trends are the sum of millions of microeconomic developments of varying size and impact, although the speed, direction and results of microeconomic changes are affected by the aggregate trends that result from the adding up the parts. Considerable emphasis was therefore paid to two exercises. The first was a sectoral disaggregation of aggregate data, to examine what was happening to individual industries, and to measure the impact of their growth/decline on aggregate trends. The second was an examination of the relationships between sectoral trends and the underlying changes in/inertia of the social and territorial division of labour. In this case, attention was paid to the structure and trajectories of individual establishments and enterprises in particular industries, their changing role in European and international divisions of labour and the identification and explanation of the microeconomic strategies from which they derived. Attention was paid, in particular, to two industries (motor vehicles and chemicals) containing differing microeconomic realities (SME systems, large private corporations and state-owned companies), with a concentration on four regional economies (Lombardia, Piemonte, Basilicata and Puglia) and their Italian and international contexts.

Both sectors demonstrated ways in which the roles of establishments, and their impacts on the geography of wealth, income and employment creation, reflect an array of profit strategies. In the case of the car sector, we noted a movement into services (comprising highly qualified professional and managerial jobs as well as clerical jobs) in and outside the vehicles sector, a decline of well-paid manual jobs in the northwest and a refocusing of Italian manufacturing operations in the south (where highly qualified personnel are underrepresented, and jobs and wages are already threatened by possibilities of further offshoring to take advantage of lower costs in less developed countries). Also noted was the emergence of a new hierarchical international division of labour in FIAT Auto, with clear implications for the geography of wealth and inequality. At the top is Italy where most high value added functions are carried out. Next come a series of countries such as Poland, Brazil and Argentina, where there are integrated production complexes, followed by a number of other developing countries, such as Morocco, India and China, that are sites for assembly plants. At the root of these developments were a frequently shifting set of profit strategies: an oscillating commitment to diversification into manufacturing and service activities not strictly related to car manufacturing; a movement downstream along the manufacturing value chain towards higher value added and commercial/market-related functions; a varying degree of commitment to investment in the design and marketing of new models; supply chain reorganization; joint powertrain and global sourcing operations with GM; and a strategy of internationalization which saw the company's expansion in the CEECs and other emerging economies, especially in Latin America.

In the case of chemicals, most attention was paid to the rise and fall of the primary petrochemicals sector in Puglia and Basilicata. In its expansion phase, this sector played a major role in driving regional performance in the south. As economic and institutional circumstances and corporate strategies changed, these and other chemical industry poles were required to confront difficult and politically mediated reconversions and redundancies, which also help explain why certain regional economies lost economic momentum. The reason for the redirection of corporate strategies was that the earlier quest for economies of scale, and diversification into a range of chemical activities (extending from industrial to speciality and fine chemicals), whose justification resided in the growth of demand in the 1960s, were abandoned in the 1980s. A radically different set of international conditions and increased pressures to upgrade saw Italy's large chemical companies, along with their European counterparts, embark on a period of intense rationalization, involving concentration on core activities, and disinvestment from the least profitable businesses. ENIChem's upgrading paths consisted of technological upgrading and work reorganization, although in Italy the redefinition of large companies' profit strategies also involved securing autonomy from the political sphere which had had a major impact on economic, locational and disinvestment decisions.

Conclusions: Inequalities, Territorial Divisions of Labour and Profit Strategies

In this volume, we have argued that trends in comparative regional development are a result of the structure and evolution of the social and territorial division of labour (and the character of agglomerations), and that these divisions of labour are a consequence of institutionally-mediated profit strategies and the conditions in which they unfold. At the most general level, changing trajectories of accumulation cause changes in the volume, composition and geography of the demand for labour and the rewards paid for different kinds of work. At the root of these trajectories of accumulation are the profit and upgrading strategies of enterprises that are themselves developed in the context of one or more commodity/value chains, of relations with other actors at a range of geographical scales and of specific modes of regulation and governance. As we argued, these strategies result in in situ change in the profile of employment and income and also in the location of different but interdependent (complementary/competing) parts of value chains in different locational contexts with different costs, skills, supporting services and research infrastructures. These changes in the profile of employment and income amount to a remaking of the territorial division of labour. As a result, the characteristics of places, networks and hierarchies are changed, as is the map of economic development.

As far as the present is concerned, certain empirical generalizations are perhaps possible. New knowledge-intensive industries, new functional roles (research and development, product definition, marketing and other advanced services) and new technological and organizational models rise, providing high incomes for white-collar, symbolic analysts (Reich, 1991) most often in core countries and regions.[3] Old-established industries that paid high wages to relatively unskilled people are subjected to strong cost competition. Insofar as these industries cease to be suited to the institutional and economic environment of rich countries, they move offshore, or go to the wall. Some technology and skill intensive parts of manufacturing also move offshore, further altering the types of firm, occupational structures and skill profiles of core areas, and raising strategic industrial redeployment issues in those areas that formerly attracted substantial amounts of inward industrial investment/performance plants.

Although these generalizations summarize some features of recent trends in the territorial division of labour, the aim was not to establish a new metanarrative. In this volume we have demonstrated instead that there are significant differences in the ways in which similar generative mechanisms work their way out in different sectors, and that a model that is applicable to one part of an interdependent division of labour is not necessarily applicable to other equally vital parts of the division of labour. A similar point applies to modes of territorial organization and regional development: in this volume we have identified a multiplicity of place-specific pathways, and have argued against preoccupations with a single model. A fundamental aim of economic geography is to develop theories of generative mechanisms. Another fundamental aim is the identification of the characteristics that places share and the characteristics of their interrelationships. These theories do not require the identification of identical cases. Instead the aim is to identify mechanisms that combine with a range of more contingent factors and to explain areal differences and the variability of the map of development and underdevelopment.

References

ACC (American Chemistry Council) (2003) *Guide To The Business of Chemistry, 2002.* ACC, Arlington, VA. http://www.americanchemistry.com/

Affinito, Massimo, De Cecco, Marcello and Dringoli, Angelo (2000) *Le privatizzazioni nell'industria manifatturiera italiana.* Donzelli Editore, Roma.

Aglietta, Micheal and Brender, Anton (1984) *Les métamorphoses de la société salariale. La France en projet.* Calmann-Lévy, Paris.

Agnew, John (2002) *Place and Politics in Modern Italy.* University of Chicago Press, Chicago.

Albert, Michel (1993) *Capitalism against capitalism.* Whurr Publishers, London.

Amatori, Franco and Brioschi, Francesco (1997) Le grandi imprese private: famiglie e coalizioni. In Barca, Fabrizio (a cura di) *Storia del capitalismo italiano.* Donzelli Editore, Roma, pp. 118–53.

Amin, Ash (1983) Industrial restructuring, state intervention and regional growth. *Reading Geographical Papers,* 77. University of Reading, Reading.

Amin, Ash and Thrift, Nigel (1994) Living in the global. In Amin, Ash and Thrift, Nigel (eds) *Globalization, institutions and regional development in Europe.* Oxford University Press, Oxford, pp.1–22.

Ardigò, Achille and Donati, Pierpaolo (1979) *Famiglia e industrializzazione.* Franco Angeli, Milano.

Arlacchi, Pino (1983) *Mafia, peasants and great estates in traditional Calabria: society in traditional Calabria.* Cambridge University Press, Cambridge.

Arlacchi, Pino (1986) *Mafia business: the Mafia ethic and the spirit of capitalism.* Verso, London.

Armstrong, Harvey and Taylor, Jim (2000) *Regional economics and policy.* Blackwell Publishers, Oxford.

Arora, Ashish, Landau, Ralph and Rosenberg, Nathan (eds) (1998) *Chemicals and long-term economic growth.* John Wiley, New York.

Arrighetti, Alessandro and Seravalli, Gilberto (1997) Istituzioni e dualismo dimensionale nell'industria italiana. In Barca, Fabrizio (a cura di) *Storia del capitalismo italiano.* Donzelli Editore, Roma, pp. 335–88.

ASTRID (Associazione per gli studi e le ricerche sulla riforma delle istituzioni democratiche e sull'innovazione nelle amministrazioni pubbliche) (2002) *La riforma del Titolo V della Costituzione e i problemi della sua attuazione*. Mimeo, Roma.

Autorità Garante della Concorrenza e del Mercato (2004) *Indagine conoscitiva sullo stato della liberalizzazione del settore del gas naturale*. In collaborazione tra l'Autorità per l'Energia Elettrica e il Gas e l'Autorità Garante della Concorrenza e del Mercato, Roma.

Aydalot, Philippe (1986) *Milieux innovateurs en Europe*. Groupe de Recherche Européen sur les Milieux Innovateurs (GREMI), Paris.

Bagnasco, Arnaldo (1977) *Tre Italie. La problematica territoriale dello sviluppo italiano*. Il Mulino, Bologna.

Bagnasco, Arnaldo (1999) Teoria del capitale sociale e political economy comparata. *Stato e Mercato*, 57 (Dicembre), 351–72.

Bagnasco, Arnaldo and Oberti, Marco (1998) Italy: 'le trompe-l'œil' of regions. In Le Galès, Patrick and Lequesne, Christian (eds) *Regions in Europe*. Routledge, London, pp.150–65.

Balcet, Giovanni and Enrietti, Aldo (2002) The impact of focused globalisation in the Italian automotive industry. *The Journal of Interdisciplinary Economics*, 13 (1–3), 97–133.

Banca d'Italia (1999) *Supplementi al Bollettino Statistico. Indicatori monetari e finanziari*. Banca d'Italia, Roma. http://www.bancaditalia.it

Banca d'Italia (2003) *Note sull'andamento dell'economia delle regioni italiane nel 2003*. Banca d'Italia, Roma.

Banca d'Italia (2004) *Supplementi al Bollettino Statistico. Indicatori monetari e finanziari*. Banca d'Italia, Roma.

Banca d'Italia (2004a) *Relazione annuale sul 2003*. Banca d'Italia, Roma.

Banfield, Edward (1958) *The moral basis of a backward society*. The Free Press, Chicago.

Barca, Fabrizio (1997) *Storia del capitalismo italiano*.Donzelli Editore, Roma.

Barca, Fabrizio (1999) Introduzione. In Ministero del Tesoro, Bilancio e Programmazione Economica, *La nuova programmazione e il Mezzogiorno*. Donzelli Editore, Roma, pp. 29–39.

Barro, Robert and Sala i Martin, Xavier (1992) Convergence. *Journal of Political Economy*, 100(2), 223–51.

Bathelt, Harald (2000) Persistent structures in a turbulent world: the division of labour in the German chemical industry. *Environment and Planning C*, 18(2), 224–47.

Baumol, William J., Panzar, John C. and Willig, Robert D (1982) *Contestable markets and the theory of industry structure*. Harcourt Brace Jovanovich Inc., New York.

Becattini, Giacomo (ed.) (1987) *Mercato e forze locali: il distretto industriale*. Il Mulino, Bologna.

Becattini, Giacomo (1990) The Marshallian industrial district as a socio-economic notion. In Pyke, Frank, Becattini, Giacomo and Sengenberger, Werner (eds) *Industrial districts and inter-firm cooperation in Italy*. Geneva, ILO, pp. 37–51.

Becattini, Giacomo (1991) The industrial district as a creative milieu. In Benko, Georges and Dunford, Michael (eds) *Industrial change and regional development. The transformation of new industrial spaces*. Belhaven, London, pp. 102–14.

Bélis-Bergouignan, Marie-Claude, Bordenave, Gérard and Lung, Yannick (2000) Global strategies in the automobile industry. *Regional Studies*, 34(1), 41–53.

Berger, Suzanne, Sturgeon, Timothy, Kurz, Constanze, Voskamp, Ulrich and Wittke, Volker (1999) *Globalization, value networks, and national models*. MIT Industrial Performance Centre, Globalization Working Paper, 99–100. MIT, Massachusetts.

Bersani, Pierluigi and Letta, Enrico (2004) *Viaggio nell'economia italiana*. Donzelli Editore, Roma.

Beynon, Huw, Sadler, David, Lewis, Jim and Hudson, Ray (1986) *The growth and internationalisation of Teesside's chemicals industry*. University of Durham, Department of Geography Working Paper 3. University of Durham, Durham, UK.

Borts, George and Stein, Jerome (1964) *Economic growth in a free market*. Columbia University Press, New York.

Boudeville, Jacques (1966) *Problems of regional economic planning*. Edinburgh University Press, Edinburgh.

Boyer, Robert and Freyssenet, Michel (2000) *Les modèles productifs*. Editions La Découverte, Paris.

Brancati, Raffaele (1995) *La questione regionale: federalismo, Mezzogiorno e politica economica*. Donzelli Editore, Roma.

Brenner, Robert (2002) *The boom and the bubble. The U.S. in the world economy*. Verso, London.

Bunel, Jacqueline and Duclos, Constant (2001) *La chimie de base*. Service des Etudes et des Statistiques Industrielles, Direction Générale de l'Industrie, des Technologies de l'Information et des Postes, Paris.

Cafiero, Salvatore and Padovani, Fabio (1999) Problemi di coerenza tra geografia della disoccupazione e delimitazione delle aree incentivabili. In Giannola, Adriano (ed) *Mezzogiorno tra Stato e mercato*. Il Mulino, Bologna, pp. 171–95.

Camera dei Deputati (2000) *Le privatizzazioni in Italia dal 1992*. Studio presentato il 10 ottobre alla Commissione Bilancio, Camera dei Deputati, Roma.

Camera dei Deputati (2004) *Indagine conoscitiva sulle politiche di privatizzazione*. Audizione del Presidente della Consob, Lamberto Cardia. V Commissione Bilancio, Tesoro e Programmazione, 5 maggio 2004, Camera dei Deputati, Roma.

Cameron, Gavin, Muelbauer, John and Snicker, Jonathan (2002) A study in structural change: relative earnings in Wales since the 1970s. *Regional Studies*, 36 (1), 1–11.

Camuffo, Arnaldo and Volpato, Giuseppe (2002) Global sourcing in the automotive supply chain: The case of Fiat Auto 'Project 178' world car. *International motor vehicle program*. MIT, Cambridge, Mass.

Castells, Manuel and Hall, Peter (1994) *Technopoles of the world: the making of twenty-first-century industrial complexes*. Routledge, London.

Castronovo, Valerio (1995) *Storia economica d'Italia*. Einaudi, Torino.

CCFA (Comité des Constructeurs Français d'Automobiles) (1999) *L'industrie automobile française. Analyse et statistiques.* Édition 1999. CCFA, Paris.

CCFA (Comité des Constructeurs Français d'Automobiles) (2003) *L'industrie automobile française. Analyse et statistiques.* Édition 2003. CCFA, Paris.

CEC (2004) *A new partnership for cohesion: convergence, competitiveness, cooperation.* Third report on Economic and Social Cohesion, Office for Official Publications of the European Communities, Luxembourg.

CEFIC (European Chemical Industry Council) (2004) *Facts and figures.* CEFIC, Brussels. http://www.cefic.be/factsandfigures/

Cella, Gian Primo (1997) *Le tre forme dello scambio.* Il Mulino, Bologna.

Censis (Centro Studi Investimenti Sociali) (2003) Notes on an '*Indagine Censis sul sommerso*'. Mimeo, 20 January 2003, Censis Roma.

Censis and Fondazione BNC (2003) *Impresa e criminalità nel Mezzogiorno: meccanismi di distorsione del mercato.* Censis, Roma.

CEPII (Centre d'Etudes Prospectives et d'Informations Internationales) (2002) Banque de données CHELEM (Comptes Harmonisés sur les Echanges et l'Economie Mondiale). CEPII, Paris. http://www.cepii.fr/

CERPEM (Centro Ricerche per il Mezzogiorno) (1990) *Dall'emergenza allo sviluppo.* Rapporto per l'Associazione degli Industriali della provincia di Brindisi. Mimeo, Bari.

Cersosimo, Domenico and Nisticò, Rosanna (2001) *Il distretto dell'intimo.* Donzelli Editore, Roma.

Chandler, Alfred Dupont (1977) *The visible hand: the managerial revolution in American business.* Belknap Press, Cambridge, Mass.

Chapman, Keith (1986) *Chemicals and the Cleveland economy.* Department of Geography, University of Aberdeen, UK.

Chapman, Keith (1991) *The international petrochemical industry.* Blackwell, Oxford.

Chapman, Keith and Edmond, Helen (2000) Mergers/acquisition and restructuring in the EU chemical industry: patterns and implications. *Regional Studies,* 34(8), 753–67.

Chemical Industries Association (1999) *Main world chemical markets by geographic area, 1996–2010.* London, UK.

Chemical Week (1981) *Italy's long road to chemical profits.* December, 2.

Cianciullo, Antonio and Fontana, Enrico (1995) *Ecomafia. I predoni dell'ambiente.* Editori Riuniti, Roma.

Clark, Gordon and Wrigley, Neil (1997) The spatial configuration of the firm and the management of sunk costs. *Economic Geography,* 73(3), 285–304.

Cleveland County Council (1994) *Changes in the chemical industry and the impact on Cleveland in the 1990s.* Cleveland County, Council Cleveland, UK.

Commissione Antimafia (Commissione Parlamentare di Inchiesta sul Fenomeno della Mafia e sulle altre Associazioni Criminali Similari) (1994) *Relazione sulla camorra.* Approvata il 21 Dicembre 1993. XI legislatura, doc. XXIII, n. 12. Tipografia del Senato, Roma.

Commissione Antimafia (Commissione Parlamentare di Inchiesta sul Fenomeno della Mafia e delle altre Associazioni Criminali Similari) (1999) *Relazione sullo stato della lotta alla criminalità organizzata nella provincia di Brindisi*. XIII legislatura, doc. XXIII, n. 31. Tipografia del Senato, Roma.

Commissione Parlamentare di Inchiesta (Commissione parlamentare d'inchiesta sulla attuazione degli interventi per la ricostruzione dei territori della Basilicata e della Campania colpiti dal terremoto del novembre 1980 e febbraio 1981) (1991) *Relazione conclusiva e relazione propositiva ed allegati, 10 volumi*. X legislatura, doc. XXIII, n. 27. Tipografia del Senato, Roma.

Cooke, Philip and Morgan, Kevin (1998) *The associational economy: firms, regions and innovation*. Oxford University Press, Oxford.

Coriat, Benjamin (1991) Technical flexibility and mass production: flexible specialisation and dynamic flexibility. In Benko, Georges and Dunford, Michael (eds) *Industrial change and regional development*. Belhaven, London, pp. 134–58.

Courlet, Claude (2001) *Territoires et régions: les grands oubliés du développement économique*. l'Harmattan, Paris.

Courlet, Claude and Pecqueur, Bernard (1992) Les systèmes industriels localisés en France: un nouvel modèle de développement. In Benko, Georges and Lipietz, Alain (eds.) *Les régions qui gagnent. Districts et réseaux: les nouveaux paradigmes de la géographie économique*. Presses Universitaires de France, Paris, pp. 81–102.

Crevoisier, Olivier and Maillat, Denis (1991) Milieu, industrial organization and territorial production system–towards a new theory of spatial development. In Camagni, Roberto (ed.) *Innovation networks—spatial perspectives*. Belhaven Press, London and New York. pp.13–34.

CUM (Comunità delle Università Mediterranee) (1995) *Sistemi produttivi locali*. Mimeo, Bari.

Dancet, Geert and Rosenstock, Manfred (1995*) State aid control by the European Commission: the case of the automobile sector*. European Commission, Brussels. http://europa.eu.int/comm/ competition/speeches/text/sp1995_043_en.html

De Cecco, Marcello (2000) La politica italiana delle privatizzazioni. In Affinito, Massimo, De Cecco, Marcello and Dringoli, Angelo (2000) *Le privatizzazioni nell'industria manifatturiera italiana*. Donzelli Editore, Roma, pp. xv–xxxvii.

De Nardis, Sergio (2000) *Le privatizzazioni italiane*. Rapporti monografici del CSC. Il Mulino, Bologna.

De Vivo, Paola (2000) *Il Mezzogiorno in Europa: sviluppo e sommerso delle economie territoriali*. Novus Campus, IRES and CGIL, Napoli.

De Vivo, Paola (2001) I molti volti del sud. Fra imprenditorialità diffusa, azione pubblica e società locale. *Il Mulino*, 50 (4), 711–720.

Dei Ottati, Gabi (1995) *Tra mercato e comunità: aspetti concettuali e ricerche empiriche sul distretto industriale*. Franco Angeli, Milano.

Del Monte, Alfredo (1996) *Istituzioni, intervento pubblico e sviluppo del Mezzogiorno*. ISPE, Roma.

Del Monte, Alfredo and Giannola, Adriano (1978) *Il Mezzogiorno nell'economia italiana*. Il Mulino, Bologna.

Della Porta, Donatella (1980) *Mediatori*. Bollati Boringhieri, Torino.

Diamanti, Ilvo (1996) *Il Male del Nord: Lega, localismo, secessione.* Donzelli Editore, Roma.

Dietzenbacher, Erik and Lahr, Michael (eds) (2004) *Wassily Leontief and input-output economics.* Cambridge University Press, Cambridge.

Dixit, Avinash, and Stiglitz, Joseph (1977) Monopolistic competition and optimal product diversity. *American Economic Review,* 67(6), 297–308.

Dunford, Michael and Perrons, Diane (1983) *The arena of capital.* Macmillan, London and Basingstoke.

Dunford, Michael (1988) *Capital, the state and regional development.* Pion, London.

Dunford, Michael (1990) 'Theories of regulation', Environment and Planning D: Society and Space, 8, 297–321.

Dunford, Michael (1991) Industrial trajectories and social relations in areas of new industrial growth. In Benko, Georges and Dunford, Michael (eds) *Industrial change and regional development. The transformation of new industrial space.* Belhaven, London, pp. 51–82.

Dunford, Michael (1996) Disparities in employment, productivity and output in the EU: the roles of labour market governance and welfare regimes, *Regional Studies,* 30(4), 339–57.

Dunford, Michael (1997) Divergence, instability and exclusion: regional dynamics in Great Britain. In Lee, Roger and Wills, Jane (eds) *Geographies of economies,* Arnold, London, pp. 259–77.

Dunford, Michael (2001) Regional inequality in the EU: the roles of distribution and growth. In Zimmermann, Friedrich and Janschitz, Susanne (eds*) Regional policies in Europe: key opportunities for regions in the 21st century.* Leykam, Graz, pp. 65–92.

Dunford, Michael (2002) Italian regional evolutions. *Environment and Planning A,* 34, 657–94.

Dunford, Michael, Cho, Myung Rae, Fernandes, Ana, Musyck, Bernard, Sadowski, Bert and Tsenkova, Sasha (1993) The organisation of production and territory: small firm systems. *International Journal of Urban and Regional Research,* 17(1), 132–6.

Dunford, Michael and Hudson, Ray (1996) *Successful European regions. Northern Ireland learning from others.* Northern Ireland Economic Council, Belfast.

Dunford, Michael and Smith, Adrian (2000) Catching up or falling behind?: economic performance and the trajectories of economic development in an enlarged Europe. *Economic Geography,* 76(2), 169–95.

Dunin-Wasowicz, Stefan (2002) *Uncertainty and the localization of industrial districts.* Paper presented at a workshop on 'The emerging industrial architecture of the wider Europe: the co-evolution of industrial and political structures', School of Slavonic and East European Studies, University College London, 22 March 2002.

Economia Italiana (1997) L'industria chimica e farmaceutica. *Economia Italiana,* 3, 697–712.

Economist, The (2004) Soaring Sorin. An atypical Italian success. *The Economist,* 10–16 January, p.49.

ECOTER (1993) *Pacchetto localizzativo Brindisi. Rapporto finale,* Roma.

ENI (Various years) *Fact Book*. ENI, Roma. http://www.eni.it

ENIChem (1995) *Annual Report*. Direzione amministrativa e unità di coordinamento immagine di ENIChem, Milano.

ENIChem (1999) *Annual Report*. Direzione amministrativa e unità di coordinamento immagine di ENIChem, Milano.

Enrietti, Aldo and Lanzetti, Renato (2002) *Fiat Auto: le ragioni della crisi e gli effetti a livello locale*. Quaderni del Dipartimento di Scienze Economiche, Bergamo.

EUROSTAT (Statistical Office of the European Communities) (2000) *EUROSTAT new chronos database*. EUROSTAT, Luxembourg.

Fagerberg, Jan, Verspagen, Bart and Caniëls, Marjolein (1997) Technology, growth and unemployment across European regions. *Regional Studies*, 31, 457–66.

Fagerberg, Jan (2003) *What have we learned (before and) after Myrdal? Growth theory and disparities*. Background paper for presentation at the conference 'Cohesion reform in a larger Union', College of Europe, Bruges, 3–4 April.

Fazio, Antonio (2004) Le considerazioni finali. Relazione alla Banca d'Italia. *Il Sole-24 Ore*, 1 Giugno.

Federchimica (Federazione nazionale dell'industria chimica) (1991) *Struttura, risultati e problemi dell'industria chimica in Italia*. Federchimica, Milano.

Federchimica (2001) *Cambiamenti nella chimica mondiale ed effetti sull'industria e sulle imprese italiane*. XXV Convegno Nazionale di Economia e Politica Industriale, Bologna, Italia.

Federchimica (2003a) *L'industria chimica in cifre*. Federchimica, Milano.

Federchimica (2003b) *Rapporto Federchimica 2003*. Federchimica, Milano.

Federchimica (2004) direct data.

FIAT (2002) Third quarter 2002 results presentation: October 2002. Gruppo FIAT, Torino.

FIAT (Various years) *Bilancio consolidato e di esercizio and Relazione sulla gestione*. Various years. Gruppo FIAT, Torino.

Financial Times (1997) Survey on the Chemical Industry. *Financial Times*, 25 September.

FIOM-Piemonte (Federazione Impiegati Operai Metallurgici-Piemonte) (2000) *FIAT Auto e Aziende terziarizzate (Piemonte): Divenire organici 1989–2000*. FIOM-Piemonte, Torino. http://www.pmt.cgil.it/

FIOM-Piemonte (Federazione Impiegati Operai Metallurgici-Piemonte) (2002) *L'industria dell'auto a Torino, 1997–2003. Una strategia dell'abbandono?* Sistema Informativo della FIOM Torino/Piemonte sulla FIAT. FIOM-Piemonte, Torino. http://www.pmt.cgil.it/

Florio, Massimo (ed.) (1991) *Grande impresa e sviluppo locale*. Clua, Ancona.

Franzini, Maurizio and Giunta, Anna (1999) Grande impresa e Mezzogiorno: alcuni elementi di riflessione. In Annibaldi, Cesare e Berta, Giuseppe (a cura di), *Grande impresa e sviluppo italiano. Studi per i cento anni della FIAT*. Il Mulino, Bologna, pp. 209–263.

Fuà, Giorgio (1983) L'industrializzazione nel Nord-Est e nel Centro. In Fuà, Giorgio and Zacchia, Carlo (eds) *Industrializzazione senza fratture*. Il Mulino, Bologna, pp. 7–46.

Fujita, Masahisa, Krugman, Paul and Venables, Anthony J. (1999) *The spatial economy*. The MIT Press, Cambridge, Mass.

Galvão, Antonio Carlos E. (1998) Inovações e desenvolvimento regional: alguns elos da discussão recente. *Revista Econômica do Nordeste*, 29(4), 387–405.

Galvão, Antonio Carlos E. (2002) A guerra fiscal no Brasil, A review. *Regional Studies*, 36(9), 1109–10.

Garofoli, Gioacchino (1991) The Italian model of spatial development in the 1970s and 1980s. In Benko, Georges and Dunford, Michael (eds) *Industrial change and regional development. The transformation of new industrial spaces.* Belhaven, London, pp. 85–101.

Gereffi, Gary, Korzeniewicz, Miguel and Korzeniewicz, Roberto P. (1994) Introduction: global commodity chains. In Gereffi, Gary and Korzeniewicz, Miguel (eds.) *Commodity chains and global capitalism.* Praeger, Westport, pp. 1–14.

Giannetti, Mariassunta (2002) The effects of integration on regional disparities: convergence, divergence or both?. *European Economic Review*, 46, 539–67.

Giannetti, Renato (1996) *Radici storiche ed esperienza dell'intervento straordinario nel Mezzogiorno. Collezioni studi meridionali.* Bibliopolis, Roma.

Giannola, Adriano (ed) (1999) *Mezzogiorno tra Stato e mercato.* Il Mulino, Bologna.

Giannola Adriano (2000) La nuova programmazione. Evoluzione e restaurazione. *Rivista economica del Mezzogiorno*, 3, 403–22.

Ginsborg, Paul (1998) *L'Italia del tempo presente.* Einaudi, Torino.

Giordano, Benito (2000) Italian regionalism or 'Padanian' nationalism–the political project of the Lega Nord in Italian politics. *Political Geography*, 19 (4), 445–71.

Graziani, Augusto (a cura di) (1989) *L'economia italiana 1945–1970.* Il Mulino, Bologna.

Greco, Lidia (2002) *Industrial redundancies. A comparative analysis of the chemical and clothing industries in the UK and Italy.* Ashgate, Aldershot.

Greco, Lidia (2003) Institutional and industrial changes in the Italian South: The case of Brindisi. *Modern Italy*, 8(2), 187–201.

Greco, Lidia (2004) An institutionalist approach to redundancies in the chemical industry. The cases of Teesside (UK) and Brindisi (I). *European Urban and Regional Studies*, 11(2), 141–55.

Haggett, Peter (2001) *Geography: a global synthesis.* Prentice Hall, New York.

Hayami, Yujiro (1997) *Development economics. From the poverty to the wealth of nations.* Clarendon Press, Oxford.

Hooghe, Liesbet (1995) Subnational mobilisation in the European Union. *West European Politics*, 18 (3), 175–98.

Hudson, Ray (1983) Capital accumulation and chemicals production in Western Europe in the post-war period. *Environment and Planning A*, 15, 105–22.

Hudson, Ray (1996) Producing an industrial wasteland: capital, labour and the state in north east England. In Martin, Ron and Rowthorn, Bob (eds) *The geography of de-industrialisation.* MacMillan, London.

Hudson, Ray (1999) The learning economy, the learning firm and the learning region: a sympathetic critique of the limits to learning. *European Urban and Regional Studies*, 6(1), 59–72.

Hudson, Ray (2000) *Production, places and environment: changing perspectives in economic geography.* Longman, London.

Hudson, Ray (2002a) Restructuring industry and territory: the experience of Europe's regions. Book Review. *Regional Studies*, 36(4), 4451.

Hudson, Ray (2002b) Changing industrial production systems and regional development in the New Europe. *Transactions of the Institute of British Geographers*, 27(3), 262–81.

Hudson, Ray (2003) Fuzzy concepts and sloppy thinking: reflections on recent developments in critical regional studies. *Regional Studies*, 37(6/7), 741–6.

Humphrey, John and Schmitz, Hubert (1998) Trust and inter-firm relations in developing and transition economies. *The Journal of Development Studies* 34(4), 32–61.

Humphrey, John and Schmitz, Hubert (2002) How does insertion in global value chains affect upgrading in industrial clusters?. *Regional Studies*, 36(9), 1017–27.

Hymer, Stephen (1975) The multinational corporation and the law of uneven development. In Radice, Hugo (ed.) *International firms and modern imperialism*. Penguin, Harmondsworth, pp. 40–8.

ICE (Istituto Nazionale per il Commercio Estero) (2001) *L'Italia nell'economia internazionale*. Rapporto ICE 2000–2001. ICE, Roma. http://www.ice.gov.it

ICE (2002) *L'Italia nell'economia internazionale*. Rapporto ICE 2001–2002. ICE, Roma. http://www.ice.gov.it

ICE (2003) *L'Italia nell'economia internazionale*. Rapporto ICE 2002–2003. ICE, Roma. http://www.ice.gov.it

ICE (2003b) *Banche dati dell'ICE*. ICE, Roma. http://www.ice.gov.it

ICE (2004) *Commercio estero e attività internazionali delle imprese—2003*. Annuario ICE-ISTAT. http://www.ice.gov.it

IMF (International Monetary Fund) (2004) *Direction of trade statistics*. IMF, Washington DC.

Infocamere (2001) *Movimprese*. Infocamere, Roma. http://www.infocamere.it/movi/movi.htm

IPI (Istituto per la promozione industriale) *L'esperienza italiana dei distretti industriali. Ricerca promossa dal Ministero delle Attività Produttive e dall'IPI*. IPI, Roma.

IRES (Istituto Ricerche Economiche) (2003) *Piemonte economico sociale 2002*. IRES, Torino.

ISTAT (Istituto centrale di statistica) (1975) *Annuario di contabilità nazionale*. ISTAT, Roma.

ISTAT (1982) *Annuario di contabilità nazionale*. XI (2). ISTAT, Roma.

ISTAT (1983) *Annuario di contabilità nazionale*. ISTAT, Roma.

ISTAT (1990) *Sommario storico di statistiche sulla popolazione, 1951–1987*. ISTAT, Roma.

ISTAT (1996) *Censimento intermedio dell'industria e dei servizi*. ISTAT, Roma.

ISTAT (1998) *Conti economici regionali, 1980–1996*. ISTAT, Roma.

ISTAT (1999) *Commercio estero e attività internazionali delle imprese-1998*. ISTAT, Roma.

ISTAT (2000) *Variabili di 'rottura'*. ISTAT, Roma.

ISTAT (2002) *Conti economici territoriali-provinciali, 1995–1999*. ISTAT, Roma.

ISTAT (2002c) *Indicatori di 'contesto chiave' e variabili di 'rottura' per la valutazione del QCS 2000–06. Informazione statistica territoriale e settoriale per le politiche strutturali 2001–2008.* ISTAT, Roma.

ISTAT (2002b) *Conti economici territoriali-regionali, 1995–2000.* Anno 2000. ISTAT, Roma.

ISTAT (2003) *Indicatori di 'contesto chiave' e variabili di 'rottura' per la valutazione del QCS 2000–06. Informazione statistica territoriale e settoriale per le politiche strutturali 2001–2008.* ISTAT, Roma.

ISTAT (2003b) *Conti economici regionali, 1995–2001.* Anno 2001. ISTAT, Roma.

ISTAT (2003c) *Occupazione e valore aggiunto nelle provincie, 1995–2000.* ISTAT, Roma.

ISTAT (2003d) *Statistiche del commercio estero.* ISTAT, Roma, http://www.coeweb.istat.it

ISTAT (2004a) La misura dell'economia sommersa secondo le statistiche ufficiali, Anno 2002. *Statistiche in breve: conti nazionali.* ISTAT, Roma.

ISTAT (2004b) *Censimento generale dell'industria e dei servizi, 2001.* ISTAT, Roma. http://dwcis.istat.it/cis/index.htm

ISTAT (2004c) *Censimento generale dell'industria e dei servizi, 1971, 1981 and, 1991.* ISTAT, Roma. http://dwcis.istat.it/cis/index.htm

Istituto Giuglielmo Tagliacarne, (1961; 1962; 1963; 1964; and 1965) Reddito netto prodotto dal settore privato e dalla pubblica amministrazione, milioni di lire correnti, 1951–1963, *Moneta e Credito. Rivista Trimestrale della Banca Nazionale del Lavoro*, 1961: 521–5; *Moneta e Credito*, 1962: 415–19; *Moneta e Credito* 1963: 434–9; *Moneta e Credito*, 1964: 392–7; and *Moneta e Credito*, 1965: 386–91.

Istituto Giuglielmo Tagliacarne (1970) Reddito netto interno al costo dei fattori prodotto dal settore privato e dalla pubblica amministrazione, milioni di lire correnti, 1963–70, *Moneta e Credito*, 1970: 467–501.

Istituto Tagliacarne, (1973) *Il reddito prodotto nelle province italiane 1951–1971.* Franco Angeli, Milano.

Italdesign (2001) *Italdesign from incorporation up to today.* http://www.italdesign.it

Jeffrey, Charlie (2000) Sub-National Mobilization and European Integration: Does it make any difference?. *Journal of Common Market Studies*, 38 (1), 1–23.

Kaldor, Nicholas (1973) *Teoria dell'equilibrio e teoria della crescita.* Franco Angeli, Milano.

Krugman, Paul (1998) *The role of geography in development.* A paper prepared for the Annual World Bank Conference on Development Economics. Washington, D.C., 20–1 April.

Krugman, Paul (1999) *Some chaotic thoughts on regional dynamics.* http://web.mit.edu/krugman/

Krugman, Paul and Venables, Anthony (1996) Integration, specialization and adjustment. *European Economic Review*, 40(3–5), 959–67.

La Malfa, Giorgio and Coppola, Roberto (1974) *Il futuro dell'industria chimica.* Franco Angeli, Milano.

La Spina, Antonio (2003) *La politica per il Mezzogiorno.* Il Mulino, Bologna.

Leborgne, Daniel and Lipietz, Alain (1988) New technologies, new modes of regulation: some spatial implications. *Environment and Planning D*, 6 (3), 263–80.

Lipietz, Alain and Leborgne, Danielle (2001) Two social strategies in the production of new industrial spaces. In Benko, Georges and Dunford, Michael (eds) *Industrial change and regional development. The transformation of new industrial spaces*, Belhaven, London, pp. 27–50.

Lega Ambiente (2001) *La chimica in Italia: rischi ambientali e sanitari.* www.legambiente.it

Lega Ambiente (2001) *Rapporto Ecomafia 2001.* Lagambiente. Roma.

Lipietz, Alain (1977) *Le capital et son espace.* François Maspero, Paris.

Lovering, John (1999) Theory led by policy: the inadequacies of the 'New Regionalism' (Illustrated from the case of Wales). *International Journal of Urban and Regional Research*, 23(2), 379–95.

Magatti, Mauro (1991) *Mercato e forze sociali.* Il Mulino, Bologna.

Maglia, Vittorio (2001) *Cambiamenti nella chimica mondiale ed effetti sull'industria e sulle imprese italiane.* Federchimica, Milano.

MAP (Ministero delle Attività Produttive) and IPI (Istituto per la Promozione Industriale) (2002) *L'esperienza italiana dei distretti industriali.* IPI, Roma.

Mariotti, Sergio and Treves, Liliana (1999) Grande impresa e Mezzogiorno: la presenza della Fiat. In Annibaldi, Cesare and Berta, Giuseppe (1999) *Grande impresa e sviluppo italiano. Studi per i cento anni della Fiat.* Il Mulino, Bologna, pp. 299–337.

Marshall, Alfred (1961) *Principles of economics: an introductory volume.* Macmillan, London.

Martin, Philippe and Ottaviano, Gianmarco (2001) Growth and agglomeration. *International Economic Review*, 42(4), 947–68.

Martin, Ron (1999) The new 'geographical turn' in economics: some critical reflections. *Cambridge Journal of Economics*, 23, 63–91.

Martin, Ron and Sunley, Peter (2001), Rethinking the 'economic' in economic geography: broadening our vision or losing our focus?. *Antipode*, 33(2), 148–61.

Martinelli, Flavia (1998) *The governance of post-war development and policy in Southern Italy. Notes for a critical reappraisal.* Paper presented at the Second European Urban and Regional Studies Conference, 17–20 September 1998, Durham-UK.

Maskell, Peter and Malmberg, Anders (1999) Localised learning and industrial competitiveness. *Cambridge Journal of Economics*, 23, 167–85.

Massey, Doreen (1979) In what sense a regional problem?. *Regional Studies*, 13, 233–43.

Massey, Doreen (1995) *Spatial divisions of labour: social structures and the geography of production.* Routledge, London.

Meldolesi, Luca (1996) Al Sud qualcosa di nuovo. Piccola impresa e occupazione. *Nord e Sud*, 43(11–12), 7–42.

Meldolesi, Luca (1999) *Un balzo in avanti. Dove va l'economia italiana.* Editori Laterza, Bari.

Mele, Renato (1975) *L'industria manifatturiera della Puglia.* CESAN, Napoli.

Mingione, Enzo (1993) Italy: The resurgence of regionalism. *International Affairs*, 69, 305–18.

Ministero dell'Economia e delle Finanze (2002) *Sistema nazionale di valutazione del QCS.* Dipartimento per le politiche di sviluppo, Roma.

Ministero dell'Economia e delle Finanze (2003) *Monitoraggio degli interventi comunitari. Programmazione 2000–2006.* Ministero dell' Economia e delle Finanze. Roma.

Morgan, Kevin (1997) The learning region: Institutions, innovation and regional renewal. *Regional Studies* 31(5), 491–503.

Mutti, Antonio (1994) Il particolarismo come risorsa. Politica ed economia nello sviluppo abruzzese. *Rassegna Italiana di Sociologia*, 4, 451–518.

Myrdal, Gunnar (1957) *Economic theory and underdeveloped regions.* Methuen, London.

Neary, Peter (2001) Of hype and hyperbolas: introducing the new economic geography. *Journal of Economic Literature*, XXXIX, 536–61.

Northern Region Strategy Team (1976) *The chemicals industry in the Northern Region.* Northern Region Strategy Team, Newcastle upon Tyne.

OECD (Organisation for Economic and Cooperation Development) (2001) *Territorial Review – Italy.* OECD, Paris.

OECD (2002) *Measuring the non-observed economy: a handbook.* OECD, Paris.

OICA (Organisation Internationale des Constructeurs d'Automobiles) (2002) *Statistics.* OICA, Paris. http://www.oica.net

O'Leary, Eoin (2001) Convergence of living standards among Irish regions: the role of productivity, profit outflows and demography, 1960–1996. *Regional Studies*, 35 (3) 197–205.

Osservatorio per il Settore Chimico (2004) *Struttura territoriale.* http://www.osservatoriochimico.it

Osservatorio Provinciale sul Mercato di Lavoro della Provincia di Torino e IPL (Istituto per il Lavoro, Bologna) (2002) *La mappatura della filiera dell'auto nella provincia di Torino.* Provincia di Torino, Torino.

Osservatorio sulla Camorra (1993) *Relazione sulla camorra.* Approved by the Commissione Antimafia (Commissione Parlamentare di Inchiesta sul Fenomeno della Mafia e delle altre Associazioni Criminali Similari) on 21st December 2003. Osservatorio sulla Camorra, Napoli.

Paci, Massimo (a cura di) (1980) *Famiglia e mercato del lavoro in un'economia periferica.* Franco Angeli, Milano.

Pagano, Ugo and Trento, Sandro (2002) *Continuity and Change in Italian Corporate Governance: The Institutional Stability of One Variety of Capitalism.* Università degli Studi di Siena, Dipartimento di Economia Politica, n. 366 Settembre.

Peck, Jamie and Tickell, Adam (1995) Social regulation after-Fordism: regulation theory, neoliberalism and the global-local nexus. *Economy and Society* 24(3): 357–86

Pecqueur, Bernard (2000) *Le développement local.* Syros/Alternatives Économiques, Paris.

Penn World Tables: Heston, Alan, Summers, Robert and Aten, Bettina (2002) *Penn World Table Version 6.1,* October. Center for International Comparisons at the University of Pennsylvania (CICUP), Philadelphia.

Perna, Tonino (1994) *Lo sviluppo insostenibile.* Liguori, Napoli.

Perrone, Nico (2001) *Enrico Mattei.* Il Mulino, Bologna.

Perrons, Diane (2001) Towards a more holistic framework for economic geography. *Antipode*, 33(2), 208–15.

Perrons, Diane (2003) The new economy, labour market inequalities and the work-life balance. In Martin, Ron and Morrison, Philippe (eds) *Geographies of labour market inequality*. Routledge, London, pp. 45–61.

Perrons, Diane (2004) *Globalization and social change: people and places in a divided world*. Routledge, London, pp. 45–61.

Perroux, François (1970) Note on the concept of growth poles. English translation by Gates, Linda and McDermott, Anne Marie of Perroux (1955) Note sur la notion de 'pole de croissance', Economie Appliquée, 7, 307–320. In McKee, David L., Dean, Robert D. and Leahy, William H. (eds) *Regional economics: theory and practice*. New York: The Free Press, pp. 93–104.

Petrusewicz, Marta (2001) The Mezzogiorno: a bias for hope?. *Modern Italy*, 6(1), 63–7.

Pettigrew, Andrew (1985) *The awakening giant*. Blackwell, London.

Piattoni, Simona (1994) Regions and economic growth: the experience of Abruzzo and Puglia. In Bullmann Udo (ed) *Die Politik der dritten Ebene. Regionen im Europa der Union*. Nomos, BadenBaden, pp. 173–96.

Piattoni, Simona (1996) *Local political class and economic development. The cases of Abruzzo and Puglia in the 1970s and 1980s*. Ph.D. thesis, MIT, Cambridge, Mass.

Pichierri, Angelo (2002) La regolazione dei sistemi locali: temi e tendenze. *Sociologia del Lavoro*, 88, 32–40.

Pinch, Steven and Henry, Nick (1999) Paul Krugman's Geographical economies, industrial clustering and the British motor sport industy. *Regional Studies*, 33(9), 815–29.

Piore, Michael and Sabel, Charles (1984) *The Second Industrial Divide*. Basic Books, New York.

Pizzorno, Alessandro (1993) *Le radici della politica assoluta*. Feltrinelli, Milano.

Porter, Michael (1990) *The competitive advantage of nations*. Macmillan Press, London.

Porter, Michael (1998) *Competitive strategy: techniques for analysing industries and competitors*. Free Press, New York.

Powell, Walter and DiMaggio, Paul (eds) (1991) *The new institutionalism in organisational analysis*. Chicago University Press, Chicago.

Provincia di Brindisi, (1999) *Rapporto 1998 sull'economia e la società della provincia di Brindisi*. Provincia di Brindisi, Brindisi, Italia.

Pugliese, Enrico et al. (1973) *Mezzogiorno e classe operaia*. Coines Edizioni, Roma.

Pulignano, Valeria (1999) Gli effetti del teamwork sull'organizzazione sindacale alla Rover e alla Fiat. In Sivini, Giordano (ed.) *Oltre Melfi. La fabbrica integrata, bilancio e comparazioni*. Rubbettino, Soveria Mannelli.

Putnam Robert, Leonardi, Robert and Nanetti, Raffaella (1993) *Making democracy work: civil tradition in modern Italy*. Princeton University Press, Princeton.

Quah, Danny (1996) *The invisible hand and the weightless economy*. Centre for Economic Performance Occasional Paper, 12. London School of Economics, London.

Rampini, Federico (2004) Industria debole tra lo Stato e le banche. Intervista a Guido Rossi. *Repubblica*, 3 Giugno.

Ranci, Pippo and Vaccà, Sergio (1979) *L'industria petrolchimica in Italia: anatomia di una crisi*. Franco Angeli, Milano.

Regionale Basilicata (2000) *Documentazione e considerazioni finali sull'attuazione dell'accordo di programma della Val Basento*. Regione Basilicata, Deliberazione del Consiglio Regionale, 29 Febbraio.

Reich, Robert (1991) *The work of nations: preparing ourselves for 21st century capitalism*. Knopf, New York.

Reyneri, Emilio (1998) *Sociologia del mercato del lavoro*. Il Mulino, Bologna.

R&S (2000) *Le privatizzazioni in Italia dal 1992*. Studio commissionato dalla Commissione Bilancio, Tesoro e Programmazione a Mediobanca, Milano.

Rohmer, Paul (1990) Endogenous technological change. *Journal of Political Economy*, 98, 71–102.

Rosenstein-Rodan, Paul (1943) Problems of industrialisation in Eastern and South-Eastern Europe. *Economic Journal*, 53, 202–11.

Rossi, Ugo (2003) *New regionalism contested: Some remarks in the light of the case of the Mezzogiorno of Italy*. Paper presented at the Regional Studies Association Conference 'Reinventing regions in a global economy', Pisa, Italy, 12–15 April.

Sapir, Jacques (1999) *Understanding the shadow economy and economic criminalisation expansion in transition economies*. Document de Travail, Centre d'Etudes des Modes d'Industrialisation. Ecole des Hautes Etudes en Sciences Sociales, Paris.

Sayer, Andrew (1997) The dialectic of culture and economy. In Lee, Roger and Wills, Jane *Geographies of economies*. Arnold, London, pp. 16–26.

Sayer, Andrew and Morgan, Kevin (1985) A modern industry in a declining region: links between method, theory and policy. In Massey, Doreen and Meegan, Richard (eds) *Politics and method*. Methuen, London, pp. 147–68.

Schneider, Friedrich and Ernste, Dominik (2000) Shadow economies. Size, causes and consequences. *Journal of Economic Literature*, 38(1), 77–114.

Scott, Allen J. (1988) *New industrial spaces: flexible production organization and regional development in North America and Western Europe*. Pion, London.

Scott, Allen J. (1998) *Regions and the world economy: The coming shape of global production, competition and political order*. Oxford University Press, Oxford.

Signorelli, Amalia (1983) *Chi può e chi aspetta. Giovani e clientelismo in un'area interna del Mezzogiorno*. Liguori, Napoli.

Signorini, Luigi (ed.) (2000) *Lo sviluppo locale. Un'indagine della Banca d'Italia sui distretti industriali*. Donzelli Editore, Roma.

Sikka, Prem (2003). The role of offshore financial centres in globalization. *Accounting Forum*, 27, 365–99.

Smith Adrian, Rainnie, Al, Dunford, Michael, Hardy, Jane, Hudson, Ray and Sadler, David (2002) Networks of value, commodities and regions: reworking divisions of labour in macro-regional economies. *Progress in Human Geography*, 26(1), 41–63.

Sole 24-ore (2001) Rapporti. Mezzogiorno. Il Sole 24–ore, 25 Giugno.

Solow, Robert (1956) A Contribution to the Theory of Economic Growth. *Quarterly Journal of Economics*, 70, 65–94.

Soskice, David (1989) Perchè variano i tassi di disoccupazione: economia e istituzioni nei paesi industriali avanzati. *Stato e Mercato*, 27, 333–78.

Soskice, David and Hall, Peter (2001) *Varieties of capitalism. The challenges facing contemporary political economies*. Oxford, University Press Oxford.

Staber, Udo (2001) Spatial proximity and firm survival in a declining industrial district: the case of knitwear firms in Baden-Württemberg. *Regional Studies*, 35(4), 329–41.

Storper, Michael (1995) The resurgence of regional economies, ten years later: the region as a nexus of untraded interdependencies. *European Urban and Regional Studies*, 2(3), 191–221.

Storper, Michael and Salais, Robert (1997) *Worlds of production. The action frameworks of the economy*. Harvard University Press, Cambridge, Mass.

Streeck, Wolfgang (1989) Skills and the limits of neo-liberalism: the enterprise of the future as a place of learning. *Work, Employment and Society*, 3, 90–104.

Streeck, Wolfgang (2001) La transformation de l'organisation de l'entreprise en Europe: une vue d'ensemble. In Toffut, Jean-Philippe (ed.) *Institutions et croissance. Les chances d'un modèle économique européen*. Editions Albin-Michel, Paris, pp. 175–230.

Sturgeon, Timothy (2000) *Turn-key production networks: a new American model of industrial organization*, Industrial Performance Center, Massachusetts Institute of Technology, Cambridge, MA.

Sturgeon, Timothy and Florida, Richard (1999) *The world that changed the machine: globalization and the economic geography of the automotive industry*. Report to the Sloan Foundation, August, New York.

SVIMEZ (Società per lo Sviluppo del Mezzogiorno) (1971) *Gli investimenti industriali agevolati nel Mezzogiorno, 1951–1968*. Giuffrè Editore, Milano.

SVIMEZ, (1993) *I conti economici del Centro-Nord e del Mezzogiorno nel ventennio 1970–1989*. Il Mulino, Bologna.

SVIMEZ (1999) *Rapporto 1999 sull'economia del Mezzogiorno*.Il Mulino, Bologna.

SVIMEZ (2000) *Rapporto 2000 sull'economia del Mezzogiorno*.Il Mulino, Bologna.

SVIMEZ (2003) *Rapporto 2003 sull'economia del Mezzogiorno*. Il Mulino, Bologna.

Tecnoparco Basento (1994) direct data.

Teesside Chemical Initiative (1998) *Teesside chemical industry survey 1997*. Survey and report commissioned by Teesside TEC, Chemical Employers' Group and Teesside Chemical Initiative. Teeside Chemical Initiative, Cleveland, UK.

Tees Valley Joint Strategy Unit (1999) *Developing a sustainable, competitive, diversified economy*. Tees Valley Joint Strategy Unit, Cleveland, UK.

Terrasi, Maria (1999) Convergence and divergence across Italian regions. *The Annals of Regional Science*, 4, 491–511.

Théret, Bruno (1994) To have or to be: on the problem of the interaction between state and economy and its 'solidarist' mode of regulation. *Economy and Society*, 23(1), 1–46.

Toniolo, Gianni and Visco, Vincenzo (2004) *Il declino economico dell'Italia. Cause e rimedi*. Mondadori, Milano.

Trigilia Carlo (1992) *Sviluppo senza autonomia*. Il Mulino, Bologna.

Trigilia, Carlo (1999) Il Mezzogiorno in cammino e la politica zoppa. *Il Mulino*, 48 (5), 814–24.

Trigilia, Carlo (2002) Dalla politica alle politiche: comuni e interessi locali. In Catanzaro, Raimondo, Piselli, Fortunata, Ramella, Francesco, Trigilia, Carlo (a cura di) *Comuni nuovi. Il cambiamento nei governi locali.* Il Mulino, Bologna, pp. 579–602.

Turani, Giuseppe (2004) *Perché abbiamo il peggior capitalismo del mondo.* Sperling e Kupfer Editori, Milano.

Usui, Chikako and Colignon, Richard (1996) Corporate restructuring: converging world pattern or societally specific embeddedness?. *The Sociological Quarterly,* 37 (4), 551–78.

Viesti, Gianfranco (1996) Che succede nell'economia del Mezzogiorno? Le trasformazioni 1990–1995. *Meridiana,* 26–7, 91–130.

Viesti, Gianfranco (2000) *Il Mezzogiorno dei distretti.* Donzelli Editore, Roma.

Viesti, Gianfranco (2000a) *Come nascono i distretti industriali.* Editori Laterza, Bari.

Viesti, Gianfranco (2001) *Basilicata: Una transizione complessa. Struttura dell'economia ed evoluzione 1995–1998.* Mimeo, Bari.

Viesti, Gianfranco (2003) *Abolire il Mezzogiorno.* Editori Laterza, Bari.

Viesti, Gianfranco and Prota, Francesco (2004) *Le politiche regionali dell'Unione Europea.* Il Mulino, Bologna.

Volpato, Giuseppe (1999) Il processo di internazionalizzazione della FIAT in campo automobilistico, 1988–1999. In Annibaldi, Cesare and Berta, Giuseppe (eds) *Grande impresa e sviluppo italiano. Studi per i cento anni della FIAT, volume 1.* Il Mulino, Bologna, pp. 343–412.

Volpato, Giuseppe (2002*) Il marketing strategico di Fiat Auto. Una crisi che viene da lontano.* Paper presented at a seminar on 'La crisi Fiat. Strategie d'impresa e politiche industriali', 27 Novembre. Facoltà di Economia, Università di Urbino, Urbino.

Willan, Philip (1991) *Puppetmasters: the political use of terrorism in Italy.* Constable and Company, London.

Womack, James, Jones, Daniel and Roos, Daniel (1990) *The machine that changed the world: the story of lean production.* Harper, New York.

Zamagni, Vera (2000) *The rise and fall of the Italian chemical industry (1950s–1990s).* Paper prepared for the colloquium on 'The Global Chemical Industry since the Petrochemical Revolution'. Milan ASSI, October.

Zamagni, Vera (2003) L'ENI e la chimica. *Energia,* 2, 16–24.

Zukin, Sharon and DiMaggio, Paul (1990) *Structures of capital.* Cambridge University Press, Cambridge.

Notes

Notes to Chapter 2

1   The concept of 'economies of scope' was introduced by Baumol, Panzar and Willig (1982) in relation to the cost functions of a multi-product firm. The intention was to identify a type of scale economy found where a multiproduct firm produces joint products. In this case the source of the productivity gains lies in the fact that certain inputs can be used simultaneously to make different products. There are economies of scope if the average costs of the multiproduct firm are less than the sum of the average costs of making each of the products separately. If y_1 and y_2 are two joint products, and $c(y)$ is the general cost function, the condition of existence of economies of scope is that:

$$c(y_1, y_2) < c(y_1, 0) + c(0, y_2) \qquad (3)$$

while the relative magnitude of the economies of scope (sc) is given by the equation :

$$sc = \frac{c(y_1, 0) + c(0, y_2) - c(y_1, y_2)}{c(y_1, y_2)} \qquad (4)$$

This argument can be extended to cover not just different products (product-mix) but also different models (range-mix).

Notes to Chapter 3

1   A second set of imperatives confronts wage earners: the requirement that they acquire the skills and qualities required to make them employable, and that they accept occupational and geographical mobility.

2   A learning curve effect exists if there is an inverse relationship between average costs and the cumulative volume of output obtained from an installation whose capacity is assumed fixed. Empirical studies have established that unit costs decline at a rate that reflects increases in the efficiency of the direct labour force. Subsequently learning-curve effects were incorporated into a wider concept of experience curve effects. Experience curve effects include increases in efficiency associated not just with direct work but also with indirect work, organizational arrangements and even small technical improvements. To them can be added the effects of classical economies of scale that are associated with increases in capacity to get the long-term productivity curve.

Notes to Chapter 4

1   The G6 was subsequently enlarged, first in 1976–7 to form the G7 with the addition of Canada and the European Community, and then in 1998–2003 to form the G8 with the addition of Russia.
2   GDP at factor cost is a measure of the value added in the production of all the goods and services an economy creates. At first, this new wealth is distributed to the people/organizations involved in their production in the shape of property incomes (such as profits, interest and rent) and wages. Accordingly, GDP at factor cost is also equal to the sum of the compensation of employees and the gross operating surplus, itself made up of profits and the income, including profits, of the self-employed. At this stage, mechanisms of redistribution step in. GDP at factor cost is equal to GDP at market prices less indirect taxes net of subsidies. Household income depends on the primary distribution of income and on the way it is altered by the redistribution of income and wealth through government taxation and expenditure. The resulting distribution of income determines the distribution of claims over the goods and services produced. These claims are effected when these incomes are spent or saved. GDP can also be estimated by measuring these expenditures. More specifically, GDP is also equal to the sum of consumer expenditure, investment, government purchases of goods and services and net exports.
3   The name derives from the decision of their political leaders to coordinate their efforts aimed at achieving closer cooperation with the west and eventual integration with the EU when they met at Visegrad on the Danube in the early 1990s.
4   Value added net of imputed financial intermediation services is equal to GDP at market prices less VAT, net indirect taxes on goods and services and taxes on imports.
5   Some of of the historical data were kindly provided by Maria Provvidenza Terrasi of the University of Pisa. The GDP data are a result of changing accounting systems, though in all cases the time periods overlap: 1951–63; 1963–70; 1970–9; 1979–80; 1980–96 and 1995–2001. The indicators used in this chapter and in chapter 6 involve either a sequence of cross-sectional measurements, comparisons of pairs of estimates from within the same accounting system, or, in one case, where the levels for particular regions are examined over

time, a chained series. The Italian data sources used are: estimates of GDP for 1951–63 from Istituto Giuglielmo Tagliacarne (1961; 1962; 1963; 1964; and 1965); 1963–70 from Istituto Giuglielmo Tagliacarne (1970; and 1973: 133-47); 1970–9 (Prodotto interno lordo ai prezzi di mercato, milliardi di lire correnti) from SVIMEZ (1993); 1979–80 (Prodotto interno lordo ai prezzi di mercato) from ISTAT (1982). Estimates of the resident population for 1951–80 are from ISTAT (1990). Estimates of full-time equivalent employment for 1970–9 are from SVIMEZ (1993b). Estimates of GDP, resident population and full-time equivalent employment for 1980–96 are from ISTAT (1998). Estimates of GDP (Valore aggiunto ai prezzi base a eurolire/euro correnti, ESA95), population and full-time equivalent employment for 1996–2001 are from ISTAT (2003b). The data for the EU15 were derived for 1951–60 from Heston, Summers and Aten (2002) and for 1960–2001 from CEPII (2002).

Notes to Chapter 5

1   According to Barca (1997), the adoption of Keynesianism and state interventionism was a result of an implicit compromise, reached, for varying reasons, by a variety of political and social forces that included liberals, communists, Christian socialists and private entrepreneurs.
2   The new Italian state inherited industrial credit (IMI) and industrial interests (IRI and AGIP) from the interwar Fascist regime. IMI and IRI were established after the 1929 crisis, in 1931 and 1933 respectively, to shelter the deposit banks from the risks of industrial investment and to reorganize the supply of industrial credit (Castronovo, 1995). AGIP was involved in the energy sector and oil refining. In 1948, the DC decided against disposing of IRI's assets. In 1953, ENI was created to develop Italy's energy and chemical interests.
3   Enrico Mattei was involved in the exploitation of natural gas in Italy, the establishment of trade agreements with the USSR and the creation of an oil sector. In the case of oil, AGIP's growth entailed securing contracts with oil-rich countries through the establishment of royalty arrangements that gave the producers 75% of the revenue. As a result, Mattei earned the ire of the seven leading Anglo-American oil corporations. Mattei died in suspicious circumstances in 1962 when an explosion caused his private jet to crash.

Notes to Chapter 7

1   In the CEECs passenger car output did increase as a result of increased production by companies set up, or taken over, by major western manufacturers. The fortunes of countries with a significant foreign presence differed sharply, however, with Poland losing ground in the late 1990s as the Czech Republic forged ahead.

2   A primary chemical results from the first phase of transformation of material inputs into intermediate goods for other industries. A secondary chemical results from the transformation of intermediate chemicals into new intermediate goods for other industries or final end-user/consumer products (consumer chemicals and pharmaceuticals). A related distinction is made between commodity chemicals, characterised by low unit values, and speciality and fine chemicals, characterised by higher prices, lower volumes and greater value added.

Notes to Chapter 8

1   FIAT specialized in small cars for three reasons (Volpato, 1999). First, in the years after WWII Italy was a relatively low income country, Second, fuel taxes were comparatively high. Third, road taxes increased rapidly as engine capacity increased.
2   In 1988, FIAT Auto market shares were 39 per cent in Italy, 11 per cent in Western Europe, 30 per cent in Poland, 29 per cent in Brazil, 25 per cent in Turkey, 24 per cent in Argentina, 41 per cent in Morocco, 20 per cent in Tunisia, 19 per cent in Egypt and 13 per cent in Venezuela (FIAT, 2000).
3   General Motors also recently considered a deal to buy the bankrupt motor arm of the Korean Daewoo group which, with FIAT, has a strong presence in Poland, although finally it decided not to proceed.
4   In the 1990s, neoliberal reforms significantly increased the bargaining power of multinational corporations. In Brazil, for example, trade liberalization, privatization and the 1994 stabilization of the exchange rate in conjunction with high hopes for growth in the MERCOSUR free trade zone (comprising the markets of Brazil, Argentina, Uruguay and Paraguay) made the area increasingly attractive to inward investors. At the same time increases in state autonomy resulting from a 1988 tax reform led to a fiscal war, as intensified interstate competition for investments resulted in the concession of tax and other advantages to inward investors that almost certainly exceeded the levels required for them to invest in Brazil. The case of the Mercedes Benz investment of US$ 695 millions in the 70,000 units per year Juiz de Sora plant in Minas Gerais provides an indication of the scale of the aid. In negotiations, Mercedes Benz secured a gift of the land for the plant, the construction of social and economic infrastructure including the road surface inside the Mercedes Benz site, and the maximum-permitted tax concessions from all three levels of government (municipality, state and federal government). In addition, the state had to give Mercedes Benz shares in the state electrical company, CEMIG, as a guarantee that the country's major, publicly-owned investment bank, BNDES, would provide the low-interest loans it was required to pay (Galvão, 2002).
5   The labour conflicts in the factories have taught FIAT's management an elementary truth: large plants like Mirafiori can no longer be governed, and, therefore, the FIAT system must be reformed. What was decided was to create factories in the South, but in a certain way . . . one does not create a Mirafiori

in the South, but several small plants. What these plants have in common is that they are independent of one another' (FIAT manager, cited in Amin, 1983).

6   At first Termoli made engines for Cassino. Subsequently, it manufactured and assembled gearboxes. In 1983, it was restructured and equipped with highly automated equipment to make engines.

7   ELASIS-Sistema di ricerca FIAT nel Mezzogiorno was established in 1988 to support innovation in the FIAT Group's southern manufacturing facilities. Consisting of six Research Centres and one Laboratory, it employs 506 researchers and 486 operational technicians. In Turin, FIAT has its main research centre with more than 900 employees and a network of over 1,000 outside researchers.

8   Automated workshops such as ROBOGATE and DIGITRON and the semi-automated LAM (Lavorazione Asincrona Motori—Asynchronic Engine Manufacturing) engine assembly system were designed to eliminate manual workers from the most troublesome sectors, and to re-establish managerial authority. The cost of automation far exceeded the investment needed for optimal economic efficiency. After the early 1980s' defeat of FIAT's Fordist workers, FIAT management admitted, for example, that 'the LAM system, designed at a time when industrial relations had broken down, is an interesting but one-off development. It is very expensive, takes up a huge amount of space and breaks down more frequently than a less sophisticated installation.'

9   Another feature was the development of management-trades union relationships (local and national) characterized by the participatory principles that developed in Italy in 1988–96, first at a company, and later at a national level.

10   FIAT Auto has, for example, recently transferred responsibility for: (1) suspension units (2,160 workers) to Magneti Marelli, (2) maintenance services (3,100 employees) to Comau Service, (3) security and fire services (670 people) to Sirio Consortium, and (4) materials handling (1,670 employees) to Arcese Villanova and TNT.

Notes to Chapter 9

1   The petrochemical sector produces primary petrochemicals (olefins and aromatics), petrochemical intermediates and petrochemical products. The most important step is the conversion of petroleum feedstock into olefins (steam cracking). Although the aim of steam cracking is to maximize the production of ethylene, the other liquids and gases that result cannot be ignored, since their utilization affects the economics of ethylene production. This consideration is the main incentive to develop and market downstream derivatives of these products and helps explain the development of integrated production complexes (Chapman, 1991).

2   In just ten years from 1990 to 2000, ICI Teeside was the subject of 26 market operations. Included were the 1993 de-merger and spin off of Zeneca, the 1999 sale of ICI's petrochemical interests to Huntsman and the 1997 acquisition of four speciality chemical activities from Unilever (Greco, 2002).

3   The Hoechst of those years no longer exists. An initial decision to concentrate on life sciences involved the abandonment of its industrial chemicals activities, while in 1998 its pharmaceutical and agro-chemical interests were transferred to the Aventis joint-venture with the French company Rhône-Poulenc.

4   In the late 1960s, the state surprisingly authorized ENI to acquire Montedison shares up to the point of exceeding the shares held by private companies. The aim was to change the management of Montedison and to put the chemical industry on a reasonable track (Zamagni, 2000).

5   In the words of one paints manufacturer (cited in Maglia, 2001, our translation) 'every new regulation is an advantage for those who do not respect it'.

6   As an entity designed to provide immaterial services, the Tecnoparco Valbasento was also a proposer and contractor for the Parco Scientifico e Techologico della Basilicata Basentech (LIT 28,529 million) in March 1994 with University of Basilicata, Metapontum Agrobios, SNIA Richerche, Ergom and Ipogea. In October 1994, a company called the Parco Scientifico e Tecnologico della Basilicata was formed. The only public partner was the CSI (with a 6 per cent interest) due, it seems, to the influence of Ministero dell'Università e della Ricerca Scientifica which ended relations with Consorzio.

Notes to Chapter 10

1   Italy's' north-south divide was initially associated with the creation, in the newly unified Italian state, of a social bloc, made up of the northern industrial, commercial and financial bourgeoisie and large southern landowners, which oversaw rapid industrial growth in the north and the reproduction of a low-productivity agrarian system in the south.

2   Mirela Barbu provided this information.

3   At the same time, a raft of low-paid personal services create jobs for a large tranche of people (including immigrants) who work but are poor, while welfare professionals/public administrators suffer from an erosion of their status and rewards, although these changes are not dealt with in this volume.

Appendices

APPENDIX 4.1: DECOMPOSING THE THEIL COEFFICIENT

Suppose there are n regional entities (n provinces for example), which can be aggregated into m groups (20 regions for example) possibly of different sizes $m_1, m_2, \ldots, m_m$, and that the m groups can be aggregated into l supergroups possibly also of different sizes (4 territorial divisions for example each comprising different numbers of regions). To facilitate the exposition start at the highest level of aggregation, and assume that the aggregate income and population figures can be disaggregated first into figures for l territorial groups. The Theil coefficient for the l territorial entities is

$$T^L = \sum_{i=1}^{l} y_i^L \log \frac{y_i^L}{p_i^L} \tag{1}$$

where $y_i^L = Y_i^L / \sum_{i=1}^{l} Y_i^L$ is the ith territorial unit's share of total income and $p_i^L = P_i^L / \sum_{i=1}^{l} P_i^L$ is the ith territorial unit's share of total population, where the superscript L emphasizes that the figures relate to the territorial entities. Theil's coefficient measures the inequality between the l territorial units (interterritorial unit inequality).

If the data is disaggregated further, the Theil coefficient, computed for the finest of the levels of spatial disaggregation, can be completely and perfectly decomposed into an intergroup component and one or more intragroup components. Assume that the aggregate income and population figures for the territorial units can be disaggregated first into figures for the m regional units of which the territorial groups are made up. The decomposition is

$$T^M = T^L_{inter} + T^L_{intra} = T^L + \sum_{i=1}^{l} y_i^L T_i^M$$

$$= \sum_{i=1}^{l} y_i^L \log \frac{y_i^L}{p_i^L} + \sum_{i=1}^{l} y_i^L \left[\sum_{i=1}^{m_1} y_{ij}^{M_1} \log \frac{y_{ij}^{M_1}}{p_{ij}^{M_1}} + \sum_{j=m_1+1}^{m_2} y_{ij}^{M_2} \log \frac{y_{ij}^{M_2}}{p_{ij}^{M_2}} + \ldots + \sum_{j=m_1+1}^{m} y_{ij}^{M_1} \log \frac{y_{ij}^{M_1}}{p_{ij}^{M_1}} \right]$$

(2)

where $y_{ij}^{M_1}$ is the share of the jth region in the income of the first of the l territorial units (indexed by i) of which it is a member (where membership is denoted by M_1 and by the fact that its index j lies in the range $1 \geq j \leq m_1$.

If the m regional entities are in turn subdivided into n smaller provincial entities, the inequalities between the provincial units can be decomposed into

$$T^N = T^L_{inter} + T^L_{intra} + T^M_{intra} = T^L + \sum_{i=1}^{l} y_i^L T_i^M + \sum_{j=1}^{m} y_i^M T_j^N \qquad (3)$$

The Theil coefficient is equal in other words to the sum of the inequalities between the first tier of territorial units, the inequalities within the territorial units and the inequalities within the second tier of regions, or is equal to the inequalities between the areas identified at the first level plus a weighted sum of the Theil coefficients at each subsequent level of the hierarchy of territorial units. The Theil index for each group is the inequality between the individuals who are members of the group, while the weights are each group's share in the income of the next higher level.

APPENDIX 4.2   MEASURING REGIONAL INEQUALITY

The most common indicators of regional inequality are centred on the standard deviation, the mean absolute deviation, the Gini coefficient, and the Theil index. As regional units vary significantly in size, all of these indicators are weighted by the region's share in the total population. If p_i denotes the population of the ith region ($i = 1,2, \ldots, n$), y_i denotes its per capita GDP and q_i denotes its GDP, the population-weighted standard deviation (WSD) is given by the equation

$$WSD = \sqrt{\sum_{i=1}^{n} (y_i - \bar{y})^2 \frac{p_i}{\sum_{i=1}^{n} p_i}} \qquad (1)$$

and the weighted absolute deviation (WAD) is given by the equation

$$WAD = \sum_{i=1}^{n} |y_i - \bar{y}| \frac{p_i}{\sum_{i=1}^{n} p_i} \qquad (2)$$

To facilitate comparisons the weighted standard deviation and the weighted absolute deviation are expressed as percentages of the mean to give a coefficient of variation (CV).

The Gini coefficient (GC) is given by the equation

$$GC = \frac{1}{2n \sum_{i=1}^{n} y_i} \sum_{i=1}^{n} \sum_{j=1}^{n} |y_i - y_j| \tag{3}$$

As with the indicators based on the mean, the deviations from which the Gini coefficient was calculated were weighted by the product of the shares of the total population in each pair of regions.

Finally, the Theil coefficient (TC) is given by the equation

$$TC = \sum_{i=1}^{n} \frac{q_i}{\sum_{i=1}^{n} q_i} \log_e \left(\frac{\frac{q_i}{\sum_{i=1}^{n} q_i}}{\frac{p_i}{\sum_{i=1}^{n} p_i}} \right) \tag{4}$$

APPENDIX 5.1:   THE 1980 IRPINIA EARTHQUAKE

After the earthquake in which 2,914 people died, some 10,000 were injured and 300,000 lost their homes, LIT 60,000 billion were allocated for emergency relief and for reconstruction and development in a large area in Campania, Basilicata and Puglia that we shall consider later (Figure 5.5).

A decision was quickly made to modernize the area through the development of industrial estates, new infrastructures, new public buildings and a relocation of the population. At the end of 1990, 28,572 people still lived in temporary accommodation and 4,405 people in requisitioned lodgings. Of 10,657 jobs expected, only 3,323 existed. At the root of this failure was the fact that a large part of the activity associated with the use of the funds fell into the hands of Camorra-controlled organizations with the active collusion of politicians, entrepreneurs and administrator. In its turn, of this outcome was a result of the fact that the Italian government entrusted the management of this programme to 'a legislative framework founded on exceptionalism, excessive discretionality, a lack of control and a vagueness of decision-making arrangements' (Commissione Parlamentare d'Inchiesta (1991); see also Commissione Antimafia (1994) and Osservatorio sulla Camorra (1993)).

APPENDIX 6.1:   THE EXPONENTIAL GROWTH MODEL

Expressed algebraically, the exponential model is given by the equation:

$$y_{t+T}^i = y_t^i e^{g_{tT} T} \qquad (1)$$

where y_t^i is the value of the indicator for region i at time t, T is the length of the period measured as the number of years since t, and g_{tT} is the rate of growth. Isolating the exponential part of the equation, taking natural logarithms of both sides, and using the cancellation property for logarithms yields

$$\ln\left(\frac{y_{t+T}^i}{y_t^i}\right) = \left(\ln(y_{t+T}^i) - \ln(y_t^i)\right) = g_{tT} T \qquad (2)$$

In the case of the cumulative growth model

$$y_{t+T}^i = y_t^i e^{G_{tT}} \qquad (3)$$

where the cumulative growth rate from year $t = 0$ to T is defined as

$$G_{OT} = \sum_{t=1}^{T} g_{t-1,\,t} \qquad (4)$$

As we measure growth relative to the national average,

$$y_T^i = y_0^i e^{G_{OT}^I + G_{OT}^i} = y_0^i e^{G_{OT}^I} e^{G_{OT}^i} \qquad (5)$$

and

$$\frac{y_T^i/y_0^i}{y_T^I/y_0^I} = e^{G_{OT}^i} \qquad (6)$$

where G_{OT}^I is the cumulative national rate of growth and G_{OT}^i is the cumulative differential rate of growth of region i. Computing the exponent of the calculated regional differential growth rates gives growth of the region relative to the country, or the level a particular region achieved as a proportion of the level it would have reached at national rates of increase or decrease. Subtracting the regional exponent from one gives the increase or decrease relative to the outcome that would have resulted from the achievement of national growth rates.

APPENDIX 7.1   RELATIVE PRODUCTIVITY DECOMPOSITIONS

Arriving at Equation 1 involves recognizing that:.

$$(w^i - W) = \sum_{j=1}^{n} w_j^i e_j^i - \sum_{j=1}^{n} W_j E_j$$

$$= \sum_{j=1}^{n} (w_j^i - W_j) e_j^i + \sum_{j=1}^{n} (e_j^i - E_j) W_j$$

$$= \sum_{j=1}^{n} (w_j^i - W_j) e_j^i + \sum_{j=1}^{n} (e_j^i - E_j) w_j^i - \sum_{j=1}^{n} (w_j^i - W_j)(e_j^i - E_j), \text{ or}$$

$$= \sum_{j=1}^{n} (w_j^i - W_j) E_j + \sum_{j=1}^{n} (e_j^i - E_j) W_j + \sum_{j=1}^{n} (w_j^i - W_j)(e_j^i - E_j) \qquad (1)$$

The derivation of Equation 2 is similar to that of Equation 1:

$$(w_t^i - w_{t-1}^i) = \sum_{j=1}^{n} w_{jt}^i e_{jt}^i - \sum_{j=1}^{n} w_{jt-1}^i e_{jt-1}^i$$

$$= \sum_{j=1}^{n} (w_{jt}^i - w_{jt-1}^i) e_{jt-1}^i + \sum_{j=1}^{n} (e_{jt}^i - e_{jt-1}^i) w_{jt}^i \qquad (2)$$

$$= \sum_{j=1}^{n} (w_{jt}^i - w_{jt-1}^i) e_{jt-1}^i + \sum_{j=1}^{n} (e_{jt}^i - e_{jt-1}^i) w_{jt-1}^i + \sum_{j=1}^{n} (w_{jt}^i - w_{jt-1}^i)(e_{jt}^i - e_{jt-1}^i)$$

APPENDIX 7.2   A DECOMPOSITION OF REGIONAL PRODUCTIVITY BY INDUSTRY

This Appendix tabulates the results of the decomposition of regional productivity by industry in 1980–95 and 1995–2000 for the four regions of Piemonte, Lombardia, Basilicata and Puglia. The results were derived from ISTAT data (ISTAT, 1998; 2003b).

The variables reported in the tables are:

RPG = actual = relative productivity shift
IES = composition = intersectoral relative productivity shift
IAS = within = intrasectoral productivity shift
COV = covariance
EI80 = share of Italian employment in 1980
EI95 = share of Italian employment in 1995
EI00 = share of Italian employment in 2000
RP80 = productivity relative to Italy in 1980
RP95 = productivity relative to Italy in 1995

RP00 = productivity relative to Italy in 2000
EM80 = share of regional employment in 1980
EM95 = share of regional employment in 1995
EM00 = share of regional employment in 2000

Piemonte (1980–95)	IES	IAS	COV	RPG	EM80	EM95	RP80	RP95
Agriculture, forestry and fishing	−0.04	0.09	−0.03	0.01	0.11	0.07	0.95	0.96
Energy producing products	−0.01	0.03	0.00	0.02	0.01	0.01	0.76	0.80
Ferrous and non-ferrous minerals and metals	−0.01	0.07	−0.02	0.03	0.01	0.01	1.02	1.01
Non-metallic minerals and mineral products	−0.01	0.03	−0.01	0.01	0.01	0.01	1.04	1.12
Chemicals, chemical products and pharmaceuticals	0.00	0.05	0.00	0.05	0.01	0.01	0.88	0.97
Metal products and machinery	−0.10	0.27	−0.08	0.08	0.12	0.08	1.05	1.02
Transport equipment	−0.09	0.16	−0.07	−0.01	0.08	0.04	1.03	1.04
Food, beverages and tobacco	0.00	0.05	0.00	0.04	0.02	0.02	0.91	0.99
Textiles, clothing, leather, leather products and footwear	−0.04	0.11	−0.03	0.03	0.06	0.04	1.17	1.27
Paper, paper products, publishing and printing	−0.02	0.03	−0.01	0.00	0.02	0.01	1.05	1.03
Word, rubber and other industrial products	−0.04	0.11	−0.04	0.03	0.04	0.03	1.12	1.39
Construction and civil engineering	0.01	0.02	0.00	0.03	0.06	0.06	1.09	1.14
Trade, hotels and public stores	0.13	0.05	0.01	0.19	0.18	0.22	1.04	1.01
Transport and commnication	0.03	0.09	0.02	0.14	0.05	0.07	0.87	0.88
Credit and insurance	0.04	0.02	0.01	0.07	0.02	0.02	0.99	0.98
Other market services	0.40	−0.08	−0.07	0.26	0.08	0.14	0.96	0.90
Non-market services	0.10	−0.02	0.00	0.08	0.12	0.16	0.93	0.89
Total	0.34	1.07	−0.34	1.07	1.00	1.00	1.00	1.02

Piemonte (1995–00)	IES	IAS	COV	RPG	EM95	EM00	RP95	RP00
Agriculture, forestry and fishing	−0.13	0.12	−0.03	−0.03	0.05	0.04	1.22	1.15
Mining and quarrying	0.00	0.00	0.00	0.00	0.00	0.00	0.58	0.65
Food products; beverages and tobacco	−0.06	0.08	−0.01	0.01	0.02	0.02	1.14	1.27
Textiles and textile products	−0.08	0.03	0.00	−0.05	0.04	0.04	1.22	1.14
Leather and leather products	0.00	−0.01	0.00	−0.01	0.00	0.00	1.16	0.91
Pulp, paper and paper products; publishing and printing	−0.01	0.02	0.00	0.02	0.02	0.02	1.08	1.08
Coke, refined petroleum products, chemicals and chemical products	0.01	−0.03	0.00	−0.02	0.01	0.01	1.02	1.01
Non-metallic mineral products	0.01	0.00	0.00	0.01	0.01	0.01	1.06	1.04
Basic metals and fabricated metal products	−0.01	−0.04	0.00	−0.05	0.05	0.05	1.05	1.01
Machinery and mechanical, electrical and optical equipment; transport equipment	−0.06	0.16	0.00	0.09	0.11	0.10	1.01	1.02
Wood and wood products, rubber and plastic products and other manufacturing	−0.02	0.01	0.00	−0.01	0.04	0.03	1.26	1.14
Electricity, gas and water supply	−0.08	0.14	−0.02	0.04	0.01	0.01	0.94	0.99
Construction	−0.01	−0.01	0.00	−0.02	0.06	0.06	1.09	1.06
Wholesale and retail trade; repair of motor vehicles, motorcycles and personal and household goods	−0.01	0.14	0.00	0.13	0.16	0.16	1.01	1.03
Hotels and restaurants	0.05	−0.11	−0.01	−0.07	0.04	0.04	1.09	0.84
Transport, storage and communication	0.07	0.10	0.00	0.17	0.06	0.06	1.06	1.06
Financial intermediation	−0.05	0.14	−0.01	0.09	0.03	0.03	0.95	0.91
Real estate, renting and business activities	1.04	−0.55	−0.15	0.34	0.09	0.11	1.00	0.98
Public administration and defence; compulsory social security	−0.08	0.03	0.00	−0.05	0.04	0.04	0.98	0.98
Education	−0.02	−0.02	0.00	−0.04	0.06	0.05	0.96	0.94
Health and social work	0.00	0.03	0.00	0.02	0.05	0.05	1.04	1.02

Other community, social and personal service activities	0.04	0.01	0.00	0.06	0.04	0.04	1.00	1.00
Private households with employed persons	0.00	0.00	0.00	0.00	0.03	0.03	0.78	0.79
Total	0.60	0.27	−0.24	0.64	1.00	1.00	1.07	1.05

Lombardia (1980–95)	IES	IAS	COV	RPG	EM80	EM95	RP80	RP95
Agriculture, forestry and fishing	−0.02	0.08	−0.03	0.03	0.04	0.03	1.39	1.71
Energy producing products	−0.03	0.03	−0.01	−0.01	0.01	0.01	1.42	1.33
Ferrous and non-ferrous minerals and metals	−0.03	0.10	−0.05	0.02	0.02	0.01	1.05	1.14
Non-metallic minerals and mineral products	−0.02	0.03	−0.01	0.01	0.01	0.01	1.07	1.22
Chemicals, chemical products and pharmaceuticals	−0.02	0.14	−0.04	0.08	0.03	0.02	1.07	1.09
Metal products and machinery	−0.14	0.37	−0.13	0.10	0.15	0.10	1.03	1.07
Transport equipment	−0.03	0.04	−0.02	−0.01	0.02	0.01	1.08	1.04
Food, beverages and tobacco	−0.01	0.06	−0.01	0.04	0.02	0.02	1.07	1.06
Textiles, clothing, leather, leather products and footwear	−0.03	0.13	−0.02	0.08	0.09	0.08	1.10	1.14
Paper, paper products, publishing and printing	−0.02	0.05	−0.01	0.02	0.03	0.02	1.02	0.98
Word, rubber and other industrial products	−0.03	0.08	−0.02	0.04	0.05	0.04	1.09	1.07
Construction and civil engineering	0.03	−0.01	0.00	0.01	0.07	0.08	1.12	1.00
Trade, hotels and public stores	0.07	0.12	0.01	0.21	0.19	0.21	1.09	1.17
Transport and commnication	0.02	0.08	0.01	0.11	0.05	0.06	0.81	0.81
Credit and insurance	0.05	0.03	0.01	0.09	0.02	0.03	1.02	1.01
Other market services	0.51	−0.08	−0.07	0.36	0.08	0.16	1.09	1.06
Non-market services	0.04	0.00	0.00	0.04	0.12	0.13	0.97	0.96
Total	0.34	1.24	−0.37	1.21	1.00	1.00	1.12	1.14

Lombardia (1995–00)	IES	IAS	COV	RPG	EM95	EM00	RP95	RP00
Agriculture, forestry and fishing	−0.07	0.13	−0.02	0.04	0.03	0.03	1.38	1.43
Mining and quarrying	−0.03	−0.05	0.01	−0.08	0.00	0.00	2.33	1.95
Food products; beverages and tobacco	−0.05	0.04	0.00	−0.01	0.02	0.02	1.16	1.22
Textiles and textile products	−0.14	0.07	−0.01	−0.09	0.05	0.04	1.13	1.10
Leather and leather products	−0.02	−0.01	0.00	−0.03	0.01	0.00	1.07	0.94
Pulp, paper and paper products; publishing and printing	−0.03	0.02	0.00	−0.01	0.02	0.02	1.00	0.98
Coke, refined petroleum products, chemicals and chemical products	−0.02	0.06	0.00	0.04	0.02	0.02	0.93	1.06
Non-metallic mineral products	0.00	0.01	0.00	0.01	0.01	0.01	1.13	1.12
Basic metals and fabricated metal products	−0.05	0.03	0.00	−0.02	0.06	0.05	1.06	1.09
Machinery and mechanical, electrical and optical equipment; transport equipment	−0.08	0.14	−0.01	0.06	0.09	0.08	1.04	1.06
Wood and wood products, rubber and plastic products and other manufacturing	−0.06	0.10	−0.01	0.03	0.04	0.04	1.10	1.12
Electricity, gas and water supply	−0.04	0.09	−0.01	0.04	0.01	0.01	1.20	1.17
Construction	−0.01	0.04	0.00	0.03	0.06	0.06	0.99	1.01
Wholesale and retail trade; repair of motor vehicles, motorcycles and personal and household goods	0.00	−0.01	0.00	0.00	0.16	0.16	1.19	1.15
Hotels and restaurants	0.02	−0.01	0.00	0.01	0.04	0.04	1.07	1.01
Transport, storage and communication	0.09	0.08	0.01	0.17	0.05	0.06	1.03	1.02
Financial intermediation	−0.04	0.47	−0.01	0.42	0.03	0.03	1.05	1.18
Real estate, renting and business activities	1.16	−0.59	−0.18	0.39	0.09	0.12	0.94	0.93
Public administration and defence; compulsory social security	−0.03	0.02	0.00	−0.01	0.03	0.03	1.11	1.11
Education	−0.03	−0.02	0.00	−0.05	0.05	0.05	0.96	0.95
Health and social work	0.01	0.03	0.00	0.04	0.05	0.05	0.98	0.97

Other community, social and personal service activities	0.06	−0.01	0.00	0.04	0.04	0.04	1.13	1.08
Private households with employed persons	−0.01	0.00	0.00	−0.01	0.03	0.03	1.13	1.14
Total	0.63	0.62	−0.24	1.01	1.00	1.00	1.13	1.13

Puglia (1980–95)	IES	IAS	COV	RPG	EM80	EM95	RP80	RP95
Agriculture, forestry and fishing	−0.15	0.36	−0.20	0.01	0.23	0.10	1.17	1.50
Energy producing products	0.00	0.03	0.00	0.04	0.01	0.01	0.67	0.85
Ferrous and non-ferrous minerals and metals	−0.02	0.12	−0.05	0.04	0.02	0.01	0.82	1.18
Non-metallic minerals and mineral products	0.00	0.01	0.00	0.01	0.02	0.01	0.83	0.76
Chemicals, chemical products and pharmaceuticals	0.00	0.03	−0.01	0.02	0.01	0.00	0.93	1.09
Metal products and machinery	−0.01	0.06	−0.01	0.05	0.04	0.03	0.80	0.77
Transport equipment	0.00	0.01	0.00	0.01	0.01	0.01	0.98	0.82
Food, beverages and tobacco	−0.02	0.05	−0.01	0.02	0.02	0.01	1.20	1.17
Textiles, clothing, leather, leather products and footwear	0.01	0.02	0.00	0.03	0.04	0.04	0.62	0.56
Paper, paper products, publishing and printing	0.00	0.01	0.00	0.00	0.00	0.00	0.74	0.74
Word, rubber and other industrial products	0.00	0.03	−0.01	0.02	0.02	0.02	0.60	0.81
Construction and civil engineering	−0.04	−0.02	0.00	−0.06	0.09	0.07	0.94	0.81
Trade, hotels and public stores	0.14	0.08	0.02	0.25	0.18	0.23	0.88	0.92
Transport and commnication	0.02	0.09	0.01	0.11	0.05	0.05	1.13	1.08
Credit and insurance	0.03	0.02	0.01	0.05	0.01	0.01	1.07	1.08
Other market services	0.24	0.00	0.00	0.25	0.09	0.15	0.67	0.77
Non-market services	0.12	0.01	0.00	0.13	0.19	0.23	1.00	1.02
Total	0.31	0.91	−0.23	0.98	1.00	1.00	0.85	0.89

Puglia (1995–00)	IES	IAS	COV	RPG	EM95	EM00	RP95	RP00
Agriculture, forestry and fishing	−0.18	0.21	−0.03	0.00	0.15	0.13	0.89	0.81
Mining and quarrying	0.00	0.01	0.00	0.01	0.00	0.00	0.33	0.42
Food products; beverages and tobacco	−0.02	0.02	0.00	0.00	0.02	0.02	0.92	0.92
Textiles and textile products	−0.02	0.08	0.00	0.06	0.03	0.03	0.54	0.63
Leather and leather products	−0.01	0.03	0.00	0.01	0.01	0.01	0.65	0.85
Pulp, paper and paper products; publishing and printing	0.00	0.03	0.00	0.03	0.01	0.01	0.67	0.86
Coke, refined petroleum products, chemicals and chemical products	0.01	−0.07	0.00	−0.07	0.00	0.00	1.28	0.94
Non-metallic mineral products	0.02	0.02	0.00	0.05	0.01	0.01	0.65	0.73
Basic metals and fabricated metal products	0.02	−0.08	0.00	−0.07	0.02	0.02	1.28	1.10
Machinery and mechanical, electrical and optical equipment; transport equipment	0.01	0.09	0.00	0.11	0.02	0.03	0.79	0.90
Wood and wood products, rubber and plastic products and other manufacturing	0.01	0.05	0.00	0.06	0.02	0.02	0.78	0.83
Electricity, gas and water supply	−0.07	0.09	−0.02	0.00	0.01	0.01	0.92	0.95
Construction	0.03	0.07	0.00	0.10	0.07	0.07	0.83	0.88
Wholesale and retail trade; repair of motor vehicles, motorcycles and personal and household goods	0.08	0.23	0.01	0.32	0.15	0.16	0.81	0.86
Hotels and restaurants	0.05	0.03	0.00	0.09	0.03	0.04	1.00	1.04
Transport, storage and communication	0.02	0.13	0.00	0.15	0.05	0.05	0.74	0.79
Financial intermediation	−0.01	0.08	0.00	0.07	0.02	0.02	0.86	0.82
Real estate, renting and business activities	0.70	−0.24	−0.06	0.40	0.07	0.09	0.92	0.98
Public administration and defence; compulsory social security	−0.10	0.08	−0.01	−0.03	0.08	0.07	0.84	0.85
Education	−0.08	0.00	0.00	−0.08	0.09	0.09	0.95	0.96
Health and social work	0.04	0.05	0.00	0.09	0.06	0.06	0.93	0.94

Other community, social and personal service activities	0.04	0.03	0.00	0.07	0.04	0.04	0.79	0.82
Private households with employed persons	0.00	0.00	0.00	0.00	0.02	0.02	0.84	0.86
Total	0.55	0.92	−0.09	1.37	1.00	1.00	0.81	0.84

Basilicata (1980–95)	IES	IAS	COV	RPG	EM80	EM95	RP80	RP95
Agriculture, forestry and fishing	−0.09	0.25	−0.10	0.06	0.34	0.20	0.60	0.73
Energy producing products	0.01	0.02	0.00	0.03	0.01	0.01	0.83	0.81
Ferrous and non-ferrous minerals and metals	0.00	0.00	0.00	0.00	0.00	0.00	0.73	0.20
Non-metallic minerals and mineral products	0.00	0.01	0.00	0.00	0.01	0.01	0.92	0.78
Chemicals, chemical products and pharmaceuticals	−0.02	0.06	−0.04	0.00	0.02	0.01	0.77	0.84
Metal products and machinery	−0.01	0.04	−0.01	0.03	0.03	0.02	0.68	0.70
Transport equipment	0.04	0.04	0.11	0.18	0.01	0.03	0.84	1.82
Food, beverages and tobacco	0.01	0.03	0.00	0.04	0.02	0.02	0.81	0.72
Textiles, clothing, leather, leather products and footwear	0.00	0.02	0.00	0.02	0.02	0.02	0.68	0.74
Paper, paper products, publishing and printing	0.00	0.00	0.00	0.00	0.00	0.00	1.00	0.54
Word, rubber and other industrial products	0.01	0.02	0.01	0.03	0.01	0.02	0.63	0.72
Construction and civil engineering	0.03	−0.11	−0.01	−0.09	0.12	0.13	1.04	0.70
Trade, hotels and public stores	0.02	0.10	0.01	0.12	0.13	0.14	0.88	1.00
Transport and commnication	0.05	0.12	0.05	0.21	0.04	0.05	1.09	1.31
Credit and insurance	0.03	0.01	0.01	0.04	0.01	0.01	1.13	1.13
Other market services	0.24	−0.05	−0.03	0.16	0.05	0.10	0.94	0.91
Non-market services	0.15	0.01	0.00	0.17	0.18	0.23	1.00	1.02
Total	0.46	0.57	0.00	1.02	1.00	1.00	0.76	0.82

Basilicata (1995–00)	IES	IAS	COV	RPG	EM95	EM00	RP95	RP00
Agriculture, forestry and fishing	−0.32	0.61	−0.17	0.13	0.15	0.11	0.86	1.03
Mining and quarrying	0.01	0.01	0.00	0.02	0.00	0.00	0.49	0.57
Food products; beverages and tobacco	0.03	0.00	0.00	0.03	0.02	0.03	0.91	0.87
Textiles and textile products	0.03	−0.02	0.00	0.01	0.01	0.02	0.74	0.57
Leather and leather products	0.01	0.00	0.00	0.00	0.00	0.00	0.97	0.52
Pulp, paper and paper products; publishing and printing	0.04	0.00	0.00	0.03	0.00	0.01	0.69	0.61
Coke, refined petroleum products, chemicals and chemical products	0.14	−0.04	−0.04	0.06	0.00	0.01	0.97	0.75
Non-metallic mineral products	0.08	−0.01	0.00	0.07	0.01	0.02	0.76	0.71
Basic metals and fabricated metal products	0.03	−0.01	0.00	0.02	0.01	0.02	0.89	0.86
Machinery and mechanical, electrical and optical equipment; transport equipment	0.08	0.08	0.01	0.17	0.05	0.05	1.19	1.20
Wood and wood products, rubber and plastic products and other manufacturing	0.10	−0.09	−0.02	−0.02	0.02	0.03	1.27	0.91
Electricity, gas and water supply	−0.09	0.13	−0.02	0.03	0.01	0.01	0.82	0.86
Construction	0.05	−0.34	−0.01	−0.30	0.10	0.10	0.99	0.75
Wholesale and retail trade; repair of motor vehicles, motorcycles and personal and household goods	0.08	0.36	0.02	0.45	0.13	0.13	0.77	0.89
Hotels and restaurants	0.07	0.04	0.01	0.12	0.04	0.04	0.87	0.91
Transport, storage and communication	0.00	0.05	0.00	0.05	0.05	0.05	0.71	0.70
Financial intermediation	0.05	0.02	0.00	0.07	0.02	0.02	0.94	0.83
Real estate, renting and business activities	1.00	−0.54	−0.18	0.28	0.07	0.09	1.02	0.97
Public administration and defence; compulsory social security	−0.08	0.10	0.00	0.01	0.08	0.07	1.22	1.23
Education	−0.27	0.00	0.00	−0.28	0.12	0.10	0.97	0.98
Health and social work	−0.01	0.05	0.00	0.04	0.06	0.06	1.00	1.00

Other community, social and personal service activities	0.02	0.00	0.00	0.02	0.03	0.03	0.99	0.96
Private households with employed persons	0.01	0.00	0.00	0.01	0.01	0.01	0.87	0.91
Total	1.05	0.39	−0.43	1.01	1.00	1.00	0.88	0.89

Subject Index

Name Index